Student Solutions Manual

for

Gustafson and Frisk's

Beginning Algebra

Seventh Edition

Michael G. Welden
Rock Valley College

THOMSON

BROOKS/COLE

Australia • Canada • Mexico • Singapore • Spain • United Kingdom • United States

Printed in Canada
1 2 3 4 5 6 7   08 07 06 05 04

Printer: Webcom Limited

ISBN: 0-534-46335-5

For more information about our products, contact us at:
Thomson Learning Academic Resource Center
1-800-423-0563

For permission to use material from this text or product, submit a request online at
http://www.thomsonrights.com.
Any additional questions about permissions can be submitted by email to **thomsonrights@thomson.com.**

Thomson Brooks/Cole
10 Davis Drive
Belmont, CA 94002-3098
USA

Asia
Thomson Learning
5 Shenton Way #01-01
UIC Building
Singapore 068808

Australia/New Zealand
Thomson Learning
102 Dodds Street
Southbank, Victoria 3006
Australia

Canada
Nelson
1120 Birchmount Road
Toronto, Ontario M1K 5G4
Canada

Europe/Middle East/South Africa
Thomson Learning
High Holborn House
50/51 Bedford Row
London WC1R 4LR
United Kingdom

Latin America
Thomson Learning
Seneca, 53
Colonia Polanco
11560 Mexico D.F.
Mexico

Spain/Portugal
Paraninfo
Calle/Magallanes, 25
28015 Madrid, Spain

Preface

This manual contains detailed solutions to all of the odd exercises of the text *Beginning Algebra*, seventh edition, by R. David Gustafson and Peter D. Frisk. It also contains solutions to all chapter summary, chapter test, and cumulative review exercises found in the text.

Many of the exercises in the text may be solved using more than one method, but it is not feasible to list all possible solutions in this manual. Also, some of the exercises may have been solved in this manual using a method that differs slightly from that presented in the text. There are a few exercises in the text whose solutions may vary from person to person. Some of these solutions may not have been included in this manual. For the solution to an exercise like this, the notation "answers may vary" has been included.

Please remember that only reading a solution does not teach you how to solve a problem. To repeat a commonly used phrase, mathematics is not a spectator sport. You MUST make an honest attempt to solve each exercise in the text without using this manual first. This manual should be viewed more or less as a last resort. Above all, DO NOT simply copy the solution from this manual onto your own paper. Doing so will not help you learn how to do the exercise, nor will it help you to do better on quizzes or tests.

I would like to thank the members of the mathematics faculty at Rock Valley College and Rebecca Subity of Brooks/Cole Publishing Company for their help and support. This solutions manual was prepared using EXP 5.1.

This book is dedicated to the Reverend Ben Loyd, the Reverend Jack Swanson, and the Reverend Pam Gregory for their friendship and inspiration throughout my life.

May your study of this material be successful and rewarding.

Michael G. Welden

Contents

Exercise 1.1 (page 10)

1. set **3.** whole **5.** subset

7. prime **9.** is not equal to **11.** is greater than or equal to

13. number **15.** natural: $1, 2, 6, 9$ **17.** positive integers: $1, 2, 6, 9$

19. integers: $-3, -1, 0, 1, 2, 6, 9$ **21.** real: $-3, -\frac{1}{2}, -1, 0, 1, 2, \frac{5}{3}, \sqrt{7}, 3.25, 6, 9$

23. odd integers: $-3, -1, 1, 9$ **25.** composite: $6, 9$

27. $4 + 5 = 9$
9: natural, odd, composite, whole

29. $15 - 15 = 0$
0: even, whole

31. $3 \cdot 8 = 24$
24: natural, even, composite, whole

33. $24 \div 8 = 3$
3: natural, odd, prime, whole

35. $5 \boxed{} 3 + 2$
$5 \boxed{=} 5$

37. $25 \boxed{<} 32$

39. $5 + 7 \boxed{} 10$
$12 \boxed{>} 10$

41. $3 + 9 \boxed{} 20 - 8$
$12 \boxed{=} 12$

43. $4 \cdot 2 \boxed{} 2 \cdot 4$
$8 \boxed{=} 8$

45. $8 \div 2 \boxed{} 4 + 2$
$4 \boxed{<} 6$

47. $3 + 2 + 5 \boxed{} 5 + 2 + 3$
$10 \boxed{=} 10$

49. $7 > 3$ **51.** $8 \leq 8$ **53.** $3 + 4 = 7$ **55.** $\sqrt{2} \approx 1.41$

57. $3 \leq 7 \Rightarrow \boxed{7 \geq 3}$ **59.** $6 > 0 \Rightarrow \boxed{0 < 6}$

61. $3 + 8 > 8 \Rightarrow \boxed{8 < 3 + 8}$ **63.** $6 - 2 < 10 - 4 \Rightarrow \boxed{10 - 4 > 6 - 2}$

65. $2 \cdot 3 < 3 \cdot 4 \Rightarrow \boxed{3 \cdot 4 > 2 \cdot 3}$ **67.** $\dfrac{12}{4} < \dfrac{24}{6} \Rightarrow \boxed{\dfrac{24}{6} > \dfrac{12}{4}}$

69.

6 is greater than 3. 6 is to the right of 3.

71.

11 is greater than 6. 11 is to the right of 6.

73.

2 is greater than 0. 2 is to the right of 0.

75.

8 is greater than 0. 8 is to the right of 0.

77.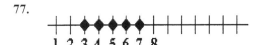

79.

81.

83.

85. ←――(――)――→
         1   5

87. ←――]――[――→
     −3   3

89. $|36| = 36$

91. $|0| = 0$

93. $|-230| = 230$

95. $|12 - 4| = |8| = 8$

97. Answers may vary.

99. Answers may vary.

101. Answers may vary.

Exercise 1.2 (page 24)

1. true

3. false; 21 has factors of 3 and 7.

5. true; 8 is to the right of -2.

7. true; $|-9| = 9$, so $9 \leq |-9|$.

9. $3 + 7 \boxed{=} 10$

11. $|-2| = 2$, so $|-2| \boxed{=} 2$

13. numerator

15. simplify

17. proper

19. 1

21. multiply

23. numerators, denominator

25. plus (or $+$)

27. repeating

29. $\dfrac{6}{12} = \dfrac{1 \cdot \cancel{6}^{1}}{2 \cdot \cancel{6}_{1}} = \dfrac{1}{2}$

31. $\dfrac{15}{20} = \dfrac{3 \cdot \cancel{5}^{1}}{4 \cdot \cancel{5}_{1}} = \dfrac{3}{4}$

33. $\dfrac{24}{18} = \dfrac{4 \cdot \cancel{6}^{1}}{3 \cdot \cancel{6}_{1}} = \dfrac{4}{3}$

35. $\dfrac{72}{64} = \dfrac{9 \cdot \cancel{8}^{1}}{8 \cdot \cancel{8}_{1}} = \dfrac{9}{8}$

37. $\dfrac{1}{2} \cdot \dfrac{3}{5} = \dfrac{1 \cdot 3}{2 \cdot 5} = \dfrac{3}{10}$

39. $\dfrac{4}{3} \cdot \dfrac{6}{5} = \dfrac{4 \cdot 6}{3 \cdot 5} = \dfrac{4 \cdot 2 \cdot \cancel{3}^{1}}{\cancel{3}_{1} \cdot 5} = \dfrac{8}{5}$

41. $\dfrac{5}{12} \cdot \dfrac{18}{5} = \dfrac{5 \cdot 18}{12 \cdot 5} = \dfrac{\cancel{5}^{1} \cdot 3 \cdot \cancel{6}^{1}}{2 \cdot \cancel{6}_{1} \cdot \cancel{5}_{1}} = \dfrac{3}{2}$

43. $\dfrac{17}{34} \cdot \dfrac{3}{6} = \dfrac{17 \cdot 3}{34 \cdot 6} = \dfrac{\cancel{17}^{1} \cdot \cancel{3}^{1}}{2 \cdot \cancel{17}_{1} \cdot 2 \cdot \cancel{3}_{1}} = \dfrac{1}{4}$

45. $12 \cdot \frac{5}{6} = \frac{12}{1} \cdot \frac{5}{6} = \frac{12 \cdot 5}{1 \cdot 6} = \frac{2 \cdot \overset{1}{\cancel{6}} \cdot 5}{1 \cdot \cancel{6}_{1}}$

$\qquad = \frac{10}{1} = 10$

47. $\frac{10}{21} \cdot 14 = \frac{10}{21} \cdot \frac{14}{1} = \frac{10 \cdot 14}{21 \cdot 1} = \frac{10 \cdot 2 \cdot \overset{1}{\cancel{7}}}{3 \cdot \cancel{7}_{1}}$

$\qquad = \frac{20}{3}$

49. $\frac{3}{5} \div \frac{2}{3} = \frac{3}{5} \cdot \frac{3}{2} = \frac{3 \cdot 3}{5 \cdot 2} = \frac{9}{10}$

51. $\frac{3}{4} \div \frac{6}{5} = \frac{3}{4} \cdot \frac{5}{6} = \frac{3 \cdot 5}{4 \cdot 6} = \frac{\overset{1}{\cancel{3}} \cdot 5}{4 \cdot 2 \cdot \cancel{3}_{1}} = \frac{5}{8}$

53. $\frac{2}{13} \div \frac{8}{13} = \frac{2}{13} \cdot \frac{13}{8} = \frac{2 \cdot 13}{13 \cdot 8} = \frac{\overset{1}{\cancel{2}} \cdot \overset{1}{\cancel{13}}}{\cancel{13}_{1} \cdot 4 \cdot \cancel{2}_{1}}$

$\qquad = \frac{1}{4}$

55. $\frac{21}{35} \div \frac{3}{14} = \frac{21}{35} \cdot \frac{14}{3} = \frac{21 \cdot 14}{35 \cdot 3}$

$\qquad = \frac{\overset{1}{\cancel{7}} \cdot \overset{1}{\cancel{3}} \cdot 14}{5 \cdot \cancel{7}_{1} \cdot \cancel{3}_{1}} = \frac{14}{5}$

57. $6 \div \frac{3}{14} = \frac{6}{1} \div \frac{3}{14} = \frac{6}{1} \cdot \frac{14}{3} = \frac{6 \cdot 14}{1 \cdot 3} = \frac{2 \cdot \overset{1}{\cancel{3}} \cdot 14}{1 \cdot \cancel{3}_{1}} = \frac{28}{1} = 28$

59. $\frac{42}{30} \div 7 = \frac{42}{30} \div \frac{7}{1} = \frac{42}{30} \cdot \frac{1}{7} = \frac{42 \cdot 1}{30 \cdot 7} = \frac{\overset{1}{\cancel{6}} \cdot \overset{1}{\cancel{7}} \cdot 1}{5 \cdot \cancel{6}_{1} \cdot \cancel{7}_{1}} = \frac{1}{5}$

61. $\frac{3}{5} + \frac{3}{5} = \frac{3+3}{5} = \frac{6}{5}$

63. $\frac{4}{13} - \frac{3}{13} = \frac{4-3}{13} = \frac{1}{13}$

65. $\frac{1}{6} + \frac{1}{24} = \frac{1 \cdot \mathbf{4}}{6 \cdot \mathbf{4}} + \frac{1}{24} = \frac{4}{24} + \frac{1}{24}$

$\qquad = \frac{4+1}{24} = \frac{5}{24}$

67. $\frac{3}{5} + \frac{2}{3} = \frac{3 \cdot \mathbf{3}}{5 \cdot \mathbf{3}} + \frac{2 \cdot \mathbf{5}}{3 \cdot \mathbf{5}} = \frac{9}{15} + \frac{10}{15}$

$\qquad = \frac{9+10}{15} = \frac{19}{15}$

69. $\frac{9}{4} - \frac{5}{6} = \frac{9 \cdot \mathbf{3}}{4 \cdot \mathbf{3}} - \frac{5 \cdot \mathbf{2}}{6 \cdot \mathbf{2}} = \frac{27}{12} - \frac{10}{12} = \frac{27-10}{12} = \frac{17}{12}$

71. $\frac{7}{10} - \frac{1}{14} = \frac{7 \cdot \mathbf{7}}{10 \cdot \mathbf{7}} - \frac{1 \cdot \mathbf{5}}{14 \cdot \mathbf{5}} = \frac{49}{70} - \frac{5}{70} = \frac{49-5}{70} = \frac{44}{70} = \frac{22 \cdot \overset{1}{\cancel{2}}}{35 \cdot \cancel{2}_{1}} = \frac{22}{35}$

73. $3 - \frac{3}{4} = \frac{3}{1} - \frac{3}{4} = \frac{3 \cdot \mathbf{4}}{1 \cdot \mathbf{4}} - \frac{3}{4} = \frac{12}{4} - \frac{3}{4} = \frac{12-3}{4} = \frac{9}{4}$

75. $\frac{17}{3} + 4 = \frac{17}{3} + \frac{4}{1} = \frac{17}{3} + \frac{4 \cdot \mathbf{3}}{1 \cdot \mathbf{3}} = \frac{17}{3} + \frac{12}{3} = \frac{17+12}{3} = \frac{29}{3}$

77. $4\frac{3}{5} + \frac{3}{5} = \left(4 + \frac{3}{5}\right) + \frac{3}{5} = \left(\frac{20}{5} + \frac{3}{5}\right) + \frac{3}{5} = \frac{23}{5} + \frac{3}{5} = \frac{26}{5} = 5\frac{1}{5}$

79. $3\frac{1}{3} - 1\frac{2}{3} = \left(3 + \frac{1}{3}\right) - \left(1 + \frac{2}{3}\right) = \left(\frac{9}{3} + \frac{1}{3}\right) - \left(\frac{3}{3} + \frac{2}{3}\right) = \frac{10}{3} - \frac{5}{3} = \frac{5}{3} = 1\frac{2}{3}$

81. $3\frac{3}{4} - 2\frac{1}{2} = \left(3 + \frac{3}{4}\right) - \left(2 + \frac{1}{2}\right) = \left(\frac{12}{4} + \frac{3}{4}\right) - \left(\frac{8}{4} + \frac{2}{4}\right) = \frac{15}{4} - \frac{10}{4} = \frac{5}{4} = 1\frac{1}{4}$

83. $8\frac{2}{9} - 7\frac{2}{3} = \left(8 + \frac{2}{9}\right) - \left(7 + \frac{2}{3}\right) = \left(\frac{72}{9} + \frac{2}{9}\right) - \left(\frac{63}{9} + \frac{6}{9}\right) = \frac{74}{9} - \frac{69}{9} = \frac{5}{9}$

85.
```
      2  3  .  4  5
  +   1  3  5  .  2
  ─────────────────
      1  5  8  .  6  5
```

87.
```
                11
              6  1  13
        6  7  .  2  3  5
    -   2  2  .  4  5
  ──────────────────────
        4  4  .  7  8  5
```

89.
```
                3  .  4
    ×        1  3  .  2
  ─────────────────────
                6  8
          1  0  2
          3  4
  ─────────────────────
          4  4  8  8
```
Put two digits to the right of the decimal
point. Answer: 44.88

91.
```
  0 . 2  3 | 1 . 0  4  6  5
```
Move decimal points 2 places right.
```
                    4 .  5  5
    2  3 . | 1  0  4 .  6  5
                9  2
            ─────────
                1  2  6
                1  1  5
            ──────────
                   1  1  5
                   1  1  5
                ──────────
                        0
```

Problems 93-99 are to be solved using a calculator. The keystrokes needed to solve each problem using a TI-83 graphing calculator appear in each solution. There may be other solutions. Keystrokes for other calculators may be slightly different.

93. $\boxed{3}\ \boxed{2}\ \boxed{3}\ \boxed{.}\ \boxed{2}\ \boxed{4}\ \boxed{+}\ \boxed{2}\ \boxed{7}\ \boxed{.}\ \boxed{2}\ \boxed{5}\ \boxed{4}\ \boxed{3}\ \boxed{\text{ENTER}}$ $\{350.4943\} \Rightarrow 350.49$

95. $\boxed{5}\ \boxed{5}\ \boxed{.}\ \boxed{7}\ \boxed{7}\ \boxed{4}\ \boxed{4}\ \boxed{3}\ \boxed{-}\ \boxed{.}\ \boxed{5}\ \boxed{6}\ \boxed{8}\ \boxed{2}\ \boxed{4}\ \boxed{5}\ \boxed{\text{ENTER}}$
$\{55.206185\} \Rightarrow 55.21$

97. $\boxed{2}\ \boxed{5}\ \boxed{.}\ \boxed{2}\ \boxed{5}\ \boxed{\times}\ \boxed{1}\ \boxed{3}\ \boxed{2}\ \boxed{.}\ \boxed{1}\ \boxed{7}\ \boxed{9}\ \boxed{\text{ENTER}}$ $\{3337.51975\} \Rightarrow 3,337.52$

99. $\boxed{4}\ \boxed{.}\ \boxed{5}\ \boxed{6}\ \boxed{9}\ \boxed{4}\ \boxed{3}\ \boxed{2}\ \boxed{3}\ \boxed{\div}\ \boxed{.}\ \boxed{4}\ \boxed{5}\ \boxed{6}\ \boxed{\text{ENTER}}$ $\{10.02068487\} \Rightarrow 10.02$

101. $43\frac{1}{2} - 12\frac{1}{3} = 43 + \frac{1}{2} - 12 - \frac{1}{3} = \frac{258}{6} + \frac{3}{6} - \frac{72}{6} - \frac{2}{6} = \frac{187}{6} = 31\frac{1}{6}$ acres

103. $14 \cdot 3\frac{1}{4} = \frac{14}{1} \cdot \frac{13}{4} = \frac{14 \cdot 13}{1 \cdot 4} = \frac{7 \cdot \cancel{2}^{1} \cdot 13}{1 \cdot 2 \cdot \cancel{2}_{1}} = \frac{91}{2} = 45\frac{1}{2}$ yd

105. 24% of $42,500 = 0.24(42,500) = \$10,200$

107. $0.23(17,500) = 4,025$ defective $\Rightarrow 17,500 - 4025 = 13,475$ acceptable units

109. $0.12(18,700,000) = 2,244,000$ increase $\Rightarrow$ sales $= 18,700,000 + 2,244,000 = \$20,944,000$

111. # gallons $= 15,675.2 \div 25.5 = 614.7137255 \Rightarrow$ cost $= 614.7137255(1.67) \approx \$1,026.57$

113. Area $=$ length $\cdot$ width $= (253.5 \text{ ft})(178.5 \text{ ft}) = 45,249.75 \text{ ft}^2$
Drums of sealer $= 45,249.75 \div 4,000 \approx 11.3 \Rightarrow$ needs 12 drums; Cost $= 12(97.50) = \$1,170$

115. Standard $= 37.50(2,530) = \$94,875$; High-capacity $= 57.35(1,670) = \$95,774.50$
The high-capacity order will produce the greater profit.

117. Silage per cow $= 0.57(12,000) = 6840$ pounds; $30(6,840) = 205,200$ lb of silage

119. Regular $= 1,730 + 36(107.75) = 1,730 + 3,879 = 5,609$
High $= 4,170 + 36(57.50) = 4,170 + 2,070 = 6,240$
The high-efficiency furnace will be more expensive after 3 years.

121. Answers may vary.   **123. Answers may vary.**   **125. Answers may vary.**

127. No. Each proper fraction is less than 1. When a number is multiplied by a number less than 1, the result is smaller than the original number.

Exercise 1.3 (page 35)

1.

3. 17 is a prime number.

5. exponent   **7.** multiplication   **9.** $P = 4s$ **11.** $P = 2l + 2w$

13. $P = a + b + c$ **15.** $P = a + b + c + d$ **17.** $C = \pi D$, or $C = 2\pi r$ **19.** $V = lwh$

21. $V = \frac{1}{3}Bh$ **23.** $V = \frac{4}{3}\pi r^3$ **25.** $4^2 = 4 \cdot 4 = 16$ **27.** $6^2 = 6 \cdot 6 = 36$

29. $\left(\dfrac{1}{10}\right)^4 = \dfrac{1}{10} \cdot \dfrac{1}{10} \cdot \dfrac{1}{10} \cdot \dfrac{1}{10} = \dfrac{1}{10,000}$

Problems 31-33 are to be solved using a calculator. The keystrokes needed to solve each problem using a TI-83 graphing calculator appear in each solution. There may be other solutions. Keystrokes for other calculators may be slightly different.

31. $\boxed{7}\,\boxed{.}\,\boxed{9}\,\boxed{\wedge}\,\boxed{3}\,\boxed{\text{ENTER}}$
$\{493.039\}$

33. $\boxed{2}\,\boxed{5}\,\boxed{.}\,\boxed{3}\,\boxed{\wedge}\,\boxed{2}\,\boxed{\text{ENTER}}$
$\{640.09\}$

35. $x^2 = x \cdot x$ **37.** $3z^4 = 3 \cdot z \cdot z \cdot z \cdot z$ **39.** $(5t)^2 = 5t \cdot 5t$

41. $5(2x)^3 = 5 \cdot 2x \cdot 2x \cdot 2x$ **43.** $4(3)^2 = 4 \cdot 9 = 36$ **45.** $(5 \cdot 2)^3 = 10^3 = 1{,}000$

47. $2(3^2) = 2 \cdot 9 = 18$ **49.** $(3 \cdot 2)^3 = 6^3 = 216$ **51.** $3 \cdot 5 - 4 = 15 - 4 = 11$

53. $3(5 - 4) = 3(1) = 3$ **55.** $3 + 5^2 = 3 + 25 = 28$ **57.** $(3 + 5)^2 = (8)^2 = 64$

59. $2 + 3 \cdot 5 - 4 = 2 + 15 - 4 = 17 - 4 = 13$ **61.** $64 \div (3 + 1) = 64 \div 4 = 16$

63. $(7 + 9) \div (2 \cdot 4) = 16 \div 8 = 2$ **65.** $(5 + 7) \div 3 \cdot 4 = 12 \div 3 \cdot 4 = 4 \cdot 4 = 16$

67. $24 \div 4 \cdot 3 + 3 = 6 \cdot 3 + 3 = 18 + 3 = 21$

69. $3^2 + 2(1 + 4) - 2 = 9 + 2(5) - 2 = 9 + 10 - 2 = 19 - 2 = 17$

71. $5^2 - (7 - 3)^2 = 5^2 - 4^2 = 25 - 16 = 9$ **73.** $(2 \cdot 3 - 4)^3 = (6 - 4)^3 = 2^3 = 8$

75. $\dfrac{3}{5} \cdot \dfrac{10}{3} + \dfrac{1}{2} \cdot 12 = \dfrac{3}{5} \cdot \dfrac{10}{3} + \dfrac{1}{2} \cdot \dfrac{12}{1} = \dfrac{\cancel{3}}{\cancel{5}} \cdot \dfrac{2 \cdot \cancel{5}}{\cancel{3}} + \dfrac{1}{\cancel{2}} \cdot \dfrac{6 \cdot \cancel{2}}{1} = \dfrac{2}{1} + \dfrac{6}{1} = 2 + 6 = 8$

77. $\left[\dfrac{1}{3} - \left(\dfrac{1}{2}\right)^2\right]^2 = \left[\dfrac{1}{3} - \dfrac{1}{4}\right]^2 = \left[\dfrac{1 \cdot \mathbf{4}}{3 \cdot \mathbf{4}} - \dfrac{1 \cdot \mathbf{3}}{4 \cdot \mathbf{3}}\right]^2 = \left[\dfrac{4}{12} - \dfrac{3}{12}\right]^2 = \left[\dfrac{1}{12}\right]^2 = \dfrac{1}{144}$

Problems 79-85 are to be solved using a calculator. The keystrokes needed to solve each problem using a TI-83 graphing calculator appear in each solution. There may be other solutions. Keystrokes for other calculators may be slightly different.

79. numerator: $\boxed{(}\ \boxed{3}\ \boxed{+}\ \boxed{5}\ \boxed{)}\ \boxed{x^2}\ \boxed{+}\ \boxed{2}\ \boxed{\text{ENTER}}$ {66}   fraction $= \dfrac{66}{6} = 11$

     denominator: $\boxed{2}\ \boxed{\times}\ \boxed{(}\ \boxed{8}\ \boxed{-}\ \boxed{5}\ \boxed{)}\ \boxed{\text{ENTER}}$ {6}

81. numerator: $\boxed{(}\ \boxed{5}\ \boxed{-}\ \boxed{3}\ \boxed{)}\ \boxed{x^2}\ \boxed{+}\ \boxed{2}\ \boxed{\text{ENTER}}$ {6}   fraction $= \dfrac{6}{6} = 1$

     denominator: $\boxed{4}\ \boxed{x^2}\ \boxed{-}\ \boxed{(}\ \boxed{8}\ \boxed{+}\ \boxed{2}\ \boxed{)}\ \boxed{\text{ENTER}}$ {6}

83. numerator: $\boxed{2}\ \boxed{\times}\ \boxed{(}\ \boxed{4}\ \boxed{+}\ \boxed{2}\ \boxed{\times}\ \boxed{(}\ \boxed{3}\ \boxed{-}\ \boxed{1}\ \boxed{)}\ \boxed{)}\ \boxed{\text{ENTER}}$ {16}

     denominator: $\boxed{3}\ \boxed{\times}\ \boxed{(}\ \boxed{3}\ \boxed{\times}\ \boxed{(}\ \boxed{2}\ \boxed{\times}\ \boxed{3}\ \boxed{-}\ \boxed{4}\ \boxed{)}\ \boxed{)}\ \boxed{\text{ENTER}}$ {18}

     fraction $= \dfrac{16}{18} = \dfrac{8}{9}$

85. numerator: $\boxed{3}\ \boxed{\times}\ \boxed{7}\ \boxed{-}\ \boxed{5}\ \boxed{\times}\ \boxed{(}\ \boxed{3}\ \boxed{\times}\ \boxed{4}\ \boxed{-}\ \boxed{1}\ \boxed{1}\ \boxed{)}\ \boxed{\text{ENTER}}$ {16}

     denominator: $\boxed{4}\ \boxed{\times}\ \boxed{(}\ \boxed{3}\ \boxed{+}\ \boxed{2}\ \boxed{)}\ \boxed{-}\ \boxed{3}\ \boxed{x^2}\ \boxed{+}\ \boxed{5}\ \boxed{\text{ENTER}}$ {16}

     fraction $= \dfrac{16}{16} = 1$

87. $39 = (3 \cdot 8) + (5 \cdot 3)$

89. $87 = (3 \cdot 8 + 5) \cdot 3$

91. $P = 4s = 4(4 \text{ in.}) = 16 \text{ in.}$

93. $P = a + b + c = 3 \text{ m} + 5 \text{ m} + 7 \text{ m}$
$$= 15 \text{ m}$$

95. $A = s^2 = (5 \text{ m})^2 = 25 \text{ m}^2$

97. $A = bh = (6 \text{ ft})(10 \text{ ft}) = 60 \text{ ft}^2$

99. $C = \pi D \approx \dfrac{22}{7}(28 \text{ m}) = \dfrac{22}{7} \cdot \dfrac{28}{1} \text{ m} = \dfrac{22}{\underset{1}{7}} \cdot \dfrac{4 \cdot \overset{1}{7}}{1} \text{ m} = 88 \text{ m}$

101. $A = \pi r^2 \approx \dfrac{22}{7}(21 \text{ ft})^2 = \dfrac{22}{7}\left(441 \text{ ft}^2\right) = \dfrac{22}{7} \cdot \dfrac{441}{1} \text{ ft}^2 = \dfrac{22}{\underset{1}{7}} \cdot \dfrac{63 \cdot \overset{1}{7}}{1} \text{ ft}^2 = 1{,}386 \text{ ft}^2$

103. $V = \frac{1}{3}Bh = \frac{1}{3}(3 \text{ cm})^2(2 \text{ cm}) = \frac{1}{3}(9 \text{ cm}^2)(2 \text{ cm}) = (3 \text{ cm}^2)(2 \text{ cm}) = 6 \text{ cm}^3$

105. $V = \dfrac{4}{3}\pi r^3 \approx \dfrac{4}{3} \cdot \dfrac{22}{7}(6 \text{ m})^3 = \dfrac{88}{21}\left(216 \text{ m}^3\right) = \dfrac{88}{21} \cdot \dfrac{216}{1} \text{ m}^3 \approx 905 \text{ m}^3$

107. Cylinder: $V = Bh = \pi(4 \text{ cm})^2(14 \text{ cm}) \approx \dfrac{22}{7} \cdot \dfrac{16}{1} \text{ cm}^2 \cdot \dfrac{14}{1} \text{ cm} = \dfrac{22 \cdot 16 \cdot 2 \cdot \overset{1}{7}}{\underset{1}{7} \cdot 1 \cdot 1} \text{ cm}^3 = 704 \text{ cm}^3$

Cone: $V = \frac{1}{3}Bh = \frac{1}{3}\pi(4 \text{ cm})^2(21 \text{ cm}) \approx \dfrac{1}{3} \cdot \dfrac{22}{7} \cdot \dfrac{16}{1} \text{ cm}^2 \cdot \dfrac{21}{1} \text{ cm} = \dfrac{22 \cdot 16 \cdot \overset{1}{\cancel{3}} \cdot \overset{1}{\cancel{7}}}{\underset{1}{\cancel{3}} \cdot \underset{1}{\cancel{7}} \cdot 1 \cdot 1} \text{ cm}^3 = 352 \text{ cm}^3$

Total $= 704 \text{ cm}^3 + 352 \text{ cm}^3 = 1{,}056 \text{ cm}^3$

109. $P = 4s = 4\left(30\frac{2}{5} \text{ m}\right) = 4\left(30 + \frac{2}{5} \text{ m}\right) = 4\left(\dfrac{150}{5} + \dfrac{2}{5} \text{ m}\right) = \dfrac{4}{1}\left(\dfrac{152}{5} \text{ m}\right)$
$$= \dfrac{608}{5} \text{ m} = 121\frac{3}{5} \text{ m}$$

111. $V = \frac{4}{3}\pi r^3 = \frac{4}{3}\pi(21.35 \text{ ft})^3 \approx 40{,}764.51 \text{ ft}^3$

113. $V = lwh = (40 \text{ ft})(40 \text{ ft})(9 \text{ ft}) = 14{,}400 \text{ ft}^3$; Per student $= 14{,}400 \text{ ft}^3 \div 30 = 480 \text{ ft}^3$ per student

115. $f = \dfrac{rs}{(r+s)(n-1)} = \dfrac{(8)(12)}{(8+12)(1.6-1)} = \dfrac{96}{(20)(0.6)} = \dfrac{96}{12} = 8$

117. Answers may vary.

119. Increasing powers produce larger numbers.

Exercise 1.4 (page 45)

1. $5 + 3(7 - 2) = 5 + 3(5) = 5 + 15 = 20$

3. $5 + 3(7) - 2 = 5 + 21 - 2 = 26 - 2 = 24$

5. arrows   **7.** subtract, greater   **9.** add, opposite

11. $4 + 8 = +(4 + 8) = +12$ **13.** $(-3) + (-7) = -(3 + 7)$ **15.** $6 + (-4) = +(6 - 4)$
$$= -10$$ $$= +2$$

17. $9 + (-11) = -(11 - 9) = -2$ **19.** $(-0.4) + 0.9 = +(0.9 - 0.4) = +0.55$

21. $\dfrac{1}{5} + \left(+\dfrac{1}{7}\right) = \dfrac{7}{35} + \left(+\dfrac{5}{35}\right) = +\left(\dfrac{7}{35} + \dfrac{5}{35}\right) = +\dfrac{12}{35}$

23. $5 + (-4) = +(5 - 4) = +1$ **25.** $-1.3 + 3.5 = +(3.5 - 1.3) = +2.2$

27. $5 + [4 + (-2)] = 5 + [2] = 7$ **29.** $-2 + (-4 + 5) = -2 + 1 = -1$

31. $(-3 + 5) + 2 = 2 + 2 = 4$ **33.** $-15 + (-4 + 12) = -15 + 8 = -7$

35. $[-4 + (-3)] + [2 + (-2)] = [-7] + [0] = -7$

37. $-4 + (-3 + 2) + (-3) = -4 + (-1) + (-3) = -5 + (-3) = -8$

39. $-|8 + (-4)| + 7 = -|4| + 7 = -4 + 7 = 3$ **41.** $-5.2 + |-2.5 + (-4)| = -5.2 + |-6.5|$
$$= -5.2 + 6.5 = 1.3$$

43. $8 - 4 = 8 + (-4) = 4$ **45.** $8 - (-4) = 8 + (+4) = 12$

47. $0 - (-5) = 0 + (+5) = 5$ **49.** $\dfrac{5}{3} - \dfrac{7}{6} = \dfrac{10}{6} - \dfrac{7}{6} = \dfrac{10}{6} + \left(-\dfrac{7}{6}\right)$
$$= \dfrac{3}{6} = \dfrac{1}{2}$$

51. $-3\dfrac{1}{2} - 5\dfrac{1}{4} = -\dfrac{7}{2} - \dfrac{21}{4} = -\dfrac{14}{4} - \dfrac{21}{4} = -\dfrac{14}{4} + \left(-\dfrac{21}{4}\right) = -\dfrac{35}{4} = -8\dfrac{3}{4}$

53. $-6.7 - (-2.5) = -6.7 + (+2.5) = -4.2$ **55.** $8 - 4 = 8 + (-4) = 4$

57. $-10 - (-3) = -10 + (+3) = -7$

59. $+3 - [(-4) - 3] = +3 - [(-4) + (-3)] = +3 - [-7] = +3 + [+7] = 10$

61. $(5 - 3) + (3 - 5) = [5 + (-3)] + [3 + (-5)] = [2] + [-2] = 0$

63. $5 - [4 + (-2) - 5] = 5 - [4 + (-2) + (-5)] = 5 - [-3] = 5 + [+3] = 8$

65. $\dfrac{5 - (-4)}{3 - (-6)} = \dfrac{5 + (+4)}{3 + (+6)} = \dfrac{9}{9} = 1$ **67.** $\dfrac{-4 - 2}{-[2 + (-3)]} = \dfrac{-4 + (-2)}{-[-1]} = \dfrac{-6}{+1} = -6$

69. $\left(\dfrac{5}{2} - 3\right) - \left(\dfrac{3}{2} - 5\right) = \left(\dfrac{5}{2} - \dfrac{6}{2}\right) - \left(\dfrac{3}{2} - \dfrac{10}{2}\right) = \left(-\dfrac{1}{2}\right) - \left(-\dfrac{7}{2}\right) = \left(-\dfrac{1}{2}\right) + \left(+\dfrac{7}{2}\right)$

$$= \dfrac{6}{2} = 3$$

71. $(5.2 - 2.5) - (5.25 - 5) = [5.2 + (-2.5)] - [5.25 + (-5)] = 2.7 - [0.25] = 2.7 + (-0.25)$

$$= 2.45$$

Problems 73-75 are to be solved using a calculator. The keystrokes needed to solve each problem using a TI-83 graphing calculator appear in each solution. There may be other solutions. Keystrokes for other calculators may be slightly different.

73. $\boxed{2}\ \boxed{.}\ \boxed{3}\ \boxed{4}\ \boxed{\wedge}\ \boxed{3}\ \boxed{-}\ \boxed{3}\ \boxed{.}\ \boxed{4}\ \boxed{7}\ \boxed{+}\ \boxed{.}\ \boxed{7}\ \boxed{2}\ \boxed{x^2}\ \boxed{\text{ENTER}}$

$\{9.861304\} \Rightarrow 9.9$

75. $\boxed{2}\ \boxed{.}\ \boxed{3}\ \boxed{4}\ \boxed{x^2}\ \boxed{-}\ \boxed{3}\ \boxed{.}\ \boxed{4}\ \boxed{7}\ \boxed{x^2}\ \boxed{-}\ \boxed{.}\ \boxed{7}\ \boxed{2}\ \boxed{x^2}\ \boxed{\text{ENTER}}$

$\{-7.0837\} \Rightarrow -7.1$

77. $(-575) + (+400) = -175$ **79.** $(+13) + (-4) = +9$ **81.** $(-14) + 10 = -4°$

She still owes $175.

83. $1700 - (-300) = 2000$ years   **85.** $(-2,300) + (1,750) + (1,875) = +1,325$ m

87. $32,000 - 28,000 = 4,000$ ft   **89.** $+32 - (+27) = 5°$

91. $12,153 - 23 + 57 = 12,187$ **93.** $500 \cdot 2 - 300 = 1000 - 300 = 700$ shares

95. $437.45 + 25.17 + 37.93 + 45.26 - 17.13 - 83.44 - 22.58 = \$422.66$

97. $115,000 - 78 - 446 - 216 - 7,612.32 - 23,445.11 + 223 = \$83,425.57$

99. **Answers may vary.**

101. The answers agree if the two numbers have the same sign. The answers do not agree if the numbers have opposite signs.

Exercise 1.5 (page 53)

1. $30 \cdot 37\dfrac{1}{2} = \dfrac{30}{1} \cdot \dfrac{75}{2} = \dfrac{15 \cdot \overset{1}{\cancel{2}} \cdot 75}{1 \cdot \underset{1}{\cancel{2}}} = 1,125$ lb   **3.** $5^3 - 8(3)^2 = 125 - 8(9) = 125 - 72 = 53$

5. positive   **7.** positive   **9.** positive   **11.** a

13. 0 **15.** $(+6)(+8) = 48$

17. $(-8)(-7) = 56$ **19.** $(+12)(-12) = -144$

21. $\left(\dfrac{1}{2}\right)(-32) = -\dfrac{1}{2} \cdot \dfrac{32}{1} = -\dfrac{32}{2} = -16$

23. $\left(-\dfrac{3}{4}\right)\left(-\dfrac{8}{3}\right) = +\dfrac{3}{4} \cdot \dfrac{8}{3} = \dfrac{24}{12} = 2$

25. $(-3)\left(-\dfrac{1}{3}\right) = +\dfrac{3}{1} \cdot \dfrac{1}{3} = \dfrac{3}{3} = 1$

27. $(3)(-4)(-6) = (-12)(-6) = 72$

29. $(-2)(3)(4) = (-6)(4) = -24$

31. $(2)(-5)(-6)(-7) = (-10)(-6)(-7)$
$= (+60)(-7) = -420$

33. $(-2)(-2)(-2)(-3)(-4) = (+4)(-2)(-3)(-4) = (-8)(-3)(-4) = (+24)(-4) = -96$

35. $(-2)^2 = (-2)(-2) = 4$

37. $(-4)^3 = (-4)(-4)(-4) = (+16)(-4)$
$= -64$

39. $(-7)^2 = (-7)(-7) = 49$

41. $-(-3)^2 = -(-3)(-3) = -(+9) = -9$

43. $(-1)(2^3) = (-1)(8) = -8$

45. $2 + (-1)(-3) = 2 + 3 = 5$

47. $(-1+2)(-3) = 1(-3) = -3$

49. $[-1-(-3)][-1+(-3)] = [-1+3][-4]$
$= [2][-4] = -8$

51. $-1(2) + (2)(-3) = -2 + (-6) = -8$

53. $(-1)(2)(-3) + 6 = (-2)(-3) + 6$
$= 6 + 6 = 12$

55. $(-1)^2[2-(-3)] = (-1)(-1)[2+(+3)] = 1[5] = 5$

57. $(-1)^3(-2)^2 + (-3)^2 = (-1)(-1)(-1)(-2)(-2) + (-3)(-3) = -4 + 9 = 5$

59. $\dfrac{80}{-20} = -4$

61. $\dfrac{-110}{-55} = 2$

63. $\dfrac{-160}{40} = -4$

65. $\dfrac{320}{-16} = -20$

67. $\dfrac{8-12}{-2} = \dfrac{-4}{-2} = 2$

69. $\dfrac{20-25}{7-12} = \dfrac{-5}{-5} = 1$

71. $\dfrac{3(4)}{-2} = \dfrac{12}{-2} = -6$

73. $\dfrac{5(-18)}{3} = \dfrac{-90}{3} = -30$

75. $\dfrac{4+(-18)}{-2} = \dfrac{-14}{-2} = 7$

77. $\dfrac{(-2)(5)(4)}{-3+1} = \dfrac{-40}{-2} = 20$

79. $\dfrac{1}{2} - \dfrac{2}{3} = \dfrac{3}{6} + \left(-\dfrac{4}{6}\right) = -\dfrac{1}{6}$

81. $\dfrac{1}{2} - \dfrac{2}{3} - \dfrac{3}{4} = \dfrac{6}{12} + \left(-\dfrac{8}{12}\right) + \left(-\dfrac{9}{12}\right)$
$= -\dfrac{11}{12}$

83. $\left(\dfrac{1}{2} - \dfrac{2}{3}\right)\left(\dfrac{1}{2} + \dfrac{2}{3}\right) = \left(\dfrac{3}{6} - \dfrac{4}{6}\right)\left(\dfrac{3}{6} + \dfrac{4}{6}\right) = \left(-\dfrac{1}{6}\right)\left(\dfrac{7}{6}\right) = -\dfrac{7}{36}$

Problems 85-87 are to be solved using a calculator. The keystrokes needed to solve each problem using a TI-83 graphing calculator appear in each solution. There may be other solutions. Keystrokes for other calculators may be slightly different.

85. (((−) 6) ^ 3 + 4 × ((−) 3) ^ 3) ÷ (4 − 6)) x^2 ENTER {−81}

87. (4 × ((−) 6)) x^2 × (−) 3 + 4 x^2 × (−) 6) ÷ (2 × (−) 6 − 2 × (−) 3) ENTER {88}

89. $(+2)(+3) = +6$

91. $(-30)(+15) = -450$

93. $(+23)(-120) = -2{,}760$

95. $\dfrac{-18}{-3} = +6$

97. $\dfrac{(+26) + (+35) + (+17) + (-25) + (-31) + (-12) + (-24)}{7} = \dfrac{-14}{7} = -2 \text{ per day}$

99. $613.50(18) = \$11{,}043 \Rightarrow$ enough $

101. **Answers may vary.**

103. If the quotient is undefined, then the denominator must equal 0, and the product of the two numbers is 0.

105. If x^5 is negative, then x must be negative.

Exercise 1.6 (page 61)

1. $0.14 \cdot 3{,}800 = 532$

3. $\dfrac{-4 + (7 - 9)}{(-9 - 7) + 4} = \dfrac{-4 + (-2)}{-16 + 4} = \dfrac{-6}{-12} = \dfrac{1}{2}$

5. sum

7. multiplication

9. algebraic

11. variables

13. $x + y$

15. $x(2y)$

17. $y - x$

19. $\dfrac{y}{x}$

21. $z + \dfrac{x}{y}$

23. $z - xy$

25. $3xy$

27. $\dfrac{x + y}{y + z}$

29. $xy + \dfrac{y}{z}$

31. the sum of x and 3

33. the quotient obtained when x is divided by y

35. the product of 2, x and y (or twice the product of x and y)

37. the quotient obtained when 5 is divided by the sum of x and y

39. the quotient obtained when the sum of 3 and x is divided by y

11

41. the product of x, y and the sum of x and y

43. $x + z = 8 + 2 = 10$

45. $y - z = 4 - 2 = 2$

47. $yz - 3 = (4)(2) - 3 = 5$

49. $\dfrac{xy}{z} = \dfrac{(8)(4)}{2} = 16$

51. $x + y = (-2) + 5 = 3$

53. $xyz = (-2)(5)(-3) = (-10)(-3) = 30$

55. $\dfrac{yz}{x} = \dfrac{(5)(-3)}{-2} = \dfrac{-15}{-2} = \dfrac{15}{2}$

57. $\dfrac{3(x+z)}{y} = \dfrac{3[(-2)+(-3)]}{5} = \dfrac{3[-5]}{5} = \dfrac{-15}{5} = -3$

59. $\dfrac{3(x+z^2)+4}{y(x-z)} = \dfrac{3\left[(-2)+(-3)^2\right]+4}{5[(-2)-(-3)]} = \dfrac{3(-2+9)+4}{5(1)} = \dfrac{3(7)+4}{5} = \dfrac{21+4}{5} = \dfrac{25}{5} = 5$

61. $6d$: 1 term; coef. $= 6$

63. $-xy - 4t + 35$: 3 terms; coef. $= -1$

65. $3ab + bc - cd - ef$: 4 terms; coef. $= 3$

67. $-4xyz + 7xy - z$: 3 terms; coef. $= -4$

69. $3x + 4y + 2z + 2$: 4 terms; coef. $= 3$

71. 3rd term: $19x$; factors: $19, x$

73. 1st term: $29xyz$; factors: $29, x, y, z$

75. 1st term: $3xyz$; factors: $3, x, y, z$

77. 3rd term: $17xz$; factors: $17, x, z$

79. coefficients: $5, 1$ and 8

81. x and y are common to the 1st and 3rd terms.

83. coefficients: $3, 1$ and 25; $3 \cdot 1 \cdot 25 = 75$

85. x and y are common to the 1st and 3rd terms.

87. $c + 4$

89. $9{,}987t$

91. $\dfrac{x}{5}$

93. $3d + 5$

95. $\dfrac{N(N-1)}{2} = \dfrac{10{,}000(10{,}000 - 1)}{2} = \dfrac{10{,}000(9{,}999)}{2} = \dfrac{99{,}990{,}000}{2} = 49{,}995{,}000$ comparisons

97. **Answers may vary.**

99. **Answers may vary.**

101. $37x \Rightarrow 37(2x)$
$37(2x) = 2(37x)$
$37x$ is doubled.

Exercise 1.7 (page 68)

1. $x + y^2 \geq z$

3. $|x| \geq \boxed{0}$

5. positive

7. real

9. $a + b = b + \underline{a}$

11. $(a + b) + c = a + \underline{(b+c)}$

13. $a(b+c) = ab + \underline{ac}$

15. $a \cdot 1 = \underline{a}$

17. element, multiplication

19. $a, \dfrac{1}{a}$

21. $x + y = 12 + (-2) = 10$

23. $xy = 12(-2) = -24$

25. $x^2 = 12^2 = 144$

27. $\dfrac{x}{y^2} = \dfrac{12}{(-2)^2} = \dfrac{12}{4} = 3$

29. $x + y = 5 + 7 = 12$
$y + x = 7 + 5 = 12$

31. $3x + 2y = 3(5) + 2(7) = 15 + 14 = 29$
$2y + 3x = 2(7) + 3(5) = 14 + 15 = 29$

33. $x(x + y) = 5(5 + 7) = 5(12) = \boxed{60}$; $(x + y)x = (5 + 7)5 = (12)5 = \boxed{60}$

35. $(x + y) + z = [2 + (-3)] + 1 = -1 + 1 = \boxed{0}$; $x + (y + z) = 2 + (-3 + 1) = 2 + (-2) = \boxed{0}$

37. $(xz)y = [2(1)](-3) = [2](-3) = \boxed{-6}$; $x(yz) = 2[-3(1)] = 2[-3] = \boxed{-6}$

39. $x^2(yz^2) = 2^2\left[-3(1)^2\right] = 4[-3(1)] = 4[-3] = \boxed{-12}$
$(x^2 y)z^2 = \left[2^2(-3)\right](1)^2 = [4(-3)](1) = [-12](1) = \boxed{-12}$

41. $3(x + y) = 3x + 3y$

43. $x(x + 3) = x \cdot x + x \cdot 3 = x^2 + 3x$

45. $-x(a + b) = (-x)a + (-x)b = -ax - bx$

47. $4(x^2 + x) = 4x^2 + 4x$

49. $-5(t + 2) = -5t + (-5)(2) = -5t - 10$

51. $-2a(x + a) = -2ax + (-2a)a$
$\qquad\qquad\qquad = -2ax - 2a^2$

53. additive inverse: -2
multiplicative inverse: $\dfrac{1}{2}$

55. additive inverse: $-\dfrac{1}{3}$
multiplicative inverse: 3

57. additive inverse: 0
multiplicative inverse: none

59. additive inverse: $\dfrac{5}{2}$
multiplicative inverse: $-\dfrac{2}{5}$

61. additive inverse: 0.2
multiplicative inverse: -5

63. additive inverse: $-\dfrac{4}{3}$
multiplicative inverse: $\dfrac{3}{4}$

65. commutative property of addition

67. commutative property of multiplication

69. distributive property

71. commutative property of addition

73. identity for multiplication

75. additive inverse

77. $3(x + 2) = 3x + 3(2) = 3x + 6$

79. $y^2 x = xy^2$

81. $(x + y)z = (y + x)z$

83. $(xy)z = x(yz)$

85. $0 + x = x$

87. **Answers may vary.**

89. Closure for addition would not be true (odd number plus odd number equals even number).
Closure for multiplication would be true (odd number times odd number equals odd number).
There would be no additive identity (0 is an even number).
There would be a multiplicative identity, 1 (1 is an odd number).

Chapter 1 Summary   (page 70)

1. natural: $1, 2, 3, 4, 5$

2. prime: $2, 3, 5$

3. odd, natural: $1, 3, 5$

4. composite: 4

5. integers: $-6, 0, 5$

6. rational: $-6, -\dfrac{2}{3}, 0, 2.6, 5$

7. prime: 5

8. real: $-6, -\dfrac{2}{3}, 0, \sqrt{2}, 2.6, \pi, 5$

9. even integers: $-6, 0$

10. odd integers: 5

11. not rational: $\sqrt{2}, \pi$

12. $-5 \boxed{\phantom{<}} 12 - 12$
$-5 \boxed{<} 0$

13. $\dfrac{24}{6} \boxed{\phantom{<}} 5$
$4 \boxed{<} 5$

14. $13 - 13 \boxed{\phantom{<}} 5 - \dfrac{25}{5}$
$0 \boxed{\phantom{<}} 5 - 5$
$0 \boxed{=} 0$

15. $\dfrac{21}{7} \boxed{} -33$
$3 \boxed{>} -33$

16. $-(-8) = +8$

17. $-(12 - 4) = -(8) = -8$

18.

19.

20.

$-3 \qquad 2$

21.

$-4 \qquad 3$

22. $|53 - 42| = |11| = 11$

23. $|-31| = 31$

24. $\dfrac{45}{27} = \dfrac{5 \cdot \cancel{9}}{3 \cdot \cancel{9}} = \dfrac{5}{3}$

25. $\dfrac{121}{11} = \dfrac{11 \cdot \cancel{11}}{1 \cdot \cancel{11}} = \dfrac{11}{1} = 11$

26. $\dfrac{31}{15} \cdot \dfrac{10}{62} = \dfrac{\cancel{31} \cdot \cancel{2} \cdot \cancel{5}}{3 \cdot \cancel{5} \cdot 2 \cdot \cancel{31}} = \dfrac{1}{3}$

27. $\dfrac{25}{36} \cdot \dfrac{12}{15} \cdot \dfrac{3}{5} = \dfrac{\cancel{5} \cdot \cancel{5} \cdot \cancel{12} \cdot \cancel{3}}{3 \cdot \cancel{12} \cdot 3 \cdot \cancel{5} \cdot \cancel{5}} = \dfrac{1}{3}$

28. $\dfrac{18}{21} \div \dfrac{6}{7} = \dfrac{18}{21} \cdot \dfrac{7}{6} = \dfrac{\overset{1}{\cancel{3}} \cdot \overset{1}{\cancel{6}} \cdot \overset{1}{\cancel{7}}}{\underset{1}{3} \cdot \underset{1}{7} \cdot \underset{1}{\cancel{6}}} = \dfrac{1}{1} = 1$

29. $\dfrac{14}{24} \div \dfrac{7}{12} \div \dfrac{2}{5} = \dfrac{14}{24} \cdot \dfrac{12}{7} \cdot \dfrac{5}{2}$

$= \dfrac{\overset{1}{\cancel{2}} \cdot \overset{1}{\cancel{7}} \cdot \overset{1}{\cancel{12}} \cdot 5}{\underset{1}{2} \cdot \underset{1}{12} \cdot \underset{1}{7} \cdot 2} = \dfrac{5}{2}$

30. $\dfrac{7}{12} + \dfrac{9}{12} = \dfrac{7+9}{12} = \dfrac{16}{12} = \dfrac{4 \cdot \overset{1}{\cancel{4}}}{3 \cdot \underset{1}{\cancel{4}}} = \dfrac{4}{3}$

31. $\dfrac{13}{24} - \dfrac{5}{24} = \dfrac{13-5}{24} = \dfrac{8}{24} = \dfrac{\overset{1}{\cancel{8}}}{3 \cdot \underset{1}{\cancel{8}}} = \dfrac{1}{3}$

32. $\dfrac{1}{3} + \dfrac{1}{7} = \dfrac{1 \cdot \mathbf{7}}{3 \cdot \mathbf{7}} + \dfrac{1 \cdot \mathbf{3}}{7 \cdot \mathbf{3}} = \dfrac{7}{21} + \dfrac{3}{21}$

$= \dfrac{7+3}{21} = \dfrac{10}{21}$

33. $\dfrac{5}{7} + \dfrac{4}{9} = \dfrac{5 \cdot \mathbf{9}}{7 \cdot \mathbf{9}} + \dfrac{4 \cdot \mathbf{7}}{9 \cdot \mathbf{7}} = \dfrac{45}{63} + \dfrac{28}{63}$

$= \dfrac{45+28}{63} = \dfrac{73}{63}$

34. $\dfrac{2}{3} - \dfrac{1}{7} = \dfrac{2 \cdot \mathbf{7}}{3 \cdot \mathbf{7}} - \dfrac{1 \cdot \mathbf{3}}{7 \cdot \mathbf{3}} = \dfrac{14}{21} - \dfrac{3}{21}$

$= \dfrac{14-3}{21} = \dfrac{11}{21}$

35. $\dfrac{4}{5} - \dfrac{2}{3} = \dfrac{4 \cdot \mathbf{3}}{5 \cdot \mathbf{3}} - \dfrac{2 \cdot \mathbf{5}}{3 \cdot \mathbf{5}} = \dfrac{12}{15} - \dfrac{10}{15}$

$= \dfrac{12-10}{15} = \dfrac{2}{15}$

36. $3\dfrac{2}{3} + 5\dfrac{1}{4} = \dfrac{11}{3} + \dfrac{21}{4} = \dfrac{11 \cdot \mathbf{4}}{3 \cdot \mathbf{4}} + \dfrac{21 \cdot \mathbf{3}}{4 \cdot \mathbf{3}} = \dfrac{44}{12} + \dfrac{63}{12} = \dfrac{44+63}{12} = \dfrac{107}{12} = 8\dfrac{11}{12}$

37. $7\dfrac{5}{12} - 4\dfrac{1}{2} = \dfrac{89}{12} - \dfrac{9}{2} = \dfrac{89}{12} - \dfrac{9 \cdot \mathbf{6}}{2 \cdot \mathbf{6}} = \dfrac{89}{12} - \dfrac{54}{12} = \dfrac{89-54}{12} = \dfrac{35}{12} = 2\dfrac{11}{12}$

38. $32.71 + 15.9 = 48.61$

39. $27.92 - 14.93 = 12.99$

40. $5.3 \cdot 3.5 = 18.55$

41. $21.83 \div 5.9 = 3.7$

42. $2.7(4.92 - 3.18) = 2.7(1.74) \approx 4.70$

43. $\dfrac{3.3 + 2.5}{0.22} = \dfrac{5.8}{0.22} \approx 26.36$

44. $\dfrac{12.5}{14.7 - 11.2} = \dfrac{12.5}{3.5} \approx 3.57$

45. $(3 - 0.7)(3.63 - 2) = (2.3)(1.63) \approx 3.75$

46. avg. $= \dfrac{5.2 + 4.7 + 9.5 + 8}{4} = \dfrac{27.4}{4}$

$= 6.85$ hours

47. $0.15(380) = 57$

48. Front/Back: $2(2.7 + 2.7 + 4.2) = 2(9.6) = 19.2$ ft. TOTAL $= 19.2 + 13.2 + 7.8 = 40.2$ ft.
Top/Bottom: $2(1.2 + 1.2 + 4.2) = 2(6.6) = 13.2$ ft.
Sides: $2(1.2 + 2.7) = 2(3.9) = 7.8$ ft.

49. $3^4 = 3 \cdot 3 \cdot 3 \cdot 3 = 81$

50. $\left(\dfrac{2}{3}\right)^2 = \dfrac{2}{3} \cdot \dfrac{2}{3} = \dfrac{4}{9}$

51. $(0.5)^2 = (0.5)(0.5) = 0.25$

52. $5^2 + 2^3 = 5 \cdot 5 + 2 \cdot 2 \cdot 2 = 25 + 8 = 33$

53. $3^2 + 4^2 = 9 + 16 = 25$

54. $(3 + 4)^2 = 7^2 = 49$

55. $V = Bh = \pi r^2 h = \pi \left(\dfrac{32.1}{2} \text{ ft}\right)^2 (18.7 \text{ ft}) = \pi (257.6025 \text{ ft}^2)(18.7 \text{ ft}) \approx 15{,}133.6 \text{ ft}^3$

56. $5 + 3^3 = 5 + 27 = 32$

57. $7 \cdot 2 - 7 = 14 - 7 = 7$

58. $4 + (8 \div 4) = 4 + 2 = 6$

59. $(4 + 8) \div 4 = 12 \div 4 = 3$

60. $5^3 - \dfrac{81}{3} = 125 - 27 = 98$

61. $(5 - 2)^2 + 5^2 + 2^2 = 3^2 + 5^2 + 2^2$
$= 9 + 25 + 4 = 38$

62. $\dfrac{4 \cdot 3 + 3^4}{31} = \dfrac{12 + 81}{31} = \dfrac{93}{31} = 3$

63. $\dfrac{4}{3} \cdot \dfrac{9}{2} + \dfrac{1}{2} \cdot 18 = \dfrac{2 \cdot \overset{1}{\cancel{2}} \cdot 3 \cdot \overset{1}{\cancel{3}}}{\underset{1}{\cancel{3}} \cdot \underset{1}{\cancel{2}}} + \dfrac{1 \cdot 9 \cdot \overset{1}{\cancel{2}}}{\underset{1}{\cancel{2}} \cdot 1}$
$= \dfrac{6}{1} + \dfrac{9}{1} = 15$

64. $8^2 - 6 = 64 - 6 = 58$

65. $(8 - 6)^2 = 2^2 = 4$

66. $\dfrac{6 + 8}{6 - 4} = \dfrac{14}{2} = 7$

67. $\dfrac{6(8) - 12}{4 + 8} = \dfrac{48 - 12}{12} = \dfrac{36}{12} = 3$

68. $2^2 + 2(3)^2 = 4 + 2(9) = 4 + 18 = 22$

69. $\dfrac{2^2 + 3}{2^3 - 1} = \dfrac{4 + 3}{8 - 1} = \dfrac{7}{7} = 1$

70. $(+7) + (+8) = +(7 + 8) = 15$

71. $(-25) + (-32) = -(25 + 32) = -57$

72. $(-2.7) + (-3.8) = -(2.7 + 3.8) = -6.5$

73. $\dfrac{1}{3} + \dfrac{1}{6} = \dfrac{2}{6} + \dfrac{1}{6} = \dfrac{3}{6} = \dfrac{1}{2}$

74. $(+12) + (-24) = -(24 - 12) = -12$

75. $(-44) + (+60) = +(60 - 44) = 16$

76. $3.7 + (-2.5) = +(3.7 - 2.5) = 1.2$

77. $-5.6 + (+2.06) = -(5.6 - 2.06)$
$= -3.54$

78. $15 - (-4) = 15 + (+4) = 19$

79. $-12 - (-13) = -12 + (+13) = 1$

80. $[-5 + (-5)] - (-5) = [-10] + (+5) = -5$

81. $1 - [5 - (-3)] = 1 - [5 + (+3)]$
$= 1 - [8] = -7$

82. $\dfrac{5}{6} - \left(-\dfrac{2}{3}\right) = \dfrac{5}{6} + \dfrac{2}{3} = \dfrac{5}{6} + \dfrac{4}{6} = \dfrac{9}{6} = \dfrac{3}{2}$

83. $\dfrac{2}{3} - \left(\dfrac{1}{3} - \dfrac{2}{3}\right) = \dfrac{2}{3} - \left(-\dfrac{1}{3}\right)$

$= \dfrac{2}{3} + \dfrac{1}{3} = \dfrac{3}{3} = 1$

84. $\left|\dfrac{3}{7} - \left(-\dfrac{4}{7}\right)\right| = \left|\dfrac{3}{7} + \dfrac{4}{7}\right| = \left|\dfrac{7}{7}\right| = |1| = 1$

85. $\dfrac{3}{7} - \left|-\dfrac{4}{7}\right| = \dfrac{3}{7} - \left(+\dfrac{4}{7}\right)$

$= \dfrac{3}{7} + \left(-\dfrac{4}{7}\right) = -\dfrac{1}{7}$

86. $(+3)(+4) = +12$

87. $(-5)(-12) = +60$

88. $\left(-\dfrac{3}{14}\right)\left(-\dfrac{7}{6}\right) = +\dfrac{3}{14} \cdot \dfrac{7}{6}$

$= \dfrac{\overset{1}{\cancel{3}} \cdot \overset{1}{\cancel{7}}}{2 \cdot 7 \cdot 2 \cdot \underset{1}{\cancel{3}}} = +\dfrac{1}{4}$

89. $(3.75)(0.37) = +1.3875$

90. $5(-7) = -35$

91. $(-15)(7) = -105$

92. $\left(-\dfrac{1}{2}\right)\left(\dfrac{4}{3}\right) = -\dfrac{1}{2} \cdot \dfrac{4}{3} = -\dfrac{1 \cdot 2 \cdot \overset{1}{\cancel{2}}}{\underset{1}{\cancel{2}} \cdot 3} = -\dfrac{2}{3}$

93. $(-12.2)(3.7) = -45.14$

94. $\dfrac{+25}{+5} = +5$

95. $\dfrac{-14}{-2} = +7$

96. $\dfrac{(-2)(-7)}{4} = \dfrac{+14}{4} = +\dfrac{7 \cdot \overset{1}{\cancel{2}}}{2 \cdot \underset{1}{\cancel{2}}} = +\dfrac{7}{2}$

97. $\dfrac{-22.5}{-3.75} = +6$

98. $\dfrac{-25}{5} = -5$

99. $\dfrac{(-3)(-4)}{(-6)} = \dfrac{+12}{-6} = -2$

100. $\left(\dfrac{-10}{2}\right)^2 - (-1)^3 = (-5)^2 - (-1)^3$

$= 25 - (-1)$

$= 25 + 1 = 26$

101. $\dfrac{[-3 + (-4)]^2}{10 + (-3)} = \dfrac{[-7]^2}{7} = \dfrac{49}{7} = 7$

102. $\left(\dfrac{-3 + (-3)}{3}\right)\left(\dfrac{-15}{5}\right) = \left(\dfrac{-6}{3}\right)\left(\dfrac{-15}{5}\right)$

$= (-2)(-3) = 6$

103. $\dfrac{-2 - (-8)}{5 + (-1)} = \dfrac{-2 + (+8)}{4} = \dfrac{6}{4} = \dfrac{3}{2}$

104. xz

105. $x + 2y$

106. $2(x + y)$

107. $x - yz$

108. the product of 3, x and y

109. 5 decreased by the product of y and z

110. 5 less than the product of y and z

111. the quotient obtained when the sum of x, y and z is divided by twice their product

112. $y + z = -3 + (-1) = -4$

113. $x + y = 2 + (-3) = -1$

114. $x + (y + z) = 2 + [-3 + (-1)]$
$= 2 + [-4] = -2$

115. $x - y = 2 - (-3) = 2 + (+3) = 5$

116. $x - (y - z) = 2 - [-3 - (-1)]$
$= 2 - [-3 + (+1)]$
$= 2 - [-2] = 2 + (+2) = 4$

117. $(x - y) - z = [2 - (-3)] - (-1)$
$= [2 + (+3)] + (+1)$
$= 5 + 1 = 6$

118. $xy = (2)(-3) = -6$

119. $yz = (-3)(-1) = 3$

120. $x(x + z) = 2[2 + (-1)] = 2[1] = 2$

121. $xyz = (2)(-3)(-1) = (-6)(-1) = 6$

122. $y^2z + x = (-3)^2(-1) + 2$
$= 9(-1) + 2 = -9 + 2 = -7$

123. $yz^3 + (xy)^2 = (-3)(-1)^3 + [2(-3)]^2$
$= (-3)(-1) + [-6]^2$
$= 3 + 36 = 39$

124. $\dfrac{xy}{z} = \dfrac{(2)(-3)}{(-1)} = \dfrac{-6}{-1} = 6$

125. $\dfrac{|xy|}{3z} = \dfrac{|2(-3)|}{3(-1)} = \dfrac{|-6|}{-3} = \dfrac{6}{-3} = -2$

126. three terms

127. 7

128. 1

129. $2 + 4 + 3 = 9$

130. closure property

131. commutative property of multiplication

132. associative property of addition

133. distributive property

134. commutative property of addition

135. associative property of multiplication

136. commutative property of addition

137. identity for multiplication

138. additive inverse

139. identity for addition

Chapter 1 Test (page 76)

1. $31, 37, 41, 43, 47$

2. 2

3.

4.

5. $-|23| = -(+23) = -23$

6. $-|7| + |-7| = -(+7) + (+7) = -7 + 7 = 0$

7.
$$3(4-2) \boxed{} -2(2-5)$$
$$3(2) \boxed{} -2(-3)$$
$$6 \boxed{=} 6$$

8.
$$1 + 4 \cdot 3 \boxed{\phantom{<}} -2(-7)$$
$$1 + 12 \boxed{\phantom{<}} +14$$
$$13 \boxed{<} 14$$

9.   25% of 136 $\boxed{}$ $\dfrac{1}{2}$ of 66
$$0.25(136) \boxed{} \dfrac{1}{2}(66)$$
$$34 \boxed{>} 33$$

10.
$$-13.7 \boxed{} -|-13.7|$$
$$-13.7 \boxed{} -(+13.7)$$
$$-13.7 \boxed{=} -13.7$$

11. $\dfrac{26}{40} = \dfrac{13 \cdot \overset{1}{\cancel{2}}}{20 \cdot \underset{1}{\cancel{2}}} = \dfrac{13}{20}$

12. $\dfrac{7}{8} \cdot \dfrac{24}{21} = \dfrac{\overset{1}{\cancel{7}} \cdot \overset{1}{\cancel{3}} \cdot \overset{1}{\cancel{8}}}{\underset{1}{\cancel{8}} \cdot \underset{1}{\cancel{3}} \cdot \underset{1}{\cancel{7}}} = \dfrac{1}{1} = 1$

13. $\dfrac{18}{35} \div \dfrac{9}{14} = \dfrac{18}{35} \cdot \dfrac{14}{9} = \dfrac{2 \cdot \overset{1}{\cancel{9}} \cdot 2 \cdot \overset{1}{\cancel{7}}}{5 \cdot \underset{1}{\cancel{7}} \cdot \underset{1}{\cancel{9}}} = \dfrac{4}{5}$

14. $\dfrac{24}{16} + 3 = \dfrac{3 \cdot \overset{1}{\cancel{8}}}{2 \cdot \underset{1}{\cancel{8}}} + \dfrac{3}{1} = \dfrac{3}{2} + \dfrac{3 \cdot \mathbf{2}}{1 \cdot \mathbf{2}} = \dfrac{3}{2} + \dfrac{6}{2} = \dfrac{3+6}{2} = \dfrac{9}{2} \left(\text{or } 4\dfrac{1}{2}\right)$

15. $\dfrac{17-5}{36} - \dfrac{2(13-5)}{12} = \dfrac{12}{36} - \dfrac{2(8)}{12} = \dfrac{12}{36} - \dfrac{16}{12} = \dfrac{\overset{1}{\cancel{12}}}{3 \cdot \underset{1}{\cancel{12}}} - \dfrac{4 \cdot \overset{1}{\cancel{4}}}{3 \cdot \underset{1}{\cancel{4}}} = \dfrac{1}{3} - \dfrac{4}{3} = \dfrac{1-4}{3} = \dfrac{-3}{3} = -1$

16. $\dfrac{|-7-(-6)|}{-7-|-6|} = \dfrac{|-7+(+6)|}{-7-(+6)} = \dfrac{|-1|}{-7+(-6)} = \dfrac{1}{-13} = -\dfrac{1}{13}$

17. $0.17(457) = 77.69 \approx 77.7$

18. $A = lw = (23.56 \text{ ft})(12.8 \text{ ft}) = 301.568 \text{ ft}^2$
$$\approx 301.57 \text{ ft}^2$$

19. $A = \frac{1}{2}bh = \frac{1}{2}(16 \text{ cm})(8 \text{ cm}) = \frac{1}{2}(128 \text{ cm}^2)$
$$= 64 \text{ cm}^2$$

20. $V = Bh = \pi r^2 h = \pi(7 \text{ in.})^2(10 \text{ in.})$
$$= \pi(49 \text{ in.}^2)(10 \text{ in.})$$
$$= \pi(490 \text{ in.}^3) \approx 1{,}539 \text{ in.}^3$$

21. $xy + z = (-2)(3) + 4 = -6 + 4 = -2$

22. $x(y+z) = -2(3+4) = -2(7) = -14$

23. $\dfrac{z+4y}{2x} = \dfrac{4+4(3)}{2(-2)} = \dfrac{4+12}{-4} = \dfrac{16}{-4} = -4$

24. $|x^3 - z| = |(-2)^3 - 4| = |-8 - 4| = |-12|$
$$= 12$$

25. $x^3 + y^2 + z = (-2)^3 + (3)^2 + 4$
$$= -8 + 9 + 4 = 5$$

26. $|x| - 3|y| - 4|z| = |-2| - 3|3| - 4|4|$
$$= 2 - 3(3) - 4(4)$$
$$= 2 - 9 - 16 = -23$$

27. $\dfrac{xy}{x+y}$

28. $5y - (x + y)$

29. $x(12 + 12) + y(7 + 7) = 24x + 14y$

30. $12a + 8b$

31. 3

32. 4 terms

33. 0

34. 5

35. commutative property of multiplication

36. distributive property

37. commutative property of addition

38. multiplicative inverse property

Exercise 2.1 (page 89)

1. $\dfrac{4}{5} + \dfrac{2}{3} = \dfrac{4 \cdot 3}{5 \cdot 3} + \dfrac{2 \cdot 5}{3 \cdot 5} = \dfrac{12}{15} + \dfrac{10}{15} = \dfrac{22}{15}$

3. $\dfrac{5}{9} \div \dfrac{3}{5} = \dfrac{5}{9} \cdot \dfrac{5}{3} = \dfrac{25}{27}$

5. $2 + 3 \cdot 4 = 2 + 12 = 14$

7. $3 + 4^3(-5) = 3 + 64(-5) = 3 + (-320)$
$\qquad\qquad = -317$

9. equation

11. equivalent

13. equal

15. equal

17. regular price

19. 100

21. $x = 2$
equation

23. $7x < 8$
not an equation

25. $x + 7 = 0$
equation

27. $1 + 1 = 3$
equation

29. $x + 2 = 3$
$1 + 2 \overset{?}{=} 3$
$3 = 3$
1 is a solution.

31. $a - 7 = 0$
$-7 - 7 \overset{?}{=} 0$
$-14 \neq 0$
-7 is not a solution.

33. $\dfrac{y}{7} = 4$
$\dfrac{28}{7} \overset{?}{=} 4$
$4 = 4$
28 is a solution.

35. $\dfrac{x}{5} = x$
$\dfrac{0}{5} \overset{?}{=} 0$
$0 = 0$
0 is a solution.

37. $3k + 5 = 5k - 1$
$3(3) + 5 \overset{?}{=} 5(3) - 1$
$9 + 5 \overset{?}{=} 15 - 1$
$14 = 14$
3 is a solution.

39. $\dfrac{5 + x}{10} - x = \dfrac{1}{2}$
$\dfrac{5 + 0}{10} - 0 \overset{?}{=} \dfrac{1}{2}$
$\dfrac{5}{10} - 0 \overset{?}{=} \dfrac{1}{2}$
$\dfrac{1}{2} = \dfrac{1}{2}$
0 is a solution.

41.
$$x + 7 = 13$$
$$x + 7 - \mathbf{7} = 13 - \mathbf{7}$$
$$x = 6$$

$$x + 7 = 13$$
$$6 + 7 \stackrel{?}{=} 13$$
$$13 = 13$$

43.
$$y - 7 = 12$$
$$y - 7 + \mathbf{7} = 12 + \mathbf{7}$$
$$y = 19$$

$$y - 7 = 12$$
$$19 - 7 \stackrel{?}{=} 12$$
$$12 = 12$$

45.
$$1 = y - 5$$
$$1 + \mathbf{5} = y - 5 + \mathbf{5}$$
$$6 = y$$

$$1 = y - 5$$
$$1 \stackrel{?}{=} 6 - 5$$
$$1 = 1$$

47.
$$p - 404 = 115$$
$$p - 404 + \mathbf{404} = 115 + \mathbf{404}$$
$$p = 519$$

Check:
$$p - 404 = 115$$
$$519 - 404 \stackrel{?}{=} 115$$
$$115 = 115$$

49.
$$-37 + z = 37$$
$$-37 + \mathbf{37} + z = 37 + \mathbf{37}$$
$$z = 74$$

Check:
$$-37 + z = 37$$
$$-37 + 74 \stackrel{?}{=} 37$$
$$37 = 37$$

51.
$$-57 = b - 29$$
$$-57 + \mathbf{29} = b - 29 + \mathbf{29}$$
$$-28 = b$$

Check:
$$-57 = b - 29$$
$$-57 \stackrel{?}{=} -28 - 29$$
$$-57 = -57$$

53.
$$\frac{4}{3} = -\frac{2}{3} + x$$
$$\frac{4}{3} + \mathbf{\frac{2}{3}} = -\frac{2}{3} + \mathbf{\frac{2}{3}} + x$$
$$\frac{6}{3} = x, \text{ or } x = 2$$

Check:
$$\frac{4}{3} = -\frac{2}{3} + x$$
$$\frac{4}{3} \stackrel{?}{=} -\frac{2}{3} + 2$$
$$\frac{4}{3} \stackrel{?}{=} -\frac{2}{3} + \frac{6}{3}$$
$$\frac{4}{3} = \frac{4}{3}$$

55.
$$d + \frac{2}{3} = \frac{3}{2}$$
$$d + \frac{2}{3} - \mathbf{\frac{2}{3}} = \frac{3}{2} - \mathbf{\frac{2}{3}}$$
$$d = \frac{9}{6} - \frac{4}{6} = \frac{5}{6}$$

Check:
$$d + \frac{2}{3} = \frac{3}{2}$$
$$\frac{5}{6} + \frac{2}{3} \stackrel{?}{=} \frac{3}{2}$$
$$\frac{5}{6} + \frac{4}{6} \stackrel{?}{=} \frac{3}{2}$$
$$\frac{9}{6} \stackrel{?}{=} \frac{3}{2}$$
$$\frac{3}{2} = \frac{3}{2}$$

57.
$$-\frac{3}{5} = x - \frac{2}{5}$$
$$-\frac{3}{5} + \frac{2}{5} = x - \frac{2}{5} + \frac{2}{5}$$
$$-\frac{1}{5} = x$$

Check: $\quad -\frac{3}{5} = x - \frac{2}{5}$
$$-\frac{3}{5} \stackrel{?}{=} -\frac{1}{5} - \frac{2}{5}$$
$$-\frac{3}{5} = -\frac{3}{5}$$

59.
$$r - \frac{1}{5} = \frac{3}{10}$$
$$r - \frac{1}{5} + \frac{1}{5} = \frac{3}{10} + \frac{1}{5}$$
$$r = \frac{3}{10} + \frac{2}{10} = \frac{5}{10} = \frac{1}{2}$$

Check: $\quad r - \frac{1}{5} = \frac{3}{10}$
$$\frac{1}{2} - \frac{1}{5} \stackrel{?}{=} \frac{3}{10}$$
$$\frac{5}{10} - \frac{2}{10} \stackrel{?}{=} \frac{3}{10}$$
$$\frac{3}{10} = \frac{3}{10}$$

61.
$$\frac{x}{5} = 5 \qquad \frac{x}{5} = 5$$
$$5 \cdot \frac{x}{5} = 5 \cdot 5 \qquad \frac{25}{5} \stackrel{?}{=} 5$$
$$x = 25 \qquad 5 = 5$$

63.
$$\frac{x}{32} = -2 \qquad \frac{x}{32} = -2$$
$$32 \cdot \frac{x}{32} = 32 \cdot (-2) \qquad \frac{-64}{32} \stackrel{?}{=} -2$$
$$x = -64 \qquad -2 = -2$$

65.
$$\frac{b}{3} = 5 \qquad \frac{b}{3} = 5$$
$$3 \cdot \frac{b}{3} = 3 \cdot 5 \qquad \frac{15}{3} \stackrel{?}{=} 5$$
$$b = 15 \qquad 5 = 5$$

67.
$$-3 = \frac{s}{11} \qquad -3 = \frac{s}{11}$$
$$11 \cdot (-3) = 11 \cdot \frac{s}{11} \qquad -3 \stackrel{?}{=} \frac{-33}{11}$$
$$-33 = s \qquad -3 = -3$$

69.
$$6x = 18 \qquad 6x = 18$$
$$\frac{6x}{6} = \frac{18}{6} \qquad 6(3) \stackrel{?}{=} 18$$
$$x = 3 \qquad 18 = 18$$

71.
$$-4x = 36 \qquad -4x = 36$$
$$\frac{-4x}{-4} = \frac{36}{-4} \qquad -4(-9) \stackrel{?}{=} 36$$
$$x = -9 \qquad 36 = 36$$

73.
$$4t = 108 \qquad 4t = 108$$
$$\frac{4t}{4} = \frac{108}{4} \qquad 4(27) \stackrel{?}{=} 108$$
$$t = 27 \qquad 108 = 108$$

75.
$$11x = -121 \qquad 11x = -121$$
$$\frac{11x}{11} = \frac{-121}{11} \qquad 11(-11) \stackrel{?}{=} -121$$
$$x = -11 \qquad -121 = -121$$

77.
$$2x = \frac{1}{7} \qquad 2x = \frac{1}{7}$$
$$\frac{1}{2} \cdot 2x = \frac{1}{2} \cdot \frac{1}{7} \qquad 2 \cdot \frac{1}{14} \stackrel{?}{=} \frac{1}{7}$$
$$x = \frac{1}{14} \qquad \frac{2}{14} \stackrel{?}{=} \frac{1}{7}$$
$$\frac{1}{7} = \frac{1}{7}$$

79.
$$5x = \frac{5}{8} \qquad 5x = \frac{5}{8}$$
$$\frac{1}{5} \cdot 5x = \frac{1}{5} \cdot \frac{5}{8} \qquad 5 \cdot \frac{1}{8} \stackrel{?}{=} \frac{5}{8}$$
$$x = \frac{1}{8} \qquad \frac{5}{8} = \frac{5}{8}$$

81.
$$\frac{z}{7} = 14 \qquad \frac{z}{7} = 14$$
$$7 \cdot \frac{z}{7} = \mathbf{7} \cdot 14 \qquad \frac{98}{7} \overset{?}{=} 14$$
$$z = 98 \qquad 14 = 14$$

83.
$$\frac{w}{7} = \frac{5}{7} \qquad \frac{w}{7} = \frac{5}{7}$$
$$7 \cdot \frac{w}{7} = \mathbf{7} \cdot \frac{5}{7} \qquad \frac{5}{7} = \frac{5}{7}$$
$$w = 5$$

85.
$$\frac{s}{-3} = -\frac{5}{6} \qquad \frac{s}{-3} = -\frac{5}{6}$$
$$-6 \cdot \frac{s}{-3} = -\mathbf{6} \cdot \left(-\frac{5}{6}\right) \qquad \frac{\frac{5}{2}}{-3} \overset{?}{=} -\frac{5}{6}$$
$$2s = 5 \qquad \frac{\frac{5}{2} \cdot 2}{-3 \cdot 2} \overset{?}{=} -\frac{5}{6}$$
$$\frac{2s}{2} = \frac{5}{2} \qquad \frac{5}{-6} = -\frac{5}{6}$$
$$s = \frac{5}{2}$$

87.
$$-32z = 64 \qquad -32z = 64$$
$$\frac{-32z}{-32} = \frac{64}{-32} \qquad -32(-2) \overset{?}{=} 64$$
$$z = -2 \qquad 64 = 64$$

89.
$$18z = -9 \qquad 18z = -9$$
$$\frac{18z}{18} = \frac{-9}{18} \qquad 18\left(-\frac{1}{2}\right) \overset{?}{=} -9$$
$$z = -\frac{1}{2} \qquad -9 = -9$$

91.
$$0.25x = 1228 \qquad 0.25x = 1228$$
$$\frac{0.25x}{0.25} = \frac{1228}{0.25} \qquad 0.25(4912) \overset{?}{=} 1228$$
$$x = 4912 \qquad 4912 = 4912$$

93.
$$\frac{b}{3} = \frac{1}{3} \qquad \frac{b}{3} = \frac{1}{3}$$
$$3 \cdot \frac{b}{3} = \mathbf{3} \cdot \frac{1}{3} \qquad \frac{1}{3} = \frac{1}{3}$$
$$b = 1$$

95.
$$-1.2w = -102 \qquad -1.2w = -102$$
$$\frac{-1.2w}{-1.2} = \frac{-102}{-1.2} \qquad -1.2(85) \overset{?}{=} -102$$
$$w = 85 \qquad -102 = -102$$

97.
$$\frac{u}{5} = -\frac{3}{10} \qquad \frac{u}{5} = -\frac{3}{10}$$
$$10 \cdot \frac{u}{5} = \mathbf{10} \cdot \left(-\frac{3}{10}\right) \qquad \frac{-\frac{3}{2}}{5} \overset{?}{=} -\frac{3}{10}$$
$$2u = -3 \qquad \frac{-\frac{3}{2} \cdot 2}{5 \cdot 2} \overset{?}{=} -\frac{3}{10}$$
$$\frac{2u}{2} = \frac{-3}{2} \qquad \frac{-3}{10} = -\frac{3}{10}$$
$$u = -\frac{3}{2}$$

99.
$$\frac{p}{0.2} = 12 \qquad \frac{p}{0.2} = 12$$
$$0.2 \cdot \frac{p}{0.2} = \mathbf{0.2} \cdot 12 \qquad \frac{2.4}{0.2} \overset{?}{=} 12$$
$$p = 2.4 \qquad 12 = 12$$

101.
$$rb = a$$
$$40\% \cdot 200 = a$$
$$0.40(200) = a$$
$$80 = a$$

103.
$$rb = a$$
$$50\% \cdot 38 = a$$
$$0.50(38) = a$$
$$19 = a$$

105.
$$rb = a$$
$$15\% \cdot b = 48$$
$$0.15b = 48$$
$$\frac{0.15b}{0.15} = \frac{48}{0.15}$$
$$b = 320$$

107.
$$rb = a$$
$$35\% \cdot b = 133$$
$$0.35b = 133$$
$$\frac{0.35b}{0.35} = \frac{133}{0.35}$$
$$b = 380$$

109.
$$rb = a$$
$$28\% \cdot b = 42$$
$$0.28b = 42$$
$$\frac{0.28b}{0.28} = \frac{42}{0.28}$$
$$b = 150$$

111.
$$rb = a$$
$$r(357.5) = 71.5$$
$$\frac{r(357.5)}{357.5} = \frac{71.5}{357.5}$$
$$r = 0.20$$
$$r = 20\%$$

113.
$$rb = a$$
$$r(4) = 0.32$$
$$\frac{r(4)}{4} = \frac{0.32}{4}$$
$$r = 0.08$$
$$r = 8\%$$

115.
$$rb = a$$
$$r(17) = 34$$
$$\frac{r(17)}{17} = \frac{34}{17}$$
$$r = 2.00$$
$$r = 200\%$$

117. Let $r =$ the regular price. Then

$$\boxed{\text{Sale price}} = \boxed{\text{Regular price}} - \boxed{\text{Markdown}}$$
$$7995 = r - 1350$$
$$7995 + \mathbf{1350} = r - 1350 + \mathbf{1350}$$
$$9345 = r$$

The regular price is \$9,345.

119. Let $w =$ the wholesale price. Then

$$\boxed{\text{Retail price}} = \boxed{\text{Wholesale price}} + \boxed{\text{Markup}}$$
$$175 = w + 85$$
$$175 - \mathbf{85} = w + 85 - \mathbf{85}$$
$$90 = w$$

The wholesale price is \$90.

121.
$$A = p + i$$
$$5010 = 4750 + i$$
$$5010 - \mathbf{4750} = 4750 - \mathbf{4750} + i$$
$$260 = i$$

The deposit earned \$260 in interest.

123.
$$v = p + a$$
$$150000 = p + 57000$$
$$150000 - \mathbf{57000} = p + 57000 - \mathbf{57000}$$
$$93000 = p$$

The original price was \$93,000.

125.
$$c = p + t$$
$$512 = 317 + t$$
$$512 - \mathbf{317} = 317 - \mathbf{317} + t$$
$$195 = t$$

The carpet cost \$195 to install.

127. Let $c =$ the condominium price. Then

$$\boxed{\text{Condo price}} = \boxed{\text{House price}} - 57595$$
$$c = 202744 - 57595$$
$$c = 145149$$

The price of the condominium is \$145,149.

129. Let $x =$ the original number in the audience.

$$\frac{1}{3} \cdot \boxed{\text{Original audience}} = \boxed{\text{Number who left}}$$
$$\frac{1}{3}x = 78$$
$$\mathbf{3} \cdot \frac{1}{3}x = \mathbf{3} \cdot 78$$
$$x = 234$$

There were originally 234 in the audience.

131. Let $x =$ the number in the senior class.

$$\frac{1}{7} \cdot \boxed{\text{Number in class}} = \boxed{\text{Number off-campus}}$$
$$\frac{1}{7}x = 217$$
$$\mathbf{7} \cdot \frac{1}{7}x = \mathbf{7} \cdot 217$$
$$x = 1519$$

There are 1,519 in the senior class.

133. Let r = the percentage not pleased.
$$a = 9200 - 4140 = 5060$$
$$rb = a$$
$$r(9200) = 5060$$
$$\frac{r(9200)}{9200} = \frac{5060}{9200}$$
$$r = 0.55$$
$$r = 55\%$$
55% of those surveyed were not pleased.

135. Let x = the selling price.
$$rb = a$$
$$0.05(x) = 13.50$$
$$\frac{0.05x}{0.05} = \frac{13.50}{0.05}$$
$$x = 270$$
The selling price was \$270.

137. Let x = the total number of patients.
$$rb = a$$
$$0.18(x) = 1008$$
$$\frac{0.18x}{0.18} = \frac{1008}{0.18}$$
$$x = 5600$$
The hospital treated 5,600 patients.

139. Answers may vary.

141.
$$A_{\text{circle}} = A_{\text{square}}$$
$$\pi r^2 = s^2$$
$$\pi(4.5)^2 = 8^2$$
$$\pi(20.25) = 64$$
$$\frac{\pi(20.25)}{20.25} = \frac{64}{20.25}$$
$$\pi \approx 3.16$$

Exercise 2.2 (page 98)

1. $P = 2l + 2w = 2(8.5\,\text{cm}) + 2(16.5\,\text{cm}) = 17\,\text{cm} + 33\,\text{cm} = 50\,\text{cm}$

3. $A = \frac{1}{2}h(b + d) = \frac{1}{2}(8.5\,\text{in.})(6.7\,\text{in.} + 12.2\,\text{in.}) = \frac{1}{2}(8.5\,\text{in.})(18.9\,\text{in.}) = \frac{1}{2}(160.65\,\text{in.}^2) = 80.325\,\text{in.}^2$

5. cost

7. percent

9.
$$5x - 1 = 4$$
$$5x - 1 + 1 = 4 + 1$$
$$5x = 5$$
$$\frac{5x}{5} = \frac{5}{5}$$
$$x = 1$$

$$5x - 1 = 4$$
$$5(1) - 1 \stackrel{?}{=} 4$$
$$5 - 1 \stackrel{?}{=} 4$$
$$4 = 4$$

11.
$$6x + 2 = -4$$
$$6x + 2 - 2 = -4 - 2$$
$$6x = -6$$
$$\frac{6x}{6} = \frac{-6}{6}$$
$$x = -1$$

$$6x + 2 = -4$$
$$6(-1) + 2 \stackrel{?}{=} -4$$
$$-6 + 2 \stackrel{?}{=} -4$$
$$-4 = -4$$

13.
$$3x - 8 = 1 \qquad 3x - 8 = 1$$
$$3x - 8 + 8 = 1 + 8 \qquad 3(3) - 8 \overset{?}{=} 1$$
$$3x = 9$$
$$9 - 8 \overset{?}{=} 1$$
$$\frac{3x}{3} = \frac{9}{3}$$
$$1 = 1$$
$$x = 3$$

15.
$$11x + 17 = -5 \qquad 11x + 17 = -5$$
$$11x + 17 - 17 = -5 - 17 \qquad 11(-2) + 17 \overset{?}{=} -5$$
$$11x = -22$$
$$-22 + 17 \overset{?}{=} -5$$
$$\frac{11x}{11} = \frac{-22}{11}$$
$$-5 = -5$$
$$x = -2$$

17.
$$43t + 72 = 158 \qquad 43t + 72 = 158$$
$$43t + 72 - 72 = 158 - 72 \qquad 43(2) + 72 \overset{?}{=} 158$$
$$43t = 86$$
$$86 + 72 \overset{?}{=} 158$$
$$\frac{43t}{43} = \frac{86}{43}$$
$$158 = 158$$
$$t = 2$$

19.
$$-47 - 21s = 58 \qquad -47 - 21s = 58$$
$$-47 + 47 - 21s = 58 + 47 \qquad -47 - 21(-5) \overset{?}{=} 58$$
$$-21s = 105$$
$$-47 + 105 \overset{?}{=} 58$$
$$\frac{-21s}{-21} = \frac{105}{-21}$$
$$58 = 58$$
$$s = -5$$

21.
$$2y - \frac{5}{3} = \frac{4}{3} \qquad 2y - \frac{5}{3} = \frac{4}{3}$$
$$2y - \frac{5}{3} + \frac{5}{3} = \frac{4}{3} + \frac{5}{3} \qquad 2\left(\frac{3}{2}\right) - \frac{5}{3} \overset{?}{=} \frac{4}{3}$$
$$2y = \frac{9}{3}$$
$$3 - \frac{5}{3} \overset{?}{=} \frac{4}{3}$$
$$2y = 3$$
$$\frac{9}{3} - \frac{5}{3} \overset{?}{=} \frac{4}{3}$$
$$\frac{2y}{2} = \frac{3}{2}$$
$$\frac{4}{3} = \frac{4}{3}$$
$$y = \frac{3}{2}$$

23.
$$-0.4y - 12 = -20 \qquad -0.4y - 12 = -20$$
$$-0.4y - 12 + 12 = -20 + 12 \qquad -0.4(20) - 12 \overset{?}{=} -20$$
$$-0.4y = -8$$
$$-8 - 12 \overset{?}{=} -20$$
$$\frac{-0.4y}{-0.4} = \frac{-8}{-0.4}$$
$$-20 = -20$$
$$y = 20$$

25.
$$\frac{x}{3} - 3 = -2$$
$$\frac{x}{3} - 3 + 3 = -2 + 3$$
$$\frac{x}{3} = 1$$
$$3 \cdot \frac{x}{3} = 3 \cdot 1$$
$$x = 3$$

$$\frac{x}{3} - 3 = -2$$
$$\frac{3}{3} - 3 \stackrel{?}{=} -2$$
$$1 - 3 \stackrel{?}{=} -2$$
$$-2 = -2$$

27.
$$\frac{z}{9} + 5 = -1$$
$$\frac{z}{9} + 5 - 5 = -1 - 5$$
$$\frac{z}{9} = -6$$
$$9 \cdot \frac{z}{9} = 9(-6)$$
$$z = -54$$

$$\frac{z}{9} + 5 = -1$$
$$\frac{-54}{9} + 5 \stackrel{?}{=} -1$$
$$-6 + 5 \stackrel{?}{=} -1$$
$$-1 = -1$$

29.
$$\frac{b}{3} + 5 = 2$$
$$\frac{b}{3} + 5 - 5 = 2 - 5$$
$$\frac{b}{3} = -3$$
$$3 \cdot \frac{b}{3} = 3(-3)$$
$$b = -9$$

$$\frac{b}{3} + 5 = 2$$
$$\frac{-9}{3} + 5 \stackrel{?}{=} 2$$
$$-3 + 5 \stackrel{?}{=} 2$$
$$2 = 2$$

31.
$$\frac{s}{11} + 9 = 6$$
$$\frac{s}{11} + 9 - 9 = 6 - 9$$
$$\frac{s}{11} = -3$$
$$11 \cdot \frac{s}{11} = 11(-3)$$
$$s = -33$$

$$\frac{s}{11} + 9 = 6$$
$$\frac{-33}{11} + 9 \stackrel{?}{=} 6$$
$$-3 + 9 \stackrel{?}{=} 6$$
$$6 = 6$$

33.
$$\frac{k}{5} - \frac{1}{2} = \frac{3}{2}$$
$$\frac{k}{5} - \frac{1}{2} + \frac{1}{2} = \frac{3}{2} + \frac{1}{2}$$
$$\frac{k}{5} = \frac{4}{2}$$
$$\frac{k}{5} = 2$$
$$5 \cdot \frac{k}{5} = 5 \cdot 2$$
$$k = 10$$

$$\frac{k}{5} - \frac{1}{2} = \frac{3}{2}$$
$$\frac{10}{5} - \frac{1}{2} \stackrel{?}{=} \frac{3}{2}$$
$$2 - \frac{1}{2} \stackrel{?}{=} \frac{3}{2}$$
$$\frac{4}{2} - \frac{1}{2} \stackrel{?}{=} \frac{3}{2}$$
$$\frac{3}{2} = \frac{3}{2}$$

35.
$$\frac{w}{16} + \frac{5}{4} = 1$$
$$\frac{w}{16} + \frac{5}{4} - \frac{5}{4} = 1 - \frac{5}{4}$$
$$\frac{w}{16} = \frac{4}{4} - \frac{5}{4}$$
$$\frac{w}{16} = -\frac{1}{4}$$
$$16 \cdot \frac{w}{16} = 16\left(-\frac{1}{4}\right)$$
$$w = -4$$

$$\frac{w}{16} + \frac{5}{4} = 1$$
$$\frac{-4}{16} + \frac{5}{4} \stackrel{?}{=} 1$$
$$-\frac{1}{4} + \frac{5}{4} \stackrel{?}{=} 1$$
$$\frac{4}{4} \stackrel{?}{=} 1$$
$$1 = 1$$

37.
$$\frac{b+5}{3} = 11$$
$$3 \cdot \frac{b+5}{3} = 3 \cdot 11$$
$$b + 5 = 33$$
$$b + 5 - 5 = 33 - 5$$
$$b = 28$$

$$\frac{b+5}{3} = 11$$
$$\frac{28+5}{3} \stackrel{?}{=} 11$$
$$\frac{33}{3} \stackrel{?}{=} 11$$
$$11 = 11$$

39.
$$\frac{r+7}{3} = 4$$
$$3 \cdot \frac{r+7}{3} = 3 \cdot 4$$
$$r + 7 = 12$$
$$r + 7 - 7 = 12 - 7$$
$$r = 5$$

$$\frac{r+7}{3} = 4$$
$$\frac{5+7}{3} \stackrel{?}{=} 4$$
$$\frac{12}{3} \stackrel{?}{=} 4$$
$$4 = 4$$

41.
$$\frac{u-2}{5} = 1$$
$$5 \cdot \frac{u-2}{5} = 5 \cdot 1$$
$$u - 2 = 5$$
$$u - 2 + 2 = 5 + 2$$
$$u = 7$$

$$\frac{u-2}{5} = 1$$
$$\frac{7-2}{5} \stackrel{?}{=} 1$$
$$\frac{5}{5} \stackrel{?}{=} 1$$
$$1 = 1$$

43.
$$\frac{x-4}{4} = -3$$
$$4 \cdot \frac{x-4}{4} = 4(-3)$$
$$x - 4 = -12$$
$$x - 4 + 4 = -12 + 4$$
$$x = -8$$

$$\frac{x-4}{4} = -3$$
$$\frac{-8-4}{4} \stackrel{?}{=} -3$$
$$\frac{-12}{4} \stackrel{?}{=} -3$$
$$-3 = -3$$

45.

$$\frac{3x}{2} - 6 = 9$$

$$\frac{3x}{2} - 6 + 6 = 9 + 6$$

$$\frac{3x}{2} = 15$$

$$2 \cdot \frac{3x}{2} = 2 \cdot 15$$

$$3x = 30$$

$$\frac{3x}{3} = \frac{30}{3}$$

$$x = 10$$

$$\frac{3x}{2} - 6 = 9$$

$$\frac{3(10)}{2} - 6 \overset{?}{=} 9$$

$$\frac{30}{2} - 6 \overset{?}{=} 9$$

$$15 - 6 \overset{?}{=} 9$$

$$9 = 9$$

47.

$$\frac{3y}{2} + 5 = 11$$

$$\frac{3y}{2} + 5 - 5 = 11 - 5$$

$$\frac{3y}{2} = 6$$

$$2 \cdot \frac{3y}{2} = 2 \cdot 6$$

$$3y = 12$$

$$\frac{3y}{3} = \frac{12}{3}$$

$$y = 4$$

$$\frac{3y}{2} + 5 = 11$$

$$\frac{3(4)}{2} + 5 \overset{?}{=} 11$$

$$\frac{12}{2} + 5 \overset{?}{=} 11$$

$$6 + 5 \overset{?}{=} 11$$

$$11 = 11$$

49.

$$\frac{2x}{3} + \frac{1}{2} = 3$$

$$\frac{2x}{3} + \frac{1}{2} - \frac{1}{2} = 3 - \frac{1}{2}$$

$$\frac{2x}{3} = \frac{5}{2}$$

$$3 \cdot \frac{2x}{3} = 3 \cdot \frac{5}{2}$$

$$2x = \frac{15}{2}$$

$$\frac{1}{2} \cdot 2x = \frac{1}{2} \cdot \frac{15}{2}$$

$$x = \frac{15}{4}$$

$$\frac{2x}{3} + \frac{1}{2} = 3$$

$$\frac{2\left(\frac{15}{4}\right)}{3} + \frac{1}{2} \overset{?}{=} 3$$

$$\frac{\frac{15}{2}}{3} + \frac{1}{2} \overset{?}{=} 3$$

$$\frac{15}{6} + \frac{1}{2} \overset{?}{=} 3$$

$$\frac{5}{2} + \frac{1}{2} \overset{?}{=} 3$$

$$\frac{6}{2} \overset{?}{=} 3$$

$$3 = 3$$

51.

$$\frac{3x}{4} - \frac{2}{5} = 2$$

$$\frac{3x}{4} - \frac{2}{5} + \frac{2}{5} = 2 + \frac{2}{5}$$

$$\frac{3x}{4} = \frac{12}{5}$$

$$4 \cdot \frac{3x}{4} = 4 \cdot \frac{12}{5}$$

$$3x = \frac{48}{5}$$

$$\frac{1}{3} \cdot 3x = \frac{1}{3} \cdot \frac{48}{5}$$

$$x = \frac{16}{5}$$

$$\frac{3x}{4} - \frac{2}{5} = 2$$

$$\frac{3\left(\frac{16}{5}\right)}{4} - \frac{2}{5} \overset{?}{=} 2$$

$$\frac{\frac{48}{5}}{4} - \frac{2}{5} \overset{?}{=} 2$$

$$\frac{48}{20} - \frac{2}{5} \overset{?}{=} 2$$

$$\frac{12}{5} - \frac{2}{5} \overset{?}{=} 2$$

$$\frac{10}{5} \overset{?}{=} 2$$

$$2 = 2$$

53.

$$\frac{3x - 12}{2} = 9$$

$$2 \cdot \frac{3x - 12}{2} = 2 \cdot 9$$

$$3x - 12 = 18$$

$$3x - 12 + 12 = 18 + 12$$

$$3x = 30$$

$$\frac{3x}{3} = \frac{30}{3}$$

$$x = 10$$

$$\frac{3x - 12}{2} = 9$$

$$\frac{3(10) - 12}{2} \overset{?}{=} 9$$

$$\frac{30 - 12}{2} \overset{?}{=} 9$$

$$\frac{18}{2} \overset{?}{=} 9$$

$$9 = 9$$

55.

$$\frac{5k - 8}{9} = 1$$

$$9 \cdot \frac{5k - 8}{9} = 9 \cdot 1$$

$$5k - 8 = 9$$

$$5k - 8 + 8 = 9 + 8$$

$$5k = 17$$

$$\frac{5k}{5} = \frac{17}{5}$$

$$k = \frac{17}{5}$$

$$\frac{5k - 8}{9} = 1$$

$$\frac{5 \cdot \frac{17}{5} - 8}{9} \overset{?}{=} 1$$

$$\frac{17 - 8}{9} \overset{?}{=} 1$$

$$\frac{9}{9} \overset{?}{=} 1$$

$$1 = 1$$

57.

$$\frac{3z+2}{17} = 0 \qquad\qquad \frac{3z+2}{17} = 0$$

$$17 \cdot \frac{3z+2}{17} = 17 \cdot 0 \qquad \frac{3\left(-\frac{2}{3}\right)+2}{17} \stackrel{?}{=} 0$$

$$3z+2 = 0 \qquad\qquad \frac{-2+2}{17} \stackrel{?}{=} 0$$

$$3z+2-2 = 0-2$$

$$3z = -2 \qquad\qquad \frac{0}{17} \stackrel{?}{=} 0$$

$$\frac{3z}{3} = \frac{-2}{3} \qquad\qquad 0 = 0$$

$$z = -\frac{2}{3}$$

59.

$$\frac{17k-28}{21} + \frac{4}{3} = 0 \qquad\qquad \frac{17k-28}{21} + \frac{4}{3} = 0$$

$$\frac{17k-28}{21} + \frac{4}{3} - \frac{4}{3} = 0 - \frac{4}{3} \qquad \frac{17\cdot 0 - 28}{21} + \frac{4}{3} \stackrel{?}{=} 0$$

$$\frac{17k-28}{21} = -\frac{4}{3} \qquad\qquad \frac{0-28}{21} + \frac{4}{3} \stackrel{?}{=} 0$$

$$21 \cdot \frac{17k-28}{21} = 21\left(-\frac{4}{3}\right) \qquad -\frac{4}{3} + \frac{4}{3} \stackrel{?}{=} 0$$

$$17k-28 = -28 \qquad\qquad 0 = 0$$

$$17k-28+28 = -28+28$$

$$17k = 0$$

$$\frac{17k}{17} = \frac{0}{17}$$

$$k = 0$$

61.

$$-\frac{x}{3} - \frac{1}{2} = -\frac{5}{2} \qquad\qquad -\frac{x}{3} - \frac{1}{2} = -\frac{5}{2}$$

$$-\frac{x}{3} - \frac{1}{2} + \frac{1}{2} = -\frac{5}{2} + \frac{1}{2} \qquad -\frac{6}{3} - \frac{1}{2} \stackrel{?}{=} -\frac{5}{2}$$

$$-\frac{x}{3} = -\frac{4}{2} \qquad\qquad -2 - \frac{1}{2} \stackrel{?}{=} -\frac{5}{2}$$

$$-\frac{x}{3} = -2 \qquad\qquad -\frac{4}{2} - \frac{1}{2} \stackrel{?}{=} -\frac{5}{2}$$

$$-3\left(-\frac{x}{3}\right) = -3(-2) \qquad -\frac{5}{2} = -\frac{5}{2}$$

$$x = 6$$

63.

$$\frac{9-5w}{15} = \frac{2}{5} \qquad\qquad \frac{9-5w}{15} = \frac{2}{5}$$

$$15 \cdot \frac{9-5w}{15} = 15 \cdot \frac{2}{5} \qquad \frac{9-5\left(\frac{3}{5}\right)}{15} \stackrel{?}{=} \frac{2}{5}$$

$$9-5w = 6 \qquad\qquad \frac{9-3}{15} \stackrel{?}{=} \frac{2}{5}$$

$$9-9-5w = 6-9 \qquad\qquad \frac{6}{15} \stackrel{?}{=} \frac{2}{5}$$

$$-5w = -3 \qquad\qquad \frac{2}{5} = \frac{2}{5}$$

$$\frac{-5w}{-5} = \frac{-3}{-5}$$

$$w = \frac{3}{5}$$

65. Let x = the original number.

Then $3x - 6$ = the other number.

$$\boxed{\text{The other number}} = 9$$

$$3x - 6 = 9$$
$$3x - 6 + 6 = 9 + 6$$
$$3x = 15$$
$$\frac{3x}{3} = \frac{15}{3}$$
$$x = 5$$

The original number is 5.

67. Let x = her former rent.

Then $2x - 100$ = the new rent.

$$\boxed{\text{The new rent}} = 400$$

$$2x - 100 = 400$$
$$2x - 100 + 100 = 400 + 100$$
$$2x = 500$$
$$\frac{2x}{2} = \frac{500}{2}$$
$$x = 250$$

Her former rent was $250.

69. Let x = the number of days.

Then $16 + 12x$ = the total cost.

$$\boxed{\text{The total cost}} = 100$$

$$16 + 12x = 100$$
$$16 - 16 + 12x = 100 - 16$$
$$12x = 84$$
$$\frac{12x}{12} = \frac{84}{12}$$
$$x = 7$$

The owner was gone for 7 days.

71. Let x = the # of minutes (after the 1st).

Then $0.85 + 0.27x$ = the total cost.

$$\boxed{\text{The total cost}} = 8.50$$

$$0.85 + 0.27x = 8.50$$
$$0.85 - 0.85 + 0.27x = 8.50 - 0.85$$
$$0.27x = 7.65$$
$$\frac{0.27x}{0.27} = \frac{7.65}{0.27}$$
$$x \approx 28.3$$

She can talk for 28 minutes **after the first minute**, for a total of 29 minutes.

73. Let x = the money from ticket sales.

Then $1500 + 0.20x$ = the total income.

$$\boxed{\text{The total income}} = 2980$$

$$1500 + 0.20x = 2980$$
$$1500 - 1500 + 0.20x = 2980 - 1500$$
$$0.20x = 1480$$
$$\frac{0.20x}{0.20} = \frac{1480}{0.20}$$
$$x = 7400$$

The total ticket sales were $7,400.

75. Let x = the score on the fifth exam.

$$\boxed{\text{Average score}} = 90$$

$$\frac{85 + 80 + 95 + 78 + x}{5} = 90$$

$$\frac{338 + x}{5} = 90$$

$$5 \cdot \frac{338 + x}{5} = 5 \cdot 90$$

$$338 + x = 450$$
$$338 - 338 + x = 450 - 338$$
$$x = 112$$

It is impossible to receive an A.

77. Let $x =$ the regular price.
Then $0.80x =$ the sale price.

$$\boxed{\begin{array}{c}\text{Final} \\ \text{price}\end{array}} = 0.90 \cdot \boxed{\begin{array}{c}\text{Sale} \\ \text{price}\end{array}}$$

$$36 = 0.90(0.80x)$$
$$36 = 0.72x$$
$$\frac{36}{0.72} = \frac{0.72x}{0.72}$$
$$50 = x$$

The original price was $50.

79. For a purchase of $100:

$$\boxed{\text{Markdown}} = \boxed{\begin{array}{c}\text{Percent} \\ \text{markdown}\end{array}} \cdot \boxed{\begin{array}{c}\text{Regular} \\ \text{price}\end{array}}$$

$$15 = r \cdot 100$$
$$\frac{15}{100} = \frac{r \cdot 100}{100}$$
$$0.15 = r$$
$$15\% = r$$

For a purchase of $250:

$$\boxed{\text{Markdown}} = \boxed{\begin{array}{c}\text{Percent} \\ \text{markdown}\end{array}} \cdot \boxed{\begin{array}{c}\text{Regular} \\ \text{price}\end{array}}$$

$$15 = r \cdot 250$$
$$\frac{15}{250} = \frac{r \cdot 250}{250}$$
$$0.06 = r$$
$$6\% = r$$

The range of the percent discount is from 6% to 15%.

81. **Answers may vary.**

83.
$$\frac{7x + \#}{22} = \frac{1}{2}$$
$$\frac{7(1) + \#}{22} = \frac{1}{2}$$
$$\frac{7 + \#}{22} = \frac{1}{2}$$
$$22 \cdot \frac{7 + \#}{22} = 22 \cdot \frac{1}{2}$$
$$7 + \# = 11$$
$$7 - 7 + \# = 11 - 7$$
$$\# = 4$$

The original equation was $\dfrac{7x + 4}{22} = \dfrac{1}{2}$.

Exercise 2.3 (page 105)

1. $x^2 z(y^3 - z) = (-3)^2(0)\left[(-5)^3 - 0\right] = 0$

3. $\dfrac{x - y^2}{2y - 1 + x} = \dfrac{-3 - (-5)^2}{2(-5) - 1 + (-3)} = \dfrac{-3 - (+25)}{-10 - 1 + (-3)} = \dfrac{-28}{-14} = 2$

5. $\dfrac{6}{7} - \dfrac{5}{8} = \dfrac{6 \cdot 8}{7 \cdot 8} - \dfrac{5 \cdot 7}{8 \cdot 7} = \dfrac{48}{56} - \dfrac{35}{56}$
$$= \dfrac{48 - 35}{56} = \dfrac{13}{56}$$

7. $\dfrac{6}{7} \div \dfrac{5}{8} = \dfrac{6}{7} \cdot \dfrac{8}{5} = \dfrac{48}{35}$

9. variables, like

11. identity

13. $3x + 17x = (3 + 17)x = 20x$

15. $8x^2 - 5x^2 = (8 - 5)x^2 = 3x^2$

17. $9x + 3y \Rightarrow$ not like terms

19. $3(x + 2) + 4x = 3 \cdot x + 3 \cdot 2 + 4x$
$$= 3x + 6 + 4x = 7x + 6$$

21. $5(z - 3) + 2z = 5 \cdot z - 5 \cdot 3 + 2z$
$$= 5z - 15 + 2z = 7z - 15$$

23. $12(x + 11) - 11 = 12 \cdot x + 12 \cdot 11 - 11$
$$= 12x + 132 - 11$$
$$= 12x + 121$$

25. $8(y + 7) - 2(y - 3) = 8 \cdot y + 8 \cdot 7 + (-2) \cdot y + (-2)(-3) = 8y + 56 - 2y + 6 = 6y + 62$

27. $2x + 4(y - x) + 3y = 2x + 4 \cdot y + 4(-x) + 3y = 2x + 4y - 4x + 3y = -2x + 7y$

29. $(x + 2) - (x - y) = 1(x + 2) - 1(x - y) = 1 \cdot x + 1 \cdot 2 + (-1) \cdot x + (-1)(-y) = x + 2 - x + y$
$$= y + 2$$

31. $2\left(4x + \dfrac{9}{2}\right) - 3\left(x + \dfrac{2}{3}\right) = 2 \cdot 4x + 2 \cdot \dfrac{9}{2} + (-3) \cdot x + (-3) \cdot \dfrac{2}{3} = 8x + 9 - 3x - 2 = 5x + 7$

33. $8x(x + 3) - 3x^2 = 8x \cdot x + 8x \cdot 3 - 3x^2 = 8x^2 + 24x - 3x^2 = 5x^2 + 24x$

35.

$$3x + 2 = 2x$$
$$3x - 3x + 2 = 2x - 3x$$
$$2 = -x$$
$$-1(2) = -1(-x)$$
$$-2 = x$$

$$3x + 2 = 2x$$
$$3(-2) + 2 \overset{?}{=} 2(-2)$$
$$-6 + 2 \overset{?}{=} -4$$
$$-4 = -4$$

37.

$$5x - 3 = 4x$$
$$5x - 5x - 3 = 4x - 5x$$
$$-3 = -x$$
$$-1(-3) = -1(-x)$$
$$3 = x$$

$$5x - 3 = 4x$$
$$5(3) - 3 \overset{?}{=} 4(3)$$
$$15 - 3 \overset{?}{=} 12$$
$$12 = 12$$

39.

$$9y - 3 = 6y$$
$$9y - 9y - 3 = 6y - 9y$$
$$-3 = -3y$$
$$\dfrac{-3}{-3} = \dfrac{-3y}{-3}$$
$$1 = y$$

$$9y - 3 = 6y$$
$$9(1) - 3 \overset{?}{=} 6(1)$$
$$9 - 3 \overset{?}{=} 6$$
$$6 = 6$$

41.

$$8y - 7 = y$$
$$8y - 8y - 7 = y - 8y$$
$$-7 = -7y$$
$$\dfrac{-7}{-7} = \dfrac{-7y}{-7}$$
$$1 = y$$

$$8y - 7 = y$$
$$8(1) - 7 \overset{?}{=} 1$$
$$8 - 7 \overset{?}{=} 1$$
$$1 = 1$$

43.

$$9 - 23w = 4w$$
$$9 - 23w + 23w = 4w + 23w$$
$$9 = 27w$$
$$\dfrac{9}{27} = \dfrac{27w}{27}$$
$$\dfrac{1}{3} = w$$

$$9 - 23w = 4w$$
$$9 - 23\left(\dfrac{1}{3}\right) \overset{?}{=} 4\left(\dfrac{1}{3}\right)$$
$$\dfrac{27}{3} - \dfrac{23}{3} \overset{?}{=} \dfrac{4}{3}$$
$$\dfrac{4}{3} = \dfrac{4}{3}$$

45.

$$22 - 3r = 8r$$
$$22 - 3r + 3r = 8r + 3r$$
$$22 = 11r$$
$$\frac{22}{11} = \frac{11r}{11}$$
$$2 = r$$

$$22 - 3r = 8r$$
$$22 - 3(2) \stackrel{?}{=} 8(2)$$
$$22 - 6 \stackrel{?}{=} 16$$
$$16 = 16$$

47.

$$3(a + 2) = 4a$$
$$3a + 6 = 4a$$
$$3a - 3a + 6 = 4a - 3a$$
$$6 = a$$

$$3(a + 2) = 4a$$
$$3(6 + 2) \stackrel{?}{=} 4(6)$$
$$3(8) \stackrel{?}{=} 24$$
$$24 = 24$$

49.

$$5(b + 7) = 6b$$
$$5b + 35 = 6b$$
$$5b - 5b + 35 = 6b - 5b$$
$$35 = b$$

$$5(b + 7) = 6b$$
$$5(35 + 7) \stackrel{?}{=} 6(35)$$
$$5(42) \stackrel{?}{=} 210$$
$$210 = 210$$

51.

$$2 + 3(x - 5) = 4(x - 1)$$
$$2 + 3x - 15 = 4x - 4$$
$$3x - 13 = 4x - 4$$
$$3x - 3x - 13 = 4x - 3x - 4$$
$$-13 = x - 4$$
$$-13 + 4 = x - 4 + 4$$
$$-9 = x$$

$$2 + 3(x - 5) = 4(x - 1)$$
$$2 + 3(-9 - 5) \stackrel{?}{=} 4(-9 - 1)$$
$$2 + 3(-14) \stackrel{?}{=} 4(-10)$$
$$2 + (-42) \stackrel{?}{=} -40$$
$$-40 = -40$$

53.

$$10x + 3(2 - x) = 5(x + 2) - 4$$
$$10x + 6 - 3x = 5x + 10 - 4$$
$$7x + 6 = 5x + 6$$
$$7x - 5x + 6 = 5x - 5x + 6$$
$$2x + 6 = 6$$
$$2x + 6 - 6 = 6 - 6$$
$$2x = 0$$
$$\frac{2x}{2} = \frac{0}{2}$$
$$x = 0$$

$$10x + 3(2 - x) = 5(x + 2) - 4$$
$$10(0) + 3(2 - 0) \stackrel{?}{=} 5(0 + 2) - 4$$
$$0 + 3(2) \stackrel{?}{=} 5(2) - 4$$
$$0 + 6 \stackrel{?}{=} 10 - 4$$
$$6 = 6$$

55.

$$3(a + 2) = 2(a - 7)$$
$$3a + 6 = 2a - 14$$
$$3a - 2a + 6 = 2a - 2a - 14$$
$$a + 6 = -14$$
$$a + 6 - 6 = -14 - 6$$
$$a = -20$$

$$3(a + 2) = 2(a - 7)$$
$$3(-20 + 2) \stackrel{?}{=} 2(-20 - 7)$$
$$3(-18) \stackrel{?}{=} 2(-27)$$
$$-54 = -54$$

57.
$$9(x + 11) + 5(13 - x) = 0$$
$$9x + 99 + 65 - 5x = 0$$
$$4x + 164 = 0$$
$$4x + 164 - 164 = 0 - 164$$
$$4x = -164$$
$$\frac{4x}{4} = \frac{-164}{4}$$
$$x = -41$$

$$9(x + 11) + 5(13 - x) = 0$$
$$9(-41 + 11) + 5[13 - (-41)] \stackrel{?}{=} 0$$
$$9(-30) + 5[13 + 41] \stackrel{?}{=} 0$$
$$-270 + 5(54) \stackrel{?}{=} 0$$
$$-270 + 270 \stackrel{?}{=} 0$$
$$0 = 0$$

59.
$$\frac{3(t - 7)}{2} = t - 6$$
$$2 \cdot \frac{3(t - 7)}{2} = 2(t - 6)$$
$$3(t - 7) = 2(t - 6)$$
$$3t - 21 = 2t - 12$$
$$3t - 2t - 21 = 2t - 2t - 12$$
$$t - 21 = -12$$
$$t - 21 + 21 = -12 + 21$$
$$t = 9$$

$$\frac{3(t - 7)}{2} = t - 6$$
$$\frac{3(9 - 7)}{2} \stackrel{?}{=} 9 - 6$$
$$\frac{3(2)}{2} \stackrel{?}{=} 9 - 6$$
$$\frac{6}{2} \stackrel{?}{=} 3$$
$$3 = 3$$

61.
$$\frac{5(2 - s)}{3} = s + 6$$
$$3 \cdot \frac{5(2 - s)}{3} = 3(s + 6)$$
$$5(2 - s) = 3(s + 6)$$
$$10 - 5s = 3s + 18$$
$$10 - 5s + 5s = 3s + 5s + 18$$
$$10 = 8s + 18$$
$$10 - 18 = 8s + 18 - 18$$
$$-8 = 8s$$
$$\frac{-8}{8} = \frac{8s}{8}$$
$$-1 = s$$

$$\frac{5(2 - s)}{3} = s + 6$$
$$\frac{5[2 - (-1)]}{3} \stackrel{?}{=} -1 + 6$$
$$\frac{5(2 + 1)}{3} \stackrel{?}{=} 5$$
$$\frac{5(3)}{3} \stackrel{?}{=} 5$$
$$\frac{15}{3} \stackrel{?}{=} 5$$
$$5 = 5$$

63.
$$\frac{4(2x - 10)}{3} = 2(x - 4)$$
$$3 \cdot \frac{4(2x - 10)}{3} = 3 \cdot 2(x - 4)$$
$$4(2x - 10) = 6(x - 4)$$
$$8x - 40 = 6x - 24$$
$$8x - 6x - 40 = 6x - 6x - 24$$
$$2x - 40 = -24$$
$$2x - 40 + 40 = -24 + 40$$
$$2x = 16$$
$$\frac{2x}{2} = \frac{16}{2}$$
$$x = 8$$

$$\frac{4(2x - 10)}{3} = 2(x - 4)$$
$$\frac{4[2(8) - 10]}{3} \stackrel{?}{=} 2(8 - 4)$$
$$\frac{4(16 - 10)}{3} \stackrel{?}{=} 2(4)$$
$$\frac{4(6)}{3} \stackrel{?}{=} 8$$
$$\frac{24}{3} \stackrel{?}{=} 8$$
$$8 = 8$$

65.
$$3.1(x - 2) = 1.3x + 2.8$$
$$3.1x - 6.2 = 1.3x + 2.8$$
$$3.1x - 1.3x - 6.2 = 1.3x - 1.3x + 2.8$$
$$1.8x - 6.2 = 2.8$$
$$1.8x - 6.2 + 6.2 = 2.8 + 6.2$$
$$1.8x = 9.0$$
$$\frac{1.8x}{1.8} = \frac{9.0}{1.8}$$
$$x = 5$$

$$3.1(x - 2) = 1.3x + 2.8$$
$$3.1(5 - 2) \overset{?}{=} 1.3(5) + 2.8$$
$$3.1(3) \overset{?}{=} 6.5 + 2.8$$
$$9.3 = 9.3$$

67.
$$2.7(y + 1) = 0.3(3y + 33)$$
$$2.7y + 2.7 = 0.9y + 9.9$$
$$2.7y - 0.9y + 2.7 = 0.9y - 0.9y + 9.9$$
$$1.8y + 2.7 = 9.9$$
$$1.8y + 2.7 - 2.7 = 9.9 - 2.7$$
$$1.8y = 7.2$$
$$\frac{1.8y}{1.8} = \frac{7.2}{1.8}$$
$$y = 4$$

$$2.7(y + 1) = 0.3(3y + 33)$$
$$2.7(4 + 1) = 0.3[3(4) + 33]$$
$$2.7(5) \overset{?}{=} 0.3(12 + 33)$$
$$13.5 \overset{?}{=} 0.3(45)$$
$$13.5 = 13.5$$

69.
$$19.1x - 4(x + 0.3) = -46.5$$
$$19.1x - 4x - 1.2 = -46.5$$
$$15.1x - 1.2 = -46.5$$
$$15.1x - 1.2 + 1.2 = -46.5 + 1.2$$
$$15.1x = -45.3$$
$$\frac{15.1x}{15.1} = \frac{-45.3}{15.1}$$
$$x = -3$$

$$19.1x - 4(x + 0.3) = -46.5$$
$$19.1(-3) - 4(-3 + 0.3) \overset{?}{=} -46.5$$
$$-57.3 - 4(-2.7) \overset{?}{=} -46.5$$
$$-57.3 + 10.8 \overset{?}{=} -46.5$$
$$-46.5 = -46.5$$

71.
$$14.3(x + 2) + 13.7(x - 3) = 15.5$$
$$14.3x + 28.6 + 13.7x - 41.1 = 15.5$$
$$28.0x - 12.5 = 15.5$$
$$28x - 12.5 + 12.5 = 15.5 + 12.5$$
$$28x = 28$$
$$\frac{28x}{28} = \frac{28}{28}$$
$$x = 1$$

$$14.3(x + 2) + 13.7(x - 3) = 15.5$$
$$14.3(1 + 2) + 13.7(1 - 3) \overset{?}{=} 15.5$$
$$14.3(3) + 13.7(-2) \overset{?}{=} 15.5$$
$$42.9 - 27.4 \overset{?}{=} 15.5$$
$$15.5 = 15.5$$

73.
$$8x + 3(2 - x) = 5(x + 2) - 4$$
$$8x + 6 - 3x = 5x + 10 - 4$$
$$5x + 6 = 5x + 6$$
$$5x - 5x + 6 = 5x - 5x + 6$$
$$6 - 6$$
Identity

75.
$$2(s + 2) = 2(s + 1) + 3$$
$$2s + 4 = 2s + 2 + 3$$
$$2s + 4 = 2s + 5$$
$$2s - 2s + 4 = 2s - 2s + 5$$
$$4 \neq 5$$
Contradiction

77.
$$\frac{2(t-1)}{6} - 2 = \frac{t+2}{6}$$
$$6\left[\frac{2(t-1)}{6} - 2\right] = 6 \cdot \frac{t+2}{6}$$
$$6 \cdot \frac{2(t-1)}{6} - 6 \cdot 2 = t+2$$
$$2(t-1) - 12 = t+2$$
$$2t - 2 - 12 = t+2$$
$$2t - 14 = t+2$$
$$2t - t - 14 = t - t + 2$$
$$t - 14 = 2$$
$$t - 14 + 14 = 2 + 14$$
$$t = 16$$

79.
$$2(3z+4) = 2(3z-2) + 13$$
$$6z + 8 = 6z - 4 + 13$$
$$6z + 8 = 6z + 9$$
$$6z - 6z + 8 = 6z - 6z + 9$$
$$8 \neq 9$$
Contradiction

81.
$$2(y-3) - \frac{y}{2} = \frac{3}{2}(y-4)$$
$$2\left[2(y-3) - \frac{y}{2}\right] = 2 \cdot \frac{3}{2}(y-4)$$
$$2 \cdot 2(y-3) - 2 \cdot \frac{y}{2} = 3(y-4)$$
$$4(y-3) - y = 3y - 12$$
$$4y - 12 - y = 3y - 12$$
$$3y - 12 = 3y - 12$$
$$3y - 3y - 12 = 3y - 3y - 12$$
$$-12 = -12$$
Identity

83.
$$\frac{3x+14}{2} = x - 2 + \frac{x+18}{2}$$
$$2 \cdot \frac{3x+14}{2} = 2\left[x - 2 + \frac{x+18}{2}\right]$$
$$3x + 14 = 2x - 2 \cdot 2 + 2 \cdot \frac{x+18}{2}$$
$$3x + 14 = 2x - 4 + x + 18$$
$$3x + 14 = 3x + 14$$
$$3x - 3x + 14 = 3x - 3x + 14$$
$$14 = 14$$
Identity

85. **Answers may vary.**

87. **Answers may vary.**

89. Let $x =$ the number.
$$x = 2x$$
$$x - x = 2x - x$$
$$0 = x$$
The number is 0.

Exercise 2.4 (page 112)

1. $V = \frac{1}{3}Bh = \frac{1}{3}s^2h = \frac{1}{3}(10\,\text{cm})^2(6\,\text{cm}) = \frac{1}{3}(100\,\text{cm}^2)(6\,\text{cm}) = \frac{1}{3}(600\,\text{cm}^3) = 200\,\text{cm}^3$

3. $3(x+2) + 4(x-3) = 3 \cdot x + 3 \cdot 2 + 4 \cdot x - 4 \cdot 3 = 3x + 6 + 4x - 12 = 7x - 6$

5. $\frac{1}{2}(x+1) - \frac{1}{2}(x+4) = \frac{1}{2} \cdot x + \frac{1}{2} \cdot 1 - \frac{1}{2} \cdot x - \frac{1}{2} \cdot 4 = \frac{1}{2}x + \frac{1}{2} - \frac{1}{2}x - \frac{4}{2} = -\frac{3}{2}$

SECTION 2.4

7. $A = P + Prt = 1200 + 1200(0.08)(3) = 1200 + 288 = \$1488$

9. $2l + 2w$ **11.** vertex   **13.** complementary   **15.** $180°$

17. Let $x =$ the length of one part.
Then $2x =$ the length of the other part.

$$\boxed{\begin{array}{c}\text{Length of}\\\text{first part}\end{array}} + \boxed{\begin{array}{c}\text{Length of}\\\text{second part}\end{array}} = 12$$

$$x + 2x = 12$$
$$3x = 12$$
$$\frac{3x}{3} = \frac{12}{3}$$
$$x = 4$$

The parts are 4 feet and 8 feet long.

19.
$$\boxed{\begin{array}{c}\text{Sum of 3}\\\text{lengths}\end{array}} = 30$$
$$x + 10 + 2x + x = 30$$
$$4x + 10 = 30$$
$$4x + 10 - 10 = 30 - 10$$
$$4x = 20$$
$$\frac{4x}{4} = \frac{20}{4}$$
$$x = 5$$

The sections are 15, 10, and 5 feet long.

21.
$$\boxed{\begin{array}{c}\text{Sum of 3}\\\text{lengths}\end{array}} = 24$$
$$x + x + 4 + x + 2 = 24$$
$$3x + 6 = 24$$
$$3x + 6 - 6 = 24 - 6$$
$$3x = 18$$
$$\frac{3x}{3} = \frac{18}{3}$$
$$x = 6$$

The sections are 6, 10, and 8 feet long.

23.
$$x + 40° = 50°$$
$$x + 40° - 40° = 50° - 40°$$
$$x = 10°$$

25.
$$x + 21° = 180°$$
$$x + 21° - 21° = 180° - 21°$$
$$x = 159°$$

27.
$$x + 12° = 59°$$
$$x + 12° - 12° = 59° - 12°$$
$$x = 47°$$

29.
$$x + 63° = 90°$$
$$x + 63° - 63° = 90° - 63°$$
$$x = 27°$$

31. Let $x =$ the measure of the complement.
$$x + 37° = 90°$$
$$x + 37° - 37° = 90° - 37°$$
$$x = 53°$$

33. Let $x =$ the measure of the complement.
$$x + 40° = 90°$$
$$x + 40° - 40° = 90° - 40°$$
$$x = 50°$$
Let $x =$ the measure of the supplement of
the complement.
$$x + 50° = 180°$$
$$x + 50° - 50° = 180° - 50°$$
$$x = 130°$$

35.
$$\boxed{\begin{array}{c}\text{Sum of}\\\text{three sides}\end{array}} = 57$$
$$x + x + x = 57$$
$$3x = 57$$
$$\frac{3x}{3} = \frac{57}{3}$$
$$x = 19$$

Each side is 19 feet long.

37. Let $L =$ the length of the pool.

Then $L - 11 =$ the width of the pool.

$$\boxed{\text{Perimeter}} = 94$$
$$2L + 2(L - 11) = 94$$
$$2L + 2L - 22 = 94$$
$$4L - 22 = 94$$
$$4L - 22 + 22 = 94 + 22$$
$$4L = 116$$
$$\frac{4L}{4} = \frac{116}{4}$$
$$L = 29$$

The dimensions are 29 m by 18 m.

39. Let $w =$ the width of the picture.

Then $2w + 5 =$ the length of the picture.

$$\boxed{\text{Perimeter}} = 112$$
$$2w + 2(2w + 5) = 112$$
$$2w + 4w + 10 = 112$$
$$6w + 10 = 112$$
$$6w + 10 - 10 = 112 - 10$$
$$6w = 102$$
$$\frac{6w}{6} = \frac{102}{6}$$
$$w = 17$$

The dimensions are 17 in. by 39 in.

41. Let $a =$ the measure of each angle.

$$\boxed{\text{Sum of angle measures}} = 180$$
$$a + a + a = 180$$
$$3a = 180$$
$$\frac{3a}{3} = \frac{180}{3}$$
$$a = 60$$

Each angle measures $60°$.

43. Let $x =$ amount invested at 5%.

$$I = PRT$$
$$300 = x(0.05)(1)$$
$$\frac{300}{0.05} = \frac{0.05x}{0.05}$$
$$6000 = x$$

He invested $6,000.

45. Let $x =$ amount in 9% fund. Then $24000 - x =$ amount in 14% fund.

$$\boxed{\begin{array}{c}\text{Interest}\\\text{at 9\%}\end{array}} + \boxed{\begin{array}{c}\text{Interest}\\\text{at 14\%}\end{array}} = \boxed{\begin{array}{c}\text{Total}\\\text{interest}\end{array}}$$
$$0.09x + 0.14(24000 - x) = 3135$$
$$9x + 14(24000 - x) = 313500$$
$$9x + 336000 - 14x = 313500$$
$$-5x + 336000 = 313500$$
$$-5x + 336000 - 336000 = 313500 - 336000$$
$$-5x = -22500$$
$$\frac{-5x}{-5} = \frac{-22500}{-5}$$
$$x = 4500 \Rightarrow \$4,500 \text{ was invested at 9\%, and \$19,500 was invested at 14\%.}$$

47. Let $x =$ amount invested in each account.

$$\boxed{\begin{array}{c}\text{Interest}\\\text{at 8\%}\end{array}} + \boxed{\begin{array}{c}\text{Interest}\\\text{at 11\%}\end{array}} = \boxed{\begin{array}{c}\text{Total}\\\text{interest}\end{array}}$$
$$0.08x + 0.11x = 712.50$$
$$8x + 11x = 71250$$
$$19x = 71250$$
$$\frac{19x}{19} = \frac{71250}{19}$$
$$x = 3750 \Rightarrow \$3,750 \text{ is invested in each account.}$$

49. Let $x =$ amount invested at 7%.

$$\boxed{\begin{smallmatrix}\text{Interest}\\\text{at 6\%}\end{smallmatrix}} + \boxed{\begin{smallmatrix}\text{Interest}\\\text{at 7\%}\end{smallmatrix}} = \boxed{\begin{smallmatrix}\text{Total}\\\text{interest}\end{smallmatrix}}$$

$$0.06(15000) + 0.07x = 1250$$
$$6(15000) + 7x = 125000$$
$$90000 + 7x = 125000$$
$$90000 - 90000 + 7x = 125000 - 90000$$
$$7x = 35000$$
$$\frac{7x}{7} = \frac{35000}{7}$$
$$x = 5000$$

$5,000 should be invested at 7%.

51. Let $r =$ the fund rate and $r + 0.01 =$ the CD rate.
The client invests $21,000 in CDs, and $10,500 in the fund.

$$\boxed{\begin{smallmatrix}\text{CD}\\\text{interest}\end{smallmatrix}} = \boxed{\begin{smallmatrix}\text{Fund}\\\text{interest}\end{smallmatrix}} + 840$$

$$(r + 0.01)21000 = r(10500) + 840$$
$$21000r + 210 = 10500r + 840$$
$$21000r - 10500r + 210 = 10500r - 10500r + 840$$
$$10500r + 210 = 840$$
$$10500r + 210 - 210 = 840 - 210$$
$$10500r = 630$$
$$\frac{10500r}{10500} = \frac{630}{10500}$$
$$r = 0.06 \Rightarrow \text{The rates are 6\% and 7\%.}$$

53. **Answers may vary.**

55. Pairs of vertical angles have equal measures.

Exercise 2.5 (page 121)

1. $3 + 4(-5) = 3 + (-20) = -17$

3. $2^3 - 3^2 = 2 \cdot 2 \cdot 2 - 3 \cdot 3 = 8 - 9 = -1$

5.
$$-2x + 3 = 9$$
$$-2x + 3 - 3 = 9 - 3$$
$$-2x = 6$$
$$\frac{-2x}{-2} = \frac{6}{-2}$$
$$x = -3$$

7.
$$\frac{2}{3}p + 1 = 5$$
$$\frac{2}{3}p + 1 - 1 = 5 - 1$$
$$\frac{2}{3}p = 4$$
$$\frac{3}{2} \cdot \frac{2}{3}p = \frac{3}{2} \cdot 4$$
$$p = \frac{12}{2}$$
$$p = 6$$

9. $d = rt$

11. $v = pn$

13. Let t = time for cars to meet.

	r	t	d
Car 1 (A to B)	50	t	$50t$
Car 2 (B to A)	55	t	$55t$

$$\boxed{\text{Distance for car 1}} + \boxed{\text{Distance for car 2}} = 315$$
$$50t + 55t = 315$$
$$105t = 315$$
$$\frac{105t}{105} = \frac{315}{105}$$
$$t = 3$$

The cars meet after 3 hours.

15. Let t = days for crews to meet.

	r	t	d
Crew 1	1.5	t	$1.5t$
Crew 2	1.2	t	$1.2t$

$$\boxed{\text{Distance for crew 1}} + \boxed{\text{Distance for crew 2}} = 9.45$$
$$1.5t + 1.2t = 9.45$$
$$2.7t = 9.45$$
$$\frac{2.7t}{2.7} = \frac{9.45}{2.7}$$
$$t = 3.5$$

The crews meet after 3.5 days.

17. Let t = time for cars to be 715 miles apart.

	r	t	d
Car 1 (going east)	60	t	$60t$
Car 2 (going west)	50	t	$50t$

$$\boxed{\text{Distance 1}} + \boxed{\text{Distance 2}} = 715$$
$$60t + 50t = 715$$
$$110t = 715$$
$$\frac{110t}{110} = \frac{715}{110}$$
$$t = 6.5$$

They will be 715 miles apart after 6.5 hours.

19. Let t = time for boys to be 2 miles apart.

	r	t	d
Boy 1	3	t	$3t$
Boy 2	4	t	$4t$

$$\boxed{\text{Distance 1}} + \boxed{\text{Distance 2}} = 2$$
$$3t + 4t = 2$$
$$7t = 2$$
$$\frac{7t}{7} = \frac{2}{7}$$
$$t = \frac{2}{7}$$

They will lose contact after $\frac{2}{7}$ hour.

21.

	r	t	d
Car	60 mph	t	$60t$
Bus	50 mph	$t + 2$	$50(t + 2)$

$$\boxed{\text{Car distance}} = \boxed{\text{Bus distance}}$$
$$60t = 50(t + 2)$$
$$60t = 50t + 100$$
$$60t - 50t = 50t - 50t + 100$$
$$10t = 100$$
$$\frac{10t}{10} = \frac{100}{10}$$
$$t = 10$$

The car will overtake the bus after 10 hours.

23. Let t = time for cars to be 82.5 miles apart.

	r	t	d
Car 1	42	t	$42t$
Car 2	53	t	$53t$

$$\boxed{\text{Distance 2}} - \boxed{\text{Distance 1}} = 82.5$$
$$53t - 42t = 82.5$$
$$11t = 82.5$$
$$\frac{11t}{11} = \frac{82.5}{11}$$
$$t = 7.5$$

They will be 82.5 miles apart after 7.5 hours.

25. Let r = rate of slow train. Then
$r + 20$ = rate of fast train.

	r	t	d
Slow train	r	3	$3r$
Fast train	$r + 20$	3	$3(r + 20)$

$\boxed{\text{Slow dist.}} + \boxed{\text{Fast dist.}} = 330$

$$3r + 3(r + 20) = 330$$
$$3r + 3r + 60 = 330$$
$$6r + 60 = 330$$
$$6r + 60 - 60 = 330 - 60$$
$$6r = 270$$
$$\frac{6r}{6} = \frac{270}{6}$$
$$r = 45$$

The rates are 45 mph and 65 mph.

27. Let t = slower time.
Then $5 - t$ = faster time.

	r	t	d
1st part	40	t	$40t$
2nd part	50	$5 - t$	$50(5 - t)$

$\boxed{\text{1st dist.}} + \boxed{\text{2nd dist.}} = 210$

$$40t + 50(5 - t) = 210$$
$$40t + 250 - 50t = 210$$
$$-10t + 250 = 210$$
$$-10t + 250 - 250 = 210 - 250$$
$$-10t = -40$$
$$\frac{-10t}{-10} = \frac{-40}{-10}$$
$$t = 4$$

The car averaged 40 mph for 4 hours.

29. Let T = total number of liters of solution.
12% of the total = liters of acid
$$0.12T = 0.3$$
$$\frac{0.12T}{0.12} = \frac{0.3}{0.12}$$
$$T = 2.5$$

There are 2.5 liters of the solution.

31. Let x = gallons of $1.15 fuel.

$\boxed{\begin{array}{c}\text{Value of}\\\text{\$1.15 fuel}\end{array}} + \boxed{\begin{array}{c}\text{Value of}\\\text{\$1.15 fuel}\end{array}} = \boxed{\begin{array}{c}\text{Value of}\\\text{mixture}\end{array}}$

$$1.15x + 0.85(20) = 1(20 + x)$$
$$115x + 85(20) = 100(20 + x)$$
$$115x + 1700 = 2000 + 100x$$
$$115x - 100x + 1700 = 2000 + 100x - 100x$$
$$15x + 1700 = 2000$$
$$15x + 1700 - 1700 = 2000 - 1700$$
$$15x = 300$$
$$\frac{15x}{15} = \frac{300}{15}$$
$$x = 20 \Rightarrow 20 \text{ gallons of the \$1.15 fuel should be used.}$$

33. Let $x =$ gallons of 3% solution used.

$$\boxed{\text{Salt in 3% solution}} + \boxed{\text{Salt in 7% solution}} = \boxed{\text{Salt in mixture}}$$

$$0.03x + 0.07(50) = 0.05(x + 50)$$
$$3x + 7(50) = 5(x + 50)$$
$$3x + 350 = 5x + 250$$
$$3x - 5x + 350 = 5x - 5x + 250$$
$$-2x + 350 = 250$$
$$-2x + 350 - 350 = 250 - 350$$
$$-2x = -100$$
$$\frac{-2x}{-2} = \frac{-100}{-2}$$
$$x = 50$$

50 gallons of the 3% mixture should be used.

35. Let $x =$ ounces of water (0%) added.

$$\boxed{\text{Amt. in 10% sol.}} + \boxed{\text{Amt. in water}} = \boxed{\text{Amt. in 8% sol.}}$$

$$0.10(30) + 0(x) = 0.08(30 + x)$$
$$10(30) + 0 = 8(30 + x)$$
$$300 = 240 + 8x$$
$$300 - 240 = 240 - 240 + 8x$$
$$60 = 8x$$
$$\frac{60}{8} = \frac{8x}{8}$$
$$7.5 = x$$

7.5 ounces of water should be added.

37. Let $x =$ pounds of lemondrops. Then $100 - x =$ pounds of jellybeans.

$$\boxed{\text{Value of lemondrops}} + \boxed{\text{Value of jellybeans}} = \boxed{\text{Value of mixture}}$$

$$1.90x + 1.20(100 - x) = 1.48(100)$$
$$190x + 120(100 - x) = 148(100)$$
$$190x + 12000 - 120x = 14800$$
$$70x + 12000 = 14800$$
$$70x + 12000 - 12000 = 14800 - 12000$$
$$70x = 2800$$
$$\frac{70x}{70} = \frac{2800}{70}$$
$$x = 40 \Rightarrow 40 \text{ lb of lemondrops and 60 lb of jellybeans should be used.}$$

39. Let $c =$ cost of cashews. Then $c - 0.30 =$ cost of peanuts. 20 pounds of each are used.

$$\boxed{\text{Value of cashews}} + \boxed{\text{Value of peanuts}} = \boxed{\text{Value of mixture}}$$

$$20c + 20(c - 0.30) = 1.05(40)$$
$$20c + 20c - 6 = 42$$
$$40c - 6 = 42$$
$$40c - 6 + 6 = 42 + 6$$
$$40c = 48$$
$$\frac{40c}{40} = \frac{48}{40}$$
$$c = 1.20$$

A bag of cashews is worth $1.20.

41. Let $x =$ pounds of regular coffee used.

$$\boxed{\text{Value of regular}} + \boxed{\text{Value of gourmet}} = \boxed{\text{Value of mixture}}$$

$$4(x) + 7(40) = 5(x + 40)$$
$$4x + 280 = 5x + 200$$
$$4x - 5x + 280 = 5x - 5x + 200$$
$$-x + 280 = 200$$
$$-x + 280 - 280 = 200 - 280$$
$$-x = -80$$
$$x = 80$$

80 pounds of regular coffee should be used.

43. Let $c =$ cost of hazelnut beans per pound.

$$\boxed{\begin{array}{c}\text{Value of}\\\text{chocolate}\end{array}} + \boxed{\begin{array}{c}\text{Value of}\\\text{hazelnut}\end{array}} = \boxed{\begin{array}{c}\text{Value of}\\\text{mixture}\end{array}}$$

$$7(2) + c(5) = 6(7)$$
$$14 + 5c = 42$$
$$14 - 14 + 5c = 42 - 14$$
$$5c = 28$$
$$c = \tfrac{28}{5} = \$5.60 \Rightarrow \text{The hazelnut beans cost } \$5.60 \text{ per pound.}$$

45. Answers may vary.

47. Answers may vary.

49. Let x and $x + 2$ represent the integers.

$$x + x + 2 = 16$$
$$2x + 2 = 16$$
$$2x = 14$$
$$x = 7$$

This says that the integers are 7 and 9, but these are not **even** integers. The equation has a solution, but not the problem itself.

51. You cannot mix a 10% solution and a 20% solution and end up with a solution that has a greater concentration (30%) than either of the original solutions.

Exercise 2.6 (page 128)

1. $2x - 5y + 3x = 2x + 3x - 5y = (2 + 3)x - 5y = 5x - 5y$

3. $\dfrac{3}{5}(x + 5) - \dfrac{8}{5}(10 + x) = \dfrac{3}{5}x + \dfrac{3}{5} \cdot 5 - \dfrac{8}{5} \cdot 10 - \dfrac{8}{5}x = \dfrac{3}{5}x + 3 - 16 - \dfrac{8}{5}x = \left(\dfrac{3}{5} - \dfrac{8}{5}\right)x - 13$

$$= -\dfrac{5}{5}x - 13$$
$$= -x - 13$$

5. literal

7. isolate

9. subtract

11. $E = IR$

$$\dfrac{E}{R} = \dfrac{IR}{R}$$
$$\dfrac{E}{R} = I, \text{ or } I = \dfrac{E}{R}$$

13. $V = lwh$

$$\dfrac{V}{lh} = \dfrac{lwh}{lh}$$
$$\dfrac{V}{lh} = w, \text{ or } w = \dfrac{V}{lh}$$

15.
$$P = a + b + c$$
$$P - a = a - a + b + c$$
$$P - a = b + c$$
$$P - a - c = b + c - c$$
$$P - a - c = b, \text{ or } b = P - a - c$$

17.
$$P = 2l + 2w$$
$$P - 2l = 2l - 2l + 2w$$
$$P - 2l = 2w$$
$$\frac{P - 2l}{2} = \frac{2w}{2}$$
$$\frac{P - 2l}{2} = w, \text{ or } w = \frac{P - 2l}{2}$$

19.
$$A = P + Prt$$
$$A - P = P - P + Prt$$
$$A - P = Prt$$
$$\frac{A - P}{Pr} = \frac{Prt}{Pr}$$
$$\frac{A - P}{Pr} = t, \text{ or } t = \frac{A - P}{Pr}$$

21.
$$C = 2\pi r$$
$$\frac{C}{2\pi} = \frac{2\pi r}{2\pi}$$
$$\frac{C}{2\pi} = r, \text{ or } r = \frac{C}{2\pi}$$

23.
$$K = \frac{wv^2}{2g}$$
$$2g \cdot K = 2g \cdot \frac{wv^2}{2g}$$
$$2gK = wv^2$$
$$\frac{2gK}{v^2} = \frac{wv^2}{v^2}$$
$$\frac{2gK}{v^2} = w, \text{ or } w = \frac{2gK}{v^2}$$

25.
$$P = I^2 R$$
$$\frac{P}{I^2} = \frac{I^2 R}{I^2}$$
$$\frac{P}{I^2} = R, \text{ or } R = \frac{P}{I^2}$$

27.
$$K = \frac{wv^2}{2g}$$
$$2g \cdot K = 2g \cdot \frac{wv^2}{2g}$$
$$2gK = wv^2$$
$$\frac{2gK}{2K} = \frac{wv^2}{2K}$$
$$g = \frac{wv^2}{2K}$$

29.
$$F = \frac{GMm}{d^2}$$
$$d^2 \cdot F = d^2 \cdot \frac{GMm}{d^2}$$
$$d^2 F = GMm$$
$$\frac{d^2 F}{Gm} = \frac{GMm}{Gm}$$
$$\frac{d^2 F}{Gm} = M, \text{ or } M = \frac{d^2 F}{Gm}$$

31.
$$F = \frac{GMm}{d^2}$$
$$d^2 \cdot F = d^2 \cdot \frac{GMm}{d^2}$$
$$d^2 F = GMm$$
$$\frac{d^2 F}{F} = \frac{GMm}{F}$$
$$d^2 = \frac{GMm}{F}$$

33.
$$G = 2(r - 1)b$$
$$\frac{G}{2b} = \frac{2(r - 1)b}{2b}$$
$$\frac{G}{2b} = r - 1$$
$$\frac{G}{2b} + 1 = r - 1 + 1$$
$$\frac{G}{2b} + 1 = r, \text{ or } r = \frac{G}{2b} + 1$$

35.
$$d = rt \qquad t = \frac{d}{r}$$
$$\frac{d}{r} = \frac{rt}{r} \qquad t = \frac{135}{45}$$
$$\frac{d}{r} = t \qquad t = 3$$

37.
$$i = prt \qquad t = \frac{i}{pr}$$
$$\frac{i}{pr} = \frac{prt}{pr} \qquad t = \frac{12}{100(0.06)}$$
$$\frac{i}{pr} = t \qquad t = \frac{12}{6} = 2$$

39.
$$P = a + b + c \qquad\qquad c = P - a - b$$
$$P - a = a - a + b + c \qquad c = 37 - 15 - 19$$
$$P - a = b + c \qquad\qquad c = 22 - 19$$
$$P - a - b = b - b + c \qquad c = 3$$
$$P - a - b = c$$

41.
$$K = \frac{1}{2}h(a+b) \qquad h = \frac{2K}{a+b}$$
$$2 \cdot K = 2 \cdot \frac{1}{2}h(a+b) \qquad h = \frac{2(48)}{7+5}$$
$$2K = h(a+b) \qquad h = \frac{96}{12} = 8$$
$$\frac{2K}{a+b} = \frac{h(a+b)}{a+b}$$
$$\frac{2K}{a+b} = h$$

43.
$$E = IR \qquad I = \frac{E}{R}$$
$$\frac{E}{R} = \frac{IR}{R} \qquad I = \frac{48}{12}$$
$$\frac{E}{R} = I \qquad I = 4 \text{ amperes}$$

45.
$$C = 2\pi r \qquad r = \frac{C}{2\pi}$$
$$\frac{C}{2\pi} = \frac{2\pi r}{2\pi} \qquad r = \frac{14.32}{2\pi}$$
$$\frac{C}{2\pi} = r \qquad r \approx 2.28 \text{ feet}$$

47.
$$P = I^2 R \qquad R = \frac{P}{I^2}$$
$$\frac{P}{I^2} = \frac{I^2 R}{I^2} \qquad R = \frac{2700}{14^2}$$
$$\frac{P}{I^2} = R \qquad R = \frac{2700}{196} = 13.78 \text{ ohms}$$

49.
$$F = \frac{GMm}{d^2}$$
$$d^2 \cdot F = d^2 \cdot \frac{GMm}{d^2}$$
$$d^2 F = GMm$$
$$\frac{Fd^2}{GM} = \frac{GMm}{GM}$$
$$\frac{Fd^2}{GM} = m, \text{ or } m = \frac{Fd^2}{GM}$$

51.

$$L = 2D + 3.25(r + R)$$
$$L - 3.25(r + R) = 2D + 3.25(r + R) - 3.25(r + R)$$
$$L - 3.25(r + R) = 2D$$
$$\frac{L - 3.25(r + R)}{2} = \frac{2D}{2}$$
$$\frac{L - 3.25r - 3.25R}{2} = D$$

$$D = \frac{L - 3.25r - 3.25R}{2}$$
$$D = \frac{25 - 3.25(1) - 3.25(3)}{2}$$
$$D = \frac{25 - 3.25 - 9.75}{2} = \frac{12}{2} = 6 \text{ ft.}$$

53.

$$C = 0.15(T - C)$$
$$C = 0.15T - 0.15C$$
$$C + 0.15C = 0.15T - 0.15C + 0.15C$$
$$1.15C = 0.15T$$
$$\frac{1.15C}{1.15} = \frac{0.15T}{1.15}$$
$$C \approx 0.1304T$$

The maximum contribution is about 13% of taxable income.

55. Answers may vary.

57. $E = mc^2$

$$E = 1(300{,}000)^2$$
$$E = 90{,}000{,}000{,}000 \text{ joules}$$

Exercise 2.7 (page 136)

1. $3x^2 - 2(y^2 - x^2) = 3x^2 + (-2)y^2 - (-2)x^2 = 3x^2 - 2y^2 + 2x^2 = 5x^2 - 2y^2$

3. $\frac{1}{3}(x + 6) - \frac{4}{3}(x - 9) = \frac{1}{3}x + \frac{1}{3} \cdot 6 - \frac{4}{3}x - \frac{4}{3}(-9) = -\frac{3}{3}x + 2 + 12 = -x + 14$

5. is less than

7. $\geq$

9. inequality

11.

$$x + 2 > 5$$
$$x + 2 - 2 > 5 - 2$$
$$x > 3$$

13.

$$-x - 3 \leq 7$$
$$-x - 3 + 3 \leq 7 + 3$$
$$-x \leq 10$$
$$-1(-x) \geq -1(10)$$
$$x \geq -10$$

15.

$$3 + x < 2$$
$$3 - 3 + x < 2 - 3$$
$$x < -1$$

17.
$$2x - 3 \le 5$$
$$2x - 3 + 3 \le 5 + 3$$
$$2x \le 8$$
$$\frac{2x}{2} \le \frac{8}{2}$$
$$x \le 4$$

19.
$$-3x - 7 > -1$$
$$-3x - 7 + 7 > -1 + 7$$
$$-3x > 6$$
$$\frac{-3x}{-3} < \frac{6}{-3}$$
$$x < -2$$

21.
$$-4x + 1 > 17$$
$$-4x + 1 - 1 > 17 - 1$$
$$-4x > 16$$
$$\frac{-4x}{-4} < \frac{16}{-4}$$
$$x < -4$$

23.
$$2x + 9 \le x + 8$$
$$2x - x + 9 \le x - x + 8$$
$$x + 9 \le 8$$
$$x + 9 - 9 \le 8 - 9$$
$$x \le -1$$

25.
$$9x + 13 \ge 8x$$
$$9x - 8x + 13 \ge 8x - 8x$$
$$x + 13 \ge 0$$
$$x + 13 - 13 \ge 0 - 13$$
$$x \ge -13$$

27.
$$8x + 4 > 6x - 2$$
$$8x - 6x + 4 > 6x - 6x - 2$$
$$2x + 4 > -2$$
$$2x + 4 - 4 > -2 - 4$$
$$2x > -6$$
$$\frac{2x}{2} > \frac{-6}{2}$$
$$x > -3$$

29.
$$5x + 7 < 2x + 1$$
$$5x - 2x + 7 < 2x - 2x + 1$$
$$3x + 7 < 1$$
$$3x + 7 - 7 < 1 - 7$$
$$3x < -6$$
$$\frac{3x}{3} < \frac{-6}{3}$$
$$x < -2$$

31.
$$7 - x \le 3x - 1$$
$$7 - x - 3x \le 3x - 3x - 1$$
$$7 - 4x \le -1$$
$$7 - 7 - 4x \le -1 - 7$$
$$-4x \le -8$$
$$\frac{-4x}{-4} \ge \frac{-8}{-4}$$
$$x \ge 2$$

33.
$$9 - 2x > 24 - 7x$$
$$9 - 2x + 7x > 24 - 7x + 7x$$
$$9 + 5x > 24$$
$$9 - 9 + 5x > 24 - 9$$
$$5x > 15$$
$$\frac{5x}{5} > \frac{15}{5}$$
$$x > 3$$

35.
$$3(x - 8) < 5x + 6$$
$$3x - 24 < 5x + 6$$
$$3x - 5x - 24 < 5x - 5x + 6$$
$$-2x - 24 < 6$$
$$-2x - 24 + 24 < 6 + 24$$
$$-2x < 30$$
$$\frac{-2x}{-2} > \frac{30}{-2}$$
$$x > -15$$

37.
$$8(5-x) \leq 10(8-x)$$
$$40 - 8x \leq 80 - 10x$$
$$40 - 8x + 10x \leq 80 - 10x + 10x$$
$$40 + 2x \leq 80$$
$$40 - 40 + 2x \leq 80 - 40$$
$$2x \leq 40$$
$$\frac{2x}{2} \leq \frac{40}{2}$$
$$x \leq 20$$

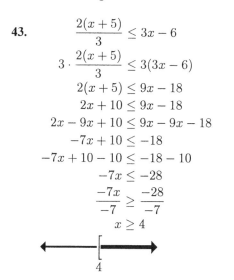

39.
$$\frac{5}{2}(7x-15) + x \geq \frac{13}{2}x - \frac{3}{2}$$
$$2\left[\frac{5}{2}(7x-15) + x\right] \geq 2\left[\frac{13}{2}x - \frac{3}{2}\right]$$
$$2 \cdot \frac{5}{2}(7x-15) + 2x \geq 2 \cdot \frac{13}{2}x - 2 \cdot \frac{3}{2}$$
$$5(7x-15) + 2x \geq 13x - 3$$
$$35x - 75 + 2x \geq 13x - 3$$
$$37x - 75 \geq 13x - 3$$
$$37x - 13x - 75 \geq 13x - 13x - 3$$
$$24x - 75 \geq -3$$
$$24x - 75 + 75 \geq -3 + 75$$
$$24x \geq 72$$
$$\frac{24x}{24} \geq \frac{72}{24}$$
$$x \geq 3$$

41.
$$\frac{3x-3}{2} < 2x + 2$$
$$2 \cdot \frac{3x-3}{2} < 2(2x+2)$$
$$3x - 3 < 4x + 4$$
$$3x - 4x - 3 < 4x - 4x + 4$$
$$-x - 3 < 4$$
$$-x - 3 + 3 < 4 + 3$$
$$-x < 7$$
$$-1(-x) > -1(7)$$
$$x > -7$$

43.
$$\frac{2(x+5)}{3} \leq 3x - 6$$
$$3 \cdot \frac{2(x+5)}{3} \leq 3(3x-6)$$
$$2(x+5) \leq 9x - 18$$
$$2x + 10 \leq 9x - 18$$
$$2x - 9x + 10 \leq 9x - 9x - 18$$
$$-7x + 10 \leq -18$$
$$-7x + 10 - 10 \leq -18 - 10$$
$$-7x \leq -28$$
$$\frac{-7x}{-7} \geq \frac{-28}{-7}$$
$$x \geq 4$$

45.
$$2 < \quad x - 5 \quad < 5$$
$$2 + 5 < x - 5 + 5 < 5 + 5$$
$$7 < \quad x \quad < 10$$

47.
$$-5 < \quad x + 4 \quad \leq 7$$
$$-5 - 4 < x + 4 - 4 \leq 7 - 4$$
$$-9 < \quad x \quad \leq 3$$

49.
$$0 \le \quad x + 10 \quad \le 10$$
$$0 - 10 \le x + 10 - 10 \le 10 - 10$$
$$-10 \le \quad x \quad \le 0$$

$$-10 \qquad 0$$

51.
$$4 < -2x < 10$$
$$\frac{4}{-2} > \frac{-2x}{2} > \frac{10}{-2}$$
$$-2 > \quad x \quad > -5$$
$$-5 < \quad x \quad < -2$$

$$-5 \qquad -2$$

53.
$$-3 \le \quad \frac{x}{2} \quad \le 5$$
$$2(-3) \le 2 \cdot \frac{x}{2} \le 2(5)$$
$$-6 \le \quad x \quad \le 10$$

$$-6 \qquad 10$$

55.
$$3 \le \quad 2x - 1 \quad < 5$$
$$3 + 1 \le 2x - 1 + 1 < 5 + 1$$
$$4 \le \quad 2x \quad < 6$$
$$\frac{4}{2} \le \quad \frac{2x}{2} \quad < \frac{6}{2}$$
$$2 \le \quad x \quad < 3$$

$$2 \qquad 3$$

57.
$$0 < \quad 10 - 5x \quad \le 15$$
$$0 - 10 < 10 - 10 - 5x \le 15 - 10$$
$$-10 < \quad -5x \quad \le 5$$
$$\frac{-10}{-5} > \quad \frac{-5x}{-5} \quad \ge \frac{5}{-5}$$
$$2 > \quad x \quad \ge -1$$
$$-1 \le \quad x \quad < 2$$

$$-1 \qquad 2$$

59.
$$-6 < \quad 3(x + 2) \quad < 9$$
$$-6 < \quad 3x + 6 \quad < 9$$
$$-6 - 6 < 3x + 6 - 6 < 9 - 6$$
$$-12 < \quad 3x \quad < 3$$
$$\frac{-12}{3} < \quad \frac{3x}{3} \quad < \frac{3}{3}$$
$$-4 < \quad x \quad < 1$$

$$-4 \qquad 1$$

61.
$$-4 < \quad \frac{x - 2}{2} \quad < 6$$
$$2(-4) < 2 \cdot \frac{x - 2}{2} < 2(6)$$
$$-8 < \quad x - 2 \quad < 12$$
$$-8 + 2 < x - 2 + 2 < 12 + 2$$
$$-6 < \quad x \quad < 14$$

$$-6 \qquad 14$$

63. Let s = score on last exam.
$$\text{Average score} \ge 80$$
$$\frac{68 + 75 + 79 + s}{4} \ge 80$$
$$4 \cdot \frac{68 + 75 + 79 + s}{4} \ge 4(80)$$
$$68 + 75 + 79 + s \ge 320$$
$$222 + s \ge 320$$
$$222 - 222 + s \ge 320 - 222$$
$$s \ge 98$$
Her last test score must be at least 98%.

65.   Let r = rating of third model.   Average rating ≥ 21

$$\frac{17 + 19 + r}{3} \geq 21$$

$$3 \cdot \frac{17 + 19 + r}{3} \geq 3(21)$$

$$17 + 19 + r \geq 63$$

$$36 + r \geq 63$$

$$36 - 36 + r \geq 63 - 36$$

$$r \geq 27 \Rightarrow \text{It must have a rating of at least 27 mpg.}$$

67.   Let s = length of each side.

$$0 < \text{Perimeter} \leq 57$$

$$0 < \quad 3s \quad \leq 57$$

$$\frac{0}{3} < \quad \frac{3s}{3} \quad \leq \frac{57}{3}$$

$$0 < \quad s \quad \leq 19$$

Each side can be from 0 feet to 19 feet long.

69. $470 \text{ ft} \leq \text{ range in feet } \leq 13{,}143 \text{ ft}$

$$\frac{470 \text{ ft}}{5280} \leq \text{ range in miles } \leq \frac{13143 \text{ ft}}{5280}$$

$0.1 \text{ miles} \leq \text{ range in miles } \leq 2.5 \text{ miles}$

The range is from 0.1 miles to 2.5 miles.

71. $17500 \text{ ft} < \text{ range in feet } < 21700 \text{ ft}$

$$\frac{17500 \text{ ft}}{5280} < \text{ range in miles } < \frac{21700 \text{ ft}}{5280}$$

$3.3 \text{ miles} < \text{ range in miles } < 4.1 \text{ miles}$

The range is between 3.3 miles and 4.1 miles.

73.

$$19° < \quad C \quad < 22°$$

$$19° < \quad \frac{5}{9}(F - 32) \quad < 22°$$

$$\frac{9}{5}(19°) < \frac{9}{5} \cdot \frac{5}{9}(F - 32) < \frac{9}{5}(22°)$$

$$34.2° < \quad F - 32 \quad < 39.6°$$

$$34.2° + 32 < F - 32 + 32 < 39.6° + 32$$

$$66.2° < \quad F \quad < 71.6°$$

The Fahrenheit temperature is between 66.2° and 71.6°.

75.

$$5.9 \text{ in.} < \quad r \quad < 6.1 \text{ in.}$$

$$2\pi(5.9 \text{ in.}) < 2\pi \cdot r < 2\pi(6.1 \text{ in})$$

$$2(3.14)(5.9 \text{ in.}) < \quad 2\pi r \quad < 2(3.14)(6.1 \text{ in.})$$

$$37.052 \text{ in.} < \quad C \quad < 38.308 \text{ in.}$$

The circumference can vary between about 37.052 inches and 38.308 inches.

77. $150 \text{ lb} < \text{ range in lbs} < 190 \text{ lb}$

$$\frac{150 \text{ lb}}{2.2} < \text{ range in kg } < \frac{190 \text{ lb}}{2.2}$$

$68.18 \text{ kg} < \text{ range in kg } < 86.36 \text{ kg}$

The weight is between 68.18 and 86.36 kg.

79. Let w = width. Then $2w - 3$ = length.

$$24 < \text{perimeter} < 48$$
$$24 < 2w + 2(2w - 3) < 48$$
$$24 < 2w + 4w - 6 < 48$$
$$24 < 6w - 6 < 48$$
$$24 + 6 < 6w - 6 + 6 < 48 + 6$$
$$30 < 6w < 54$$
$$\frac{30}{6} < \frac{6w}{6} < \frac{54}{6}$$
$$5 < w < 9$$

The width could be between 5 and 9 feet.

81. **Answers may vary.**

83. This is not correct because x could represent a negative number, and then multiplying both sides of the equation would require the inequality to change from $<$ to $>$.

Chapter 2 Summary (page 140)

1.
$$3x + 7 = 1$$
$$3(-2) + 7 \stackrel{?}{=} 1$$
$$-6 + 7 \stackrel{?}{=} 1$$
$$1 = 1$$
-2 is a solution.

2.
$$5 - 2x = 3$$
$$5 - 2(-1) \stackrel{?}{=} 3$$
$$5 + 2 \stackrel{?}{=} 3$$
$$7 \neq 3$$
-1 is not a solution.

3.
$$2(x + 3) = x$$
$$2(-3 + 3) \stackrel{?}{=} -3$$
$$2(0) \stackrel{?}{=} -3$$
$$0 \neq -3$$
-3 is not a solution.

4.
$$5(3 - x) = 2 - 4x$$
$$5(3 - 13) \stackrel{?}{=} 2 - 4(13)$$
$$5(-10) \stackrel{?}{=} 2 - 52$$
$$-50 = -50$$
13 is a solution.

5.
$$3(x + 5) = 2(x - 3)$$
$$3(-21 + 5) \stackrel{?}{=} 2(-21 - 3)$$
$$3(-16) \stackrel{?}{=} 2(-24)$$
$$-48 = -48$$
-21 is a solution.

6.
$$2(x - 7) = x + 14$$
$$2(0 - 7) \stackrel{?}{=} 0 + 14$$
$$2(-7) \stackrel{?}{=} 14$$
$$-14 \neq 14$$
0 is not a solution.

7.
$$x - 7 = -6 \qquad x - 7 = -6$$
$$x - 7 + 7 = -6 + 7 \qquad 1 - 7 \stackrel{?}{=} -6$$
$$x = 1 \qquad -6 = -6$$

8.
$$y - 4 = 5 \qquad y - 4 = 5$$
$$y - 4 + 4 = 5 + 4 \qquad 9 - 4 \stackrel{?}{=} 5$$
$$y = 9 \qquad 5 = 5$$

9.
$$p + 4 = 20 \qquad p + 4 = 20$$
$$p + 4 - 4 = 20 - 4 \qquad 16 + 4 \stackrel{?}{=} 20$$
$$p = 16 \qquad 20 = 20$$

10.
$$x + \frac{3}{5} = \frac{3}{5} \qquad x + \frac{3}{5} = \frac{3}{5}$$
$$x + \frac{3}{5} - \frac{3}{5} = \frac{3}{5} - \frac{3}{5} \qquad 0 + \frac{3}{5} \stackrel{?}{=} \frac{3}{5}$$
$$x = 0 \qquad \frac{3}{5} = \frac{3}{5}$$

11.

$$y - \frac{7}{2} = \frac{1}{2} \qquad y - \frac{7}{2} = \frac{1}{2}$$

$$y - \frac{7}{2} + \frac{7}{2} = \frac{1}{2} + \frac{7}{2} \qquad 4 - \frac{7}{2} \stackrel{?}{=} \frac{1}{2}$$

$$y = \frac{8}{2} = 4 \qquad \frac{8}{2} - \frac{7}{2} \stackrel{?}{=} \frac{1}{2}$$

$$\frac{1}{2} = \frac{1}{2}$$

12.

$$z + \frac{5}{3} = -\frac{1}{3} \qquad z + \frac{5}{3} = -\frac{1}{3}$$

$$z + \frac{5}{3} - \frac{5}{3} = -\frac{1}{3} - \frac{5}{3} \qquad -2 + \frac{5}{3} \stackrel{?}{=} -\frac{1}{3}$$

$$z = -\frac{6}{3} \qquad -\frac{6}{3} + \frac{5}{3} \stackrel{?}{=} -\frac{1}{3}$$

$$z = -2 \qquad -\frac{1}{3} = -\frac{1}{3}$$

13. Let r = the regular price. Then

$$\boxed{\text{Sale price}} = \boxed{\text{Regular price}} - \boxed{\text{Markdown}}$$

$$69.95 = r - 35.45$$

$$69.95 + 35.45 = r - 35.45 + 35.45$$

$$105.40 = r$$

The regular price is $105.40.

14. Let w = the wholesale price. Then

$$\boxed{\text{Retail price}} = \boxed{\text{Wholesale price}} + \boxed{\text{Markup}}$$

$$212.95 = w + 115.25$$

$$212.95 - 115.25 = w + 115.25 - 115.25$$

$$97.70 = w$$

The wholesale price is $97.70.

15.

$$3x = 15 \qquad 3x = 15$$

$$\frac{3x}{3} = \frac{15}{3} \qquad 3(5) \stackrel{?}{=} 15$$

$$x = 5 \qquad 15 = 15$$

16.

$$8r = -16 \qquad 8r = -16$$

$$\frac{8r}{8} = \frac{-16}{8} \qquad 8(-2) \stackrel{?}{=} -16$$

$$r = -2 \qquad -16 = -16$$

17.

$$10z = 5 \qquad 10z = 5$$

$$\frac{10z}{10} = \frac{5}{10} \qquad 10\left(\frac{1}{2}\right) \stackrel{?}{=} 5$$

$$z = \frac{1}{2} \qquad 5 = 5$$

18.

$$14s = 21 \qquad 14s = 21$$

$$\frac{14s}{14} = \frac{21}{14} \qquad 14\left(\frac{3}{2}\right) \stackrel{?}{=} 21$$

$$s = \frac{3}{2} \qquad 21 = 21$$

19.

$$\frac{y}{3} = 6 \qquad \frac{y}{3} = 6$$

$$3 \cdot \frac{y}{3} = 3 \cdot 6 \qquad \frac{18}{3} \stackrel{?}{=} 6$$

$$y = 18 \qquad 6 = 6$$

20.

$$\frac{w}{7} = -5 \qquad \frac{w}{7} = -5$$

$$7 \cdot \frac{w}{7} = 7(-5) \qquad \frac{-35}{7} \stackrel{?}{=} -5$$

$$w = -35 \qquad -5 = -5$$

21.

$$\frac{a}{-7} = \frac{1}{14} \qquad \frac{a}{-7} = \frac{1}{14}$$

$$-7 \cdot \frac{a}{-7} = -7\left(\frac{1}{14}\right) \qquad \frac{-\frac{1}{2}}{-7} \stackrel{?}{=} \frac{1}{14}$$

$$a = -\frac{7}{14} \qquad \frac{2\left(-\frac{1}{2}\right)}{2(-7)} \stackrel{?}{=} \frac{1}{14}$$

$$a = -\frac{1}{2} \qquad \frac{-1}{-14} = \frac{1}{14}$$

22.

$$\frac{t}{12} = \frac{1}{2} \qquad \frac{t}{12} = \frac{1}{2}$$

$$12 \cdot \frac{t}{12} = 12\left(\frac{1}{2}\right) \qquad \frac{6}{12} \stackrel{?}{=} \frac{1}{2}$$

$$t = \frac{12}{2} \qquad \frac{1}{2} = \frac{1}{2}$$

$$t = 6$$

23.
$$rb = a$$
$$35\% \cdot 700 = a$$
$$0.35(700) = a$$
$$245 = a$$

24.
$$rb = a$$
$$72\% \cdot b = 936$$
$$0.72b = 936$$
$$\frac{0.72b}{0.72} = \frac{936}{0.72}$$
$$b = 1,300$$

25.
$$rb = a$$
$$r \cdot 2300 = 851$$
$$\frac{r \cdot 2300}{2300} = \frac{851}{2300}$$
$$r = 0.37$$
$$r = 37\%$$

26.
$$rb = a$$
$$r \cdot 576 = 72$$
$$\frac{r \cdot 576}{576} = \frac{72}{576}$$
$$r = 0.125$$
$$r = 12.5\%$$

27.
$$5y + 6 = 21$$
$$5y + 6 - 6 = 21 - 6$$
$$5y = 15$$
$$\frac{5y}{5} = \frac{15}{5}$$
$$y = 3$$

$$5y + 6 = 21$$
$$5(3) + 6 \overset{?}{=} 21$$
$$15 + 6 \overset{?}{=} 21$$
$$21 = 21$$

28.
$$5y - 9 = 1$$
$$5y - 9 + 9 = 1 + 9$$
$$5y = 10$$
$$\frac{5y}{5} = \frac{10}{5}$$
$$y = 2$$

$$5y - 9 = 1$$
$$5(2) - 9 \overset{?}{=} 1$$
$$10 - 9 \overset{?}{=} 1$$
$$1 = 1$$

29.
$$-12z + 4 = -8$$
$$-12z + 4 - 4 = -8 - 4$$
$$-12z = -12$$
$$\frac{-12z}{-12} = \frac{-12}{-12}$$
$$z = 1$$

$$-12z + 4 = -8$$
$$-12(1) + 4 \overset{?}{=} -8$$
$$-12 + 4 \overset{?}{=} -8$$
$$-8 = -8$$

30.
$$17z + 3 = 20$$
$$17z + 3 - 3 = 20 - 3$$
$$17z = 17$$
$$\frac{17z}{17} = \frac{17}{17}$$
$$z = 1$$

$$17z + 3 = 20$$
$$17(1) + 3 \overset{?}{=} 20$$
$$17 + 3 \overset{?}{=} 20$$
$$20 = 20$$

31.
$$13 - 13t = 0$$
$$13 - 13 - 13t = 0 - 13$$
$$-13t = -13$$
$$\frac{-13t}{-13} = \frac{-13}{-13}$$
$$t = 1$$

$$13 - 13t = 0$$
$$13 - 13(1) \overset{?}{=} 0$$
$$13 - 13 \overset{?}{=} 0$$
$$0 = 0$$

32.
$$10 + 7t = -4$$
$$10 - 10 + 7t = -4 - 10$$
$$7t = -14$$
$$\frac{7t}{7} = \frac{-14}{7}$$
$$t = -2$$

$$10 + 7t = -4$$
$$10 + 7(-2) \overset{?}{=} -4$$
$$10 + (-14) \overset{?}{=} -4$$
$$-4 = -4$$

33.
$$23a - 43 = 3$$
$$23a - 43 + 43 = 3 + 43$$
$$23a = 46$$
$$\frac{23a}{23} = \frac{46}{23}$$
$$a = 2$$

$$23a - 43 = 3$$
$$23(2) - 43 \overset{?}{=} 3$$
$$46 - 43 \overset{?}{=} 3$$
$$3 = 3$$

34.
$$84 - 21a = -63$$
$$84 - 84 - 21a = -63 - 84$$
$$-21a = -147$$
$$\frac{-21a}{-21} = \frac{-147}{-21}$$
$$a = 7$$

$$84 - 21a = -63$$
$$84 - 21(7) \overset{?}{=} -63$$
$$84 - 147 \overset{?}{=} -63$$
$$-63 = -63$$

35.
$$3x + 7 = 1$$
$$3x + 7 - 7 = 1 - 7$$
$$3x = -6$$
$$\frac{3x}{3} = \frac{-6}{3}$$
$$x = -2$$

$$3x + 7 = 1$$
$$3(-2) + 7 \overset{?}{=} 1$$
$$-6 + 7 \overset{?}{=} 1$$
$$1 = 1$$

36.
$$7 - 9x = 16$$
$$7 - 7 - 9x = 16 - 7$$
$$-9x = 9$$
$$\frac{-9x}{-9} = \frac{9}{-9}$$
$$x = -1$$

$$7 - 9x = 16$$
$$7 - 9(-1) \overset{?}{=} 16$$
$$7 + 9 \overset{?}{=} 16$$
$$16 = 16$$

37.
$$\frac{b + 3}{4} = 2$$
$$4 \cdot \frac{b + 3}{4} = 4 \cdot 2$$
$$b + 3 = 8$$
$$b + 3 - 3 = 8 - 3$$
$$b = 5$$

$$\frac{b + 3}{4} = 2$$
$$\frac{5 + 3}{4} \overset{?}{=} 2$$
$$\frac{8}{4} \overset{?}{=} 2$$
$$2 = 2$$

38.
$$\frac{b - 7}{2} = -2$$
$$2 \cdot \frac{b - 7}{2} = 2(-2)$$
$$b - 7 = -4$$
$$b - 7 + 7 = -4 + 7$$
$$b = 3$$

$$\frac{b - 7}{2} = -2$$
$$\frac{3 - 7}{2} \overset{?}{=} -2$$
$$\frac{-4}{2} \overset{?}{=} -2$$
$$-2 = -2$$

39.
$$\frac{x - 8}{5} = 1$$
$$5 \cdot \frac{x - 8}{5} = 5(1)$$
$$x - 8 = 5$$
$$x - 8 + 8 = 5 + 8$$
$$x = 13$$

$$\frac{x - 8}{5} = 1$$
$$\frac{13 - 8}{5} \overset{?}{=} 1$$
$$\frac{5}{5} \overset{?}{=} 1$$
$$1 = 1$$

40.
$$\frac{x + 10}{2} = -1$$
$$2 \cdot \frac{x + 10}{2} = 2(-1)$$
$$x + 10 = -2$$
$$x + 10 - 10 = -2 - 10$$
$$x = -12$$

$$\frac{x + 10}{2} = -1$$
$$\frac{-12 + 10}{2} \overset{?}{=} -1$$
$$\frac{-2}{2} \overset{?}{=} -1$$
$$-1 = -1$$

41.
$$\frac{2y - 2}{4} = 2$$
$$4 \cdot \frac{2y - 2}{4} = 4(2)$$
$$2y - 2 = 8$$
$$2y - 2 + 2 = 8 + 2$$
$$2y = 10$$
$$\frac{2y}{2} = \frac{10}{2}$$
$$y = 5$$

$$\frac{2y - 2}{4} = 2$$
$$\frac{2(5) - 2}{4} \overset{?}{=} 2$$
$$\frac{10 - 2}{4} \overset{?}{=} 2$$
$$\frac{8}{4} \overset{?}{=} 2$$
$$2 = 2$$

42.
$$\frac{3y + 12}{11} = 3$$
$$11 \cdot \frac{3y + 12}{11} = 11(3)$$
$$3y + 12 = 33$$
$$3y + 12 - 12 = 33 - 12$$
$$3y = 21$$
$$\frac{3y}{3} = \frac{21}{3}$$
$$y = 7$$

$$\frac{3y + 12}{11} = 3$$
$$\frac{3(7) + 12}{11} \overset{?}{=} 3$$
$$\frac{21 + 12}{11} \overset{?}{=} 3$$
$$\frac{33}{11} \overset{?}{=} 3$$
$$3 = 3$$

43.
$$\frac{x}{2} + 7 = 11 \qquad \frac{x}{2} + 7 = 11$$
$$\frac{x}{2} + 7 - 7 = 11 - 7 \qquad \frac{8}{2} + 7 \overset{?}{=} 11$$
$$\frac{x}{2} = 4 \qquad 4 + 7 \overset{?}{=} 11$$
$$2 \cdot \frac{x}{2} = 2 \cdot 4 \qquad 11 = 11$$
$$x = 8$$

44.
$$\frac{r}{3} - 3 = 7 \qquad \frac{r}{3} - 3 = 7$$
$$\frac{r}{3} - 3 + 3 = 7 + 3 \qquad \frac{30}{3} - 3 \overset{?}{=} 7$$
$$\frac{r}{3} = 10 \qquad 10 - 3 \overset{?}{=} 7$$
$$3 \cdot \frac{r}{3} = 3 \cdot 10 \qquad 7 = 7$$
$$r = 30$$

45.
$$\frac{a}{2} + \frac{9}{4} = 6 \qquad \frac{a}{2} + \frac{9}{4} = 6$$
$$\frac{a}{2} + \frac{9}{4} - \frac{9}{4} = \frac{24}{4} - \frac{9}{4} \qquad \frac{\frac{15}{2}}{2} + \frac{9}{4} \overset{?}{=} 6$$
$$\frac{a}{2} = \frac{15}{4} \qquad \frac{2 \cdot \frac{15}{2}}{2 \cdot 2} + \frac{9}{4} \overset{?}{=} 6$$
$$2 \cdot \frac{a}{2} = 2 \cdot \frac{15}{4} \qquad \frac{15}{4} + \frac{9}{4} \overset{?}{=} 6$$
$$a = \frac{15}{2} \qquad \frac{24}{4} \overset{?}{=} 6$$
$$6 = 6$$

46.
$$\frac{x}{8} - 2.3 = 3.2 \qquad \frac{x}{8} - 2.3 = 3.2$$
$$\frac{x}{8} - 2.3 + 2.3 = 3.2 + 2.3 \qquad \frac{44}{8} - 2.3 \overset{?}{=} 3.2$$
$$\frac{x}{8} = 5.5 \qquad 5.5 - 2.3 \overset{?}{=} 3.2$$
$$8 \cdot \frac{x}{8} = 8(5.5) \qquad 3.2 = 3.2$$
$$x = 44$$

47. Let $x =$ the regular price.

$$\boxed{\begin{array}{c}\text{Sale}\\\text{price}\end{array}} = \boxed{\begin{array}{c}\text{Regular}\\\text{price}\end{array}} - \boxed{\text{Markdown}}$$
$$240 = x - 0.25x$$
$$240 = 0.75x$$
$$\frac{240}{0.75} = \frac{0.75x}{0.75}$$
$$320 = x$$
The regular price is \$320.

48. Let $r =$ the sales tax rate.

$$\boxed{\begin{array}{c}\text{Total}\\\text{price}\end{array}} = \boxed{\begin{array}{c}\text{Price}\\\text{before tax}\end{array}} + \boxed{\text{Sales tax}}$$
$$40.47 = 38 + 38r$$
$$40.47 - 38 = 38 - 38 + 38r$$
$$2.47 = 38r$$
$$\frac{2.47}{38} = \frac{38r}{38}$$
$$0.065 = r$$
The sales tax rate is 6.5%.

49. Let $r =$ the % increase.

$$\boxed{\begin{array}{c}\text{New}\\\text{price}\end{array}} = \boxed{\begin{array}{c}\text{Original}\\\text{price}\end{array}} + \boxed{\text{Increase}}$$
$$1100 = 560 + 560r$$
$$1100 - 560 = 560 - 560 + 560r$$
$$540 = 560r$$
$$\frac{540}{560} = \frac{560r}{560}$$
$$0.964 = r$$
The percent increase is 96.4%.

50. Let $r =$ the % discount.

$$\boxed{\text{Markdown}} = \boxed{\begin{array}{c}\text{Percent}\\\text{markdown}\end{array}} \cdot \boxed{\begin{array}{c}\text{Regular}\\\text{price}\end{array}}$$
$$465 - 215 = r \cdot 465$$
$$250 = 465r$$
$$\frac{250}{465} = \frac{465r}{465}$$
$$0.538 = r$$
The percent markdown is 53.8%.

51. $5x + 9x = (5 + 9)x = 14x$

52. $7a + 12a = (7 + 12)a = 19a$

53. $18b - 13b = (18 - 13)b = 5b$

54. $21x - 23x = (21 - 23)x = -2x$

55. $5y - 7y = (5 - 7)y = -2y$

56. $19x - 19 \Rightarrow$ not like terms

57. $7(x + 2) + 2(x - 7) = 7x + 14 + 2x - 14$
$$= 7x + 2x + 14 - 14$$
$$= (7 + 2)x + 0 = 9x$$

58. $2(3 - x) + x - 6x = 6 - 2x + x - 6x$
$$= 6 + (-2 + 1 - 6)x$$
$$= 6 - 7x$$

59. $y^2 + 3(y^2 - 2) = y^2 + 3y^2 - 6$
$$= (1 + 3)y^2 - 6$$
$$= 4y^2 - 6$$

60. $2x^2 - 2(x^2 - 2) = 2x^2 - 2x^2 + 4$
$$= (2 - 2)x^2 + 4$$
$$= 0x^2 + 4 = 4$$

61.
$$2x - 19 = 2 - x$$
$$2x + x - 19 = 2 - x + x$$
$$3x - 19 = 2$$
$$3x - 19 + 19 = 2 + 19$$
$$3x = 21$$
$$\frac{3x}{3} = \frac{21}{3}$$
$$x = 7$$

$$2x - 19 = 2 - x$$
$$2(7) - 19 \overset{?}{=} 2 - 7$$
$$14 - 19 \overset{?}{=} -5$$
$$-5 = -5$$

62.
$$5b - 19 = 2b + 20$$
$$5b - 2b - 19 = 2b - 2b + 20$$
$$3b - 19 = 20$$
$$3b - 19 + 19 = 20 + 19$$
$$3b = 39$$
$$\frac{3b}{3} = \frac{39}{3}$$
$$b = 13$$

$$5b - 19 = 2b + 20$$
$$5(13) - 19 \overset{?}{=} 2(13) + 20$$
$$65 - 19 \overset{?}{=} 26 + 20$$
$$46 = 46$$

63.
$$3x + 20 = 5 - 2x$$
$$3x + 2x + 20 = 5 - 2x + 2x$$
$$5x + 20 = 5$$
$$5x + 20 - 20 = 5 - 20$$
$$5x = -15$$
$$\frac{5x}{5} = \frac{-15}{5}$$
$$x = -3$$

$$3x + 20 = 5 - 2x$$
$$3(-3) + 20 \overset{?}{=} 5 - 2(-3)$$
$$-9 + 20 \overset{?}{=} 5 + 6$$
$$11 = 11$$

64.

$$0.9x + 10 = 0.7x + 1.8$$
$$0.9x - 0.7x + 10 = 0.7x - 0.7x + 1.8$$
$$0.2x + 10 = 1.8$$
$$0.2x + 10 - 10 = 1.8 - 10$$
$$0.2x = -8.2$$
$$\frac{0.2x}{0.2} = \frac{-8.2}{0.2}$$
$$x = -41$$

$$0.9x + 10 = 0.7x + 1.8$$
$$0.9(-41) + 10 \stackrel{?}{=} 0.7(-41) + 1.8$$
$$-36.9 + 10 \stackrel{?}{=} -28.7 + 1.8$$
$$-26.9 = -26.9$$

65.

$$10(t - 3) = 3(t + 11)$$
$$10t - 30 = 3t + 33$$
$$10t - 3t - 30 = 3t - 3t + 33$$
$$7t - 30 = 33$$
$$7t - 30 + 30 = 33 + 30$$
$$7t = 63$$
$$\frac{7t}{7} = \frac{63}{7}$$
$$t = 9$$

$$10(t - 3) = 3(t + 11)$$
$$10(9 - 3) \stackrel{?}{=} 3(9 + 11)$$
$$10(6) \stackrel{?}{=} 3(20)$$
$$60 = 60$$

66.

$$2(5x - 7) = 2(x - 35)$$
$$10x - 14 = 2x - 70$$
$$10x - 2x - 14 = 2x - 2x - 70$$
$$8x - 14 = -70$$
$$8x - 14 + 14 = -70 + 14$$
$$8x = -56$$
$$\frac{8x}{8} = \frac{-56}{8}$$
$$x = -7$$

$$2(5x - 7) = 2(x - 35)$$
$$2[5(-7) - 7] \stackrel{?}{=} 2(-7 - 35)$$
$$2[-35 - 7] \stackrel{?}{=} 2(-42)$$
$$2(-42) \stackrel{?}{=} 2(-42)$$
$$-84 = -84$$

67.

$$\frac{3u - 6}{5} = 3$$
$$5 \cdot \frac{3u - 6}{5} = 5(3)$$
$$3u - 6 = 15$$
$$3u - 6 + 6 = 15 + 6$$
$$3u = 21$$
$$\frac{3u}{3} = \frac{21}{3}$$
$$u = 7$$

$$\frac{3u - 6}{5} = 3$$
$$\frac{3(7) - 6}{5} \stackrel{?}{=} 3$$
$$\frac{21 - 6}{5} \stackrel{?}{=} 3$$
$$\frac{15}{5} \stackrel{?}{=} 3$$
$$3 = 3$$

68.

$$\frac{5v - 35}{3} = -5$$

$$3 \cdot \frac{5v - 35}{3} = 3(-5)$$

$$5v - 35 = -15$$

$$5v - 35 + 35 = -15 + 35$$

$$5v = 20$$

$$\frac{5v}{5} = \frac{20}{5}$$

$$v = 4$$

$$\frac{5v - 35}{3} = -5$$

$$\frac{5(4) - 35}{3} \stackrel{?}{=} -5$$

$$\frac{20 - 35}{3} \stackrel{?}{=} -5$$

$$\frac{-15}{3} \stackrel{?}{=} -5$$

$$-5 = -5$$

69.

$$\frac{7x - 28}{4} = -21$$

$$4 \cdot \frac{7x - 28}{4} = 4(-21)$$

$$7x - 28 = -84$$

$$7x - 28 + 28 = -84 + 28$$

$$7x = -56$$

$$\frac{7x}{7} = \frac{-56}{7}$$

$$x = -8$$

$$\frac{7x - 28}{4} = -21$$

$$\frac{7(-8) - 28}{4} \stackrel{?}{=} -21$$

$$\frac{-56 - 28}{4} \stackrel{?}{=} -21$$

$$\frac{-84}{4} \stackrel{?}{=} -21$$

$$-21 = -21$$

70.

$$\frac{27 + 9y}{5} = -27$$

$$5 \cdot \frac{27 + 9y}{5} = 5(-27)$$

$$27 + 9y = -135$$

$$27 - 27 + 9y = -135 - 27$$

$$9y = -162$$

$$\frac{9y}{9} = \frac{-162}{9}$$

$$y = -18$$

$$\frac{27 + 9y}{5} = -27$$

$$\frac{27 + 9(-18)}{5} \stackrel{?}{=} -27$$

$$\frac{27 + (-162)}{5} \stackrel{?}{=} -27$$

$$\frac{-135}{5} \stackrel{?}{=} -27$$

$$-27 = -27$$

71.

$$2x - 5 = x - 5 + x$$

$$2x - 5 = 2x - 5$$

$$2x - 2x - 5 = 2x - 2x - 5$$

$$-5 = -5$$

Identity

72.

$$-3(a + 1) - a = -4a + 3$$

$$-3a - 3 - a = -4a + 3$$

$$-4a - 3 = -4a + 3$$

$$-4a + 4a - 3 = -4a + 4a + 3$$

$$-3 \neq 3$$

Contradiction

73. $2(x-1) + 4 = 4(1+x) - (2x+2)$
$2x - 2 + 4 = 4 + 4x - 2x - 2$
$2x + 2 = 2x + 2$
$2x - 2x + 2 = 2x - 2x + 2$
$2 = 2$
Identity

74. Let $x =$ the length of one part. Then
$2x - 7 =$ the length of the other part.

$$\boxed{\begin{smallmatrix}\text{Length of}\\\text{first part}\end{smallmatrix}} + \boxed{\begin{smallmatrix}\text{Length of}\\\text{second part}\end{smallmatrix}} = 8$$

$$x + 2x - 7 = 8$$
$$3x - 7 = 8$$
$$3x - 7 + 7 = 8 + 7$$
$$3x = 15$$
$$\frac{3x}{3} = \frac{15}{3}$$
$$x = 5$$

One part should be cut 5 feet long.

75. $x + 47° = 62°$
$x + 47° - 47° = 62° - 47°$
$x = 15°$

76. $x + 135° = 180°$
$x + 135° - 135° = 180° - 135°$
$x = 45°$

77. Let $x =$ the measure of the complement.
$x + 69° = 90°$
$x + 69° - 69° = 90° - 69°$
$x = 21°$

78. Let $x =$ the measure of the supplement.
$x + 69° = 180°$
$x + 69° - 69° = 180° - 69°$
$x = 111°$

79. Let $w =$ the width of the picture. Then $2w + 3 =$ the length of the picture.

$$\boxed{\text{Perimeter}} = 84$$
$$2w + 2(2w + 3) = 84$$
$$2w + 4w + 6 = 84$$
$$6w + 6 = 84$$
$$6w + 6 - 6 = 84 - 6$$
$$6w = 78$$
$$\frac{6w}{6} = \frac{78}{6}$$
$$w = 13 \Rightarrow \text{The width is 13 inches.}$$

80. Let $x =$ amount in 7% CD. Then $27000 - x =$ amount in 9% fund.

$$\boxed{\begin{smallmatrix}\text{Interest}\\\text{at 7\%}\end{smallmatrix}} + \boxed{\begin{smallmatrix}\text{Interest}\\\text{at 9\%}\end{smallmatrix}} = \boxed{\begin{smallmatrix}\text{Total}\\\text{interest}\end{smallmatrix}}$$

$$0.07x + 0.09(27000 - x) = 2110$$
$$7x + 9(27000 - x) = 211000$$
$$7x + 243000 - 9x = 211000$$
$$-2x + 243000 = 211000$$
$$-2x + 243000 - 243000 = 211000 - 243000$$
$$-2x = -32000$$
$$\frac{-2x}{-2} = \frac{-32000}{-2}$$
$$x = 16000 \Rightarrow \$16,000 \text{ is invested at 7\%, and \$11,000 is invested at 9\%.}$$

81. Let $t =$ time for friends to meet.

	r	t	d
Walk	3	t	$3t$
Bike	12	t	$12t$

$$\boxed{\begin{array}{c}\text{Distance} \\ \text{walked}\end{array}} + \boxed{\begin{array}{c}\text{Distance} \\ \text{biked}\end{array}} = 5$$

$$3t + 12t = 5$$
$$15t = 5$$
$$\frac{15t}{15} = \frac{5}{15}$$
$$t = \frac{1}{3}$$

They meet after $\frac{1}{3}$ hour (or 20 minutes).

82. Let $x =$ liters of 1% butterfat milk used.

$$\boxed{\begin{array}{c}\text{Butterfat in} \\ \text{4\% milk}\end{array}} + \boxed{\begin{array}{c}\text{Butterfat in} \\ \text{1\% milk}\end{array}} = \boxed{\begin{array}{c}\text{Butterfat in} \\ \text{2\% mixture}\end{array}}$$

$$0.04(12) + 0.01x = 0.02(12 + x)$$
$$4(12) + 1x = 2(12 + x)$$
$$48 + x = 24 + 2x$$
$$x - 2x + 48 = 24 + 2x - 2x$$
$$-x + 48 = 24$$
$$-x + 48 - 48 = 24 - 48$$
$$-x = -24$$
$$x = 24$$

24 liters of the 1% milk should be used.

83. Let $x =$ pounds of 90¢ candy.

Then $20 - x =$ pounds of $1.50 candy.

$$\boxed{\begin{array}{c}\text{Value of} \\ \text{90¢ candy}\end{array}} + \boxed{\begin{array}{c}\text{Value of} \\ \text{\$1.50 candy}\end{array}} = \boxed{\begin{array}{c}\text{Value of} \\ \text{mixture}\end{array}}$$

$$0.90x + 1.50(20 - x) = 1.20(20)$$
$$90x + 150(20 - x) = 120(20)$$
$$90x + 3000 - 150x = 2400$$
$$-60x + 3000 = 2400$$
$$-60x + 3000 - 3000 = 2400 - 3000$$
$$-60x = -600$$
$$\frac{-60x}{-60} = \frac{-600}{-60}$$
$$x = 10$$

10 lb of each should be used.

84. Let $x =$ # of kilowatt hours used.

$$17.50 + 0.18x = 43.96$$
$$1750 + 18x = 4396$$
$$1750 - 1750 + 18x = 4396 - 1750$$
$$18x = 2646$$
$$\frac{18x}{18} = \frac{2646}{18}$$
$$x = 147$$

147 kilowatt hours were used.

85. Let $x =$ feet of gutter needed.

$$35 + 1.50x = 162.50$$
$$350 + 15x = 1625$$
$$350 - 350 + 15x = 1625 - 350$$
$$15x = 1275$$
$$\frac{15x}{15} = \frac{1275}{15}$$
$$x = 85 \Rightarrow 85 \text{ feet of gutter were needed.}$$

86.
$$E = IR$$
$$\frac{E}{I} = \frac{IR}{I}$$
$$\frac{E}{I} = R, \text{ or } R = \frac{E}{I}$$

87.
$$i = prt$$
$$\frac{i}{pr} = \frac{prt}{pr}$$
$$\frac{i}{pr} = t, \text{ or } t = \frac{i}{pr}$$

88.
$$P = I^2R$$
$$\frac{P}{I^2} = \frac{I^2R}{I^2}$$
$$\frac{P}{I^2} = R, \text{ or } R = \frac{P}{I^2}$$

89.
$$d = rt$$
$$\frac{d}{t} = \frac{rt}{t}$$
$$\frac{d}{t} = r, \text{ or } r = \frac{d}{t}$$

90.
$$V = lwh$$
$$\frac{V}{lw} = \frac{lwh}{lw}$$
$$\frac{V}{lw} = h, \text{ or } h = \frac{V}{lw}$$

91.
$$y = mx + b$$
$$y - b = mx + b - b$$
$$y - b = mx$$
$$\frac{y - b}{x} = \frac{mx}{x}$$
$$\frac{y - b}{x} = m, \text{ or } m = \frac{y - b}{x}$$

92.
$$V = \pi r^2 h$$
$$\frac{V}{\pi r^2} = \frac{\pi r^2 h}{\pi r^2}$$
$$\frac{V}{\pi r^2} = h, \text{ or } h = \frac{V}{\pi r^2}$$

93.
$$a = 2\pi rh$$
$$\frac{a}{2\pi h} = \frac{2\pi rh}{2\pi h}$$
$$\frac{a}{2\pi h} = r, \text{ or } r = \frac{a}{2\pi h}$$

94.
$$F = \frac{GMm}{d^2}$$
$$d^2 \cdot F = d^2 \cdot \frac{GMm}{d^2}$$
$$d^2 F = GMm$$
$$\frac{d^2 F}{Mm} = \frac{GMm}{Mm}$$
$$\frac{d^2 F}{Mm} = G, \text{ or } G = \frac{d^2 F}{Mm}$$

95.
$$P = \frac{RT}{mV}$$
$$mV \cdot P = mV \cdot \frac{RT}{mV}$$
$$mVP = RT$$
$$\frac{mVP}{VP} = \frac{RT}{VP}$$
$$m = \frac{RT}{VP}$$

96.
$$3x + 2 < 5$$
$$3x + 2 - 2 < 5 - 2$$
$$3x < 3$$
$$\frac{3x}{3} < \frac{3}{3}$$
$$x < 1$$

97.
$$-5x - 8 < 7$$
$$-5x - 8 + 8 < 7 + 8$$
$$-5x < 15$$
$$\frac{-5x}{-5} > \frac{15}{-5}$$
$$x > -3$$

98.
$$5x - 3 \geq 2x + 9$$
$$5x - 2x - 3 \geq 2x - 2x + 9$$
$$3x - 3 \geq 9$$
$$3x - 3 + 3 \geq 9 + 3$$
$$3x \geq 12$$
$$\frac{3x}{3} \geq \frac{12}{3}$$
$$x \geq 4$$

99.
$$7x + 1 \leq 8x - 5$$
$$7x - 8x + 1 \leq 8x - 8x - 5$$
$$-x + 1 \leq -5$$
$$-x + 1 - 1 \leq -5 - 1$$
$$-x \leq -6$$
$$-(-x) \geq -(-6)$$
$$x \geq 6$$

100.
$$5(3 - x) \le 3(x - 3)$$
$$15 - 5x \le 3x - 9$$
$$15 - 5x - 3x \le 3x - 3x - 9$$
$$15 - 8x \le -9$$
$$15 - 15 - 8x \le -9 - 15$$
$$-8x \le -24$$
$$\frac{-8x}{-8} \ge \frac{-24}{-8}$$
$$x \ge 3$$

101.
$$3(5 - x) \ge 2x$$
$$15 - 3x \ge 2x$$
$$15 - 3x + 3x \ge 2x + 3x$$
$$15 \ge 5x$$
$$\frac{15}{5} \ge \frac{5x}{5}$$
$$3 \ge x, \text{ or } x \le 3$$

102.
$$8 < \quad x + 2 \quad < 13$$
$$8 - 2 < x + 2 - 2 < 13 - 2$$
$$6 < \quad x \quad < 11$$

103.
$$0 \le \quad 2 - 2x \quad < 4$$
$$0 - 2 \le 2 - 2 - 2x < 4 - 2$$
$$-2 \le \quad -2x \quad < 2$$
$$\frac{-2}{-2} \ge \quad \frac{-2x}{-2} \quad > \frac{2}{-2}$$
$$1 \ge \quad x \quad > -1$$
$$-1 < \quad x \quad \le 1$$

104. Let l = length. Then $l - 6$ = width.
$$\text{perimeter} \quad \le 68$$
$$2l + 2(l - 6) \le 68$$
$$2l + 2l - 12 \le 68$$
$$4l - 12 \quad \le 68$$
$$4l - 12 + 12 \le 68 + 12$$
$$4l \quad \le 80$$
$$\frac{4l}{4} \quad \le \frac{80}{4}$$
$$l \quad \le 20$$
The length must be no more than 20 feet.

Chapter 2 Test (page 144)

1.
$$5x + 3 = -2$$
$$5(-1) + 3 \overset{?}{=} -2$$
$$-5 + 3 \overset{?}{=} -2$$
$$-2 = -2$$
-1 is a solution.

2.
$$3(x + 2) = 2x$$
$$3(-6 + 2) \overset{?}{=} 2(-6)$$
$$3(-4) \overset{?}{=} -12$$
$$-12 = -12$$
-6 is a solution.

3.
$$-3(2 - x) = 0$$
$$-3[2 - (-2)] \overset{?}{=} 0$$
$$-3[4] \overset{?}{=} 0$$
$$-12 \ne 0$$
-2 is not a solution.

4. $3(x+2) = 2x + 7$

$3(1+2) \overset{?}{=} 2(1) + 7$

$3(3) \overset{?}{=} 2 + 7$

$9 = 9$

1 is a solution.

5. $x + 17 = -19$

$x + 17 - 17 = -19 - 17$

$x = -36$

6. $a - 15 = 32$

$a - 15 + 15 = 32 + 15$

$a = 47$

7. $12x = -144$

$\dfrac{12x}{12} = \dfrac{-144}{12}$

$x = -12$

8. $\dfrac{x}{7} = -1$

$7 \cdot \dfrac{x}{7} = 7(-1)$

$x = -7$

9. $8x + 2 = -14$

$8x + 2 - 2 = -14 - 2$

$8x = -16$

$\dfrac{8x}{8} = \dfrac{-16}{8}$

$x = -2$

10. $3 = 5 - 2x$

$3 - 5 = 5 - 5 - 2x$

$-2 = -2x$

$\dfrac{-2}{-2} = \dfrac{-2x}{-2}$

$1 = x$

11. $\dfrac{2x - 5}{3} = 3$

$3 \cdot \dfrac{2x - 5}{3} = 3(3)$

$2x - 5 = 9$

$2x - 5 + 5 = 9 + 5$

$2x = 14$

$\frac{2x}{2} = \frac{14}{2}$

$x = 7$

12. $\dfrac{3x - 18}{2} = 6x$

$2 \cdot \dfrac{3x - 18}{2} = 2(6x)$

$3x - 18 = 12x$

$3x - 3x - 18 = 12x - 3x$

$-18 = 9x$

$\frac{-18}{9} = \frac{9x}{9}$

$-2 = x$

13. $23 - 5(x + 10) = -12$

$23 - 5x - 50 = -12$

$-5x - 27 = -12$

$-5x - 27 + 27 = -12 + 27$

$-5x = 15$

$\dfrac{-5x}{-5} = \dfrac{15}{-5}$

$x = -3$

14. $\dfrac{7}{8}(x - 4) = 5x - \dfrac{7}{2}$

$8 \cdot \dfrac{7}{8}(x - 4) = 8\left(5x - \dfrac{7}{2}\right)$

$7(x - 4) = 8(5x) - 8 \cdot \frac{7}{2}$

$7x - 28 = 40x - 28$

$7x - 40x - 28 = 40x - 40x - 28$

$-33x - 28 = -28$

$-33x - 28 + 28 = -28 + 28$

$-33x = 0$

$x = 0$

15. $x + 5(x - 3) = x + 5x - 15 = 6x - 15$

16. $3x - 5(2 - x) = 3x - 10 + 5x = 8x - 10$

17. $-3x(x + 3) + 3x(x - 3) = -3x^2 - 9x + 3x^2 - 9x = -3x^2 + 3x^2 - 9x - 9x = -18x$

18. $-4x(2x - 5) - 7x(4x + 1) = -8x^2 + 20x - 28x^2 - 7x = -36x^2 + 13x$

19.
$$x + 45° = 120°$$
$$x + 45° - 45° = 120° - 45°$$
$$x = 75°$$

20. Let x = the measure of the supplement.
$$x + 105° = 180°$$
$$x + 105° - 105° = 180° - 105°$$
$$x = 75°$$

21. Let x = amount at 6%. Then $10000 - x$ = amount at 5%.

Interest at 6%	+	Interest at 5%	=	Total interest

$$0.06x + 0.05(10000 - x) = 560$$
$$6x + 5(10000 - x) = 56000$$
$$6x + 50000 - 5x = 56000$$
$$x + 50000 = 56000$$
$$x + 50000 - 50000 = 56000 - 50000$$
$$x = 6000 \Rightarrow \$6,000 \text{ is invested at 6\%, and } \$4,000 \text{ is invested at 5\%.}$$

22. Let t = time to meet.

	r	t	d
Car	65	t	$65t$
Truck	55	t	$55t$

Car Dist	+	Truck Dist	= 72

$$65t + 55t = 72$$
$$120t = 72$$
$$\frac{120t}{120} = \frac{72}{120}$$
$$t = \frac{3}{5}$$

They meet after $\frac{3}{5}$ of an hour.

23. Let x = liters of water used.

Amt. in 10% sol.	+	Amt. in water	=	Amt. in 8% sol.

$$0.10(30) + 0(x) = 0.08(30 + x)$$
$$10(30) + 0 = 8(30 + x)$$
$$300 = 240 + 8x$$
$$300 - 240 = 8x$$
$$60 = 8x$$
$$\frac{60}{8} = \frac{8x}{8}$$
$$\frac{15}{2} = x$$

$7\frac{1}{2}$ liters of water should be used.

24.
$$d = rt$$
$$\frac{d}{r} = \frac{rt}{r}$$
$$\frac{d}{r} = t, \text{ or } t = \frac{d}{r}$$

25.
$$P = 2l + 2w$$
$$P - 2w = 2l + 2w - 2w$$
$$P - 2w = 2l$$
$$\frac{P - 2w}{2} = \frac{2l}{2}$$
$$\frac{P - 2w}{2} = l, \text{ or } l = \frac{P - 2w}{2}$$

26.
$$A = 2\pi rh$$
$$\frac{A}{2\pi r} = \frac{2\pi rh}{2\pi r}$$
$$\frac{A}{2\pi r} = h, \text{ or } h = \frac{A}{2\pi r}$$

27.
$$A = P + Prt$$
$$A - P = P - P + Prt$$
$$A - P = Prt$$
$$\frac{A - P}{Pt} = \frac{Prt}{Pt}$$
$$\frac{A - P}{Pt} = r, \text{ or } r = \frac{A - P}{Pt}$$

28.
$$8x - 20 \geq 4$$
$$8x - 20 + 20 \geq 4 + 20$$
$$8x \geq 24$$
$$\frac{8x}{8} \geq \frac{24}{8}$$
$$x \geq 3$$

3

29.
$$x - 2(x + 7) > 14$$
$$x - 2x - 14 > 14$$
$$-x - 14 > 14$$
$$-x - 14 + 14 > 14 + 14$$
$$-x > 28$$
$$-(-x) < -(28)$$
$$x < -28$$

-28

30.
$$-4 \le 2(x+1) < 10$$
$$-4 \le 2x+2 < 10$$
$$-4-2 \le 2x+2-2 < 10-2$$
$$-6 \le 2x < 8$$
$$\frac{-6}{2} \le \frac{2x}{2} < \frac{8}{2}$$
$$-3 \le x < 4$$

31.
$$-2 < 5(x-1) \le 10$$
$$-2 < 5x-5 \le 10$$
$$-2+5 < 5x-5+5 \le 10+5$$
$$3 < 5x \le 15$$
$$\frac{3}{5} < \frac{5x}{5} \le \frac{15}{5}$$
$$\frac{3}{5} < x \le 3$$

Cumulative Review Exercises (page 145)

1. $\frac{27}{9} = 3$: integer, rational number, real number, positive number

2. $-0.25 = -\frac{1}{4}$: rational number, real number, negative number

3.

2 3 4 5 6 7

4.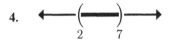

5. $\frac{|-3|-|3|}{|-3-3|} = \frac{(+3)-(+3)}{|-6|} = \frac{0}{+6} = 0$

6. $\frac{5}{7} \cdot \frac{14}{3} = \frac{5}{\underset{1}{\cancel{7}}} \cdot \frac{\overset{1}{\cancel{14}}}{3} = \frac{10}{3}$

7. $2\frac{3}{5} + 5\frac{1}{2} = \frac{13}{5} + \frac{11}{2} = \frac{13 \cdot 2}{5 \cdot 2} + \frac{11 \cdot 5}{2 \cdot 5} = \frac{26}{10} + \frac{55}{10} = \frac{81}{10} = 8\frac{1}{10}$

8. $35.7 - 0.05 = 35.65$

9. $(3x - 2y)z = [3(-5) - 2(3)]0 = 0$

10. $\frac{x - 3y + |z|}{2 - x} = \frac{-5 - 3(3) + |0|}{2 - (-5)} = \frac{-5 - 9 + 0}{2 + 5} = \frac{-14}{7} = -2$

11. $x^2 - y^2 + z^2 = (-5)^2 - (3)^2 + 0^2$
$$= 25 - 9 + 0 = 16$$

12. $\frac{x}{y} + \frac{y+2}{3-z} = \frac{-5}{3} + \frac{3+2}{3-0} = -\frac{5}{3} + \frac{5}{3} = 0$

13.
$$rb = a$$
$$0.075(330) = a$$
$$24.75 = a$$

14.
$$rb = a$$
$$0.32b = 1688$$
$$\frac{0.32b}{0.32} = \frac{1688}{0.32}$$
$$b = 5{,}275$$

15. 2nd term: $5x^2y$; coefficient: 5

16. 3rd term: $37y$; factors: 37, y

17. $3x - 5x + 2y = -2x + 2y$

18. $3(x-7) + 2(8-x) = 3x - 21 + 16 - 2x$
$$= x - 5$$

19. $2x^2y^3 - xy(xy^2) = 2x^2y^3 - x^2y^3 = x^2y^3$

20. $x^2(3 - y) + x(xy + x) = 3x^2 - x^2y + x^2y + x^2 = 4x^2$

21.
$$3(x - 5) + 2 = 2x$$
$$3x - 15 + 2 = 2x$$
$$3x - 13 = 2x$$
$$3x - 3x - 13 = 2x - 3x$$
$$-13 = -x$$
$$13 = x$$

22.
$$\frac{x-5}{3} - 5 = 7$$
$$\frac{x-5}{3} - 5 + 5 = 7 + 5$$
$$\frac{x-5}{3} = 12$$
$$3 \cdot \frac{x-5}{3} = 3(12)$$
$$x - 5 = 36$$
$$x - 5 + 5 = 36 + 5$$
$$x = 41$$

23.
$$\frac{2x-1}{5} = \frac{1}{2}$$
$$10 \cdot \frac{2x-1}{5} = 10 \cdot \frac{1}{2}$$
$$2(2x - 1) = 5$$
$$4x - 2 = 5$$
$$4x - 2 + 2 = 5 + 2$$
$$4x = 7$$
$$\frac{4x}{4} = \frac{7}{4}$$
$$x = \frac{7}{4}$$

24.
$$2(a - 5) - (3a + 1) = 0$$
$$2a - 10 - 3a - 1 = 0$$
$$-a - 11 = 0$$
$$-a + a - 11 = 0 + a$$
$$-11 = a$$

25. Let x = Dealer's invoice.

$$\boxed{\text{Price}} = \boxed{\substack{\text{Dealer's} \\ \text{invoice}}} + \boxed{\text{Markup}}$$
$$23499 = x + 0.03x$$
$$23499 = 1.03x$$
$$\frac{23499}{1.03} = \frac{1.03x}{1.03}$$
$$22814.56 = r$$

The dealer's invoice was $22,814.56.

26. Let x = original sofa price.

$$\boxed{\text{Price}} = \boxed{\substack{\text{New sofa} \\ \text{price}}} + \boxed{\substack{\text{New chair} \\ \text{price}}}$$
$$780 = x - 0.35x + 300 - 0.35(300)$$
$$780 = 0.65x + 300 - 105$$
$$780 = 0.65x + 195$$
$$780 - 195 = 0.65x + 195 - 195$$
$$585 = 0.65x$$
$$\frac{585}{0.65} = \frac{0.65x}{0.65}$$
$$900 = x$$

The original price of the sofa was $900.

27. Let x = original car price.
$$13725.25 = x + 0.085x$$
$$13725.25 = 1.085x$$
$$\frac{13725.25}{1.085} = \frac{1.085x}{1.085}$$
$$12650 = x$$

The original price was $12,650.

28. Let x and $3x$ = the unknown amounts.

$$\boxed{\text{Cement}} + \boxed{\text{Gravel}} = \boxed{\text{Concrete}}$$
$$x + 3x = 500$$
$$4x = 500$$
$$\frac{4x}{4} = \frac{500}{4}$$
$$x = 125 \text{ lbs cement}$$

29. Let x and $2x =$ the lengths.

Total length $= 35$

$14 + x + 2x = 35$

$14 + 3x = 35$

$14 - 14 + 3x = 35 - 14$

$3x = 21$

$\frac{3x}{3} = \frac{21}{3}$

$x = 7$ ft

The section will not span the doorway.

30. Let $x =$ the length of one panel.

$\boxed{\text{Total length}} = 18$

$x + x + 3.4 = 18$

$2x + 3.4 = 18$

$2x + 3.4 - 3.4 = 18 - 3.4$

$2x = 14.6$

$\frac{2x}{2} = \frac{14.6}{2}$

$x = 7.3$

The lengths are 7.3 and 10.7 feet.

31. $A = \frac{1}{2}h(b + B)$

$2A = 2 \cdot \frac{1}{2}h(b + B)$

$2A = h(b + B)$

$\frac{2A}{b+B} = \frac{h(b+B)}{b+B}$

$\frac{2A}{b+B} = h$, or $h = \frac{2A}{b+B}$

32. $y = mx + b$

$y - b = mx + b - b$

$y - b = mx$

$\frac{y-b}{m} = \frac{mx}{m}$

$\frac{y-b}{m} = x$, or $x = \frac{y-b}{m}$

33. $4^2 - 5^2 = 16 - 25 = -9$

34. $(4 - 5)^2 = (-1)^2 = 1$

35. $5(4^3 - 2^3) = 5(64 - 8) = 5(56) = 280$

36. $-2(5^4 - 7^3) = -2(625 - 343)$
$= -2(282) = -564$

37.

$8(4 + x) > 10(6 + x)$

$32 + 8x > 60 + 10x$

$32 + 8x - 10x > 60 + 10x - 10x$

$32 - 2x > 60$

$32 - 32 - 2x > 60 - 32$

$-2x > 28$

$\frac{-2x}{-2} < \frac{28}{-2}$

$x < -14$

-14

38.

$-9 < 3(x + 2) \le 3$

$-9 < 3x + 6 \le 3$

$-9 - 6 < 3x + 6 - 6 \le 3 - 6$

$-15 < 3x \le -3$

$\frac{-15}{3} < \frac{3x}{3} \le \frac{-3}{3}$

$-5 < x \le -1$

$-5 \qquad -1$

Exercise 3.1 (page 156)

1. $-3 - 3(-5) = -3 - (-15) = -3 + 15$
$$= 12$$

3. opposite of -8: 8

5.
$$-4x + 7 = -21$$
$$-4x + 7 - 7 = -21 - 7$$
$$-4x = -28$$
$$\frac{-4x}{-4} = \frac{-28}{-4}$$
$$x = 7$$

7. $(x+1)(x+y)^2 = (-2+1)[-2+(-5)]^2$
$$= (-1)(-7)^2$$
$$= -1(+49) = -49$$

9. ordered pair

11. origin

13. rectangular coordinate

15. no

17. origin, left, up

19. II

21.

x	y
4	3 or -3
0	5 or -5
-3	4 or -4
5 or -5	0
-4	-3 or 3
0	-5 or 5
3	-4 or 4

23. The point $(-10, 60)$ indicates that 10 minutes before the workout started, her heart rate was 60 beats per minute.

25. The point with an x-coordinate of 30 is $(30, 150)$, so her heart rate was 150 beats per minute one half-hour after starting.

27. The points on the graph with a y-coordinate of 100 have x-coordinates of approximately 5 and 50, so her heart rate was 100 beats per minute after about 5 and 50 minutes.

29. Before the workout, her heart rate was 60 beats per minute. After the workout, her heart rate was about 70 beats per minute, or about 10 beats per minute higher.

31.

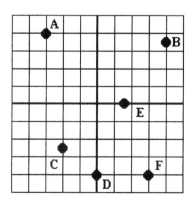

33.

City	Ordered Pair
Carbondale	(3, J)
Champaign	(4.5, D)
Chicago	(5.5, B)
Peoria	(3.5, C)
Rockford	(3.5, A)
Springfield	(2.25, E)
St. Louis	(2, H)

35. **a.** The highest point on the graph for the 60° angle is $(-2, 7)$, for a height of 7 ft. The highest point on the graph for the 30° angle is $(0, 3)$, for a height of 3 ft. The difference is 4 ft.

b. The farthest point on the graph for the 60° angle is $(3, 0)$, for a final location of 3 ft. The farthest point on the graph for the 30° angle is $(7, 0)$, for a final location of 7 ft. The difference is 4 ft.

37. **a.** To find the charge for a 1-day rental, find the y-coordinate of the point with an x-coordinate of 1. This is the point $(1, 2)$. The charge will be $2.

b. To find the charge for a 2-day rental, find the y-coordinate of the point with an x-coordinate of 2. This is the point $(2, 4)$. The charge will be $4.

c. To find the charge for a 5-day rental, find the y-coordinate of the point with an x-coordinate of 5. This is the point $(5, 7)$. The charge will be $7.

d. To find the charge for a 7-day rental, find the y-coordinate of the point with an x-coordinate of 7. This is the point $(7, 9)$. The charge will be $9.

39.

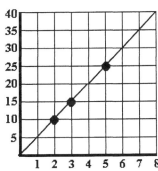

a. $(7, 35) \Rightarrow 35$ miles
b. $(4, 20) \Rightarrow 4$ gallons
c. $(6.5, 32.5) \Rightarrow 32.5$ miles

41.

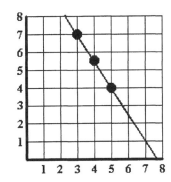

a. A 3-year old car is worth $7000.
b. $(7, 1) \Rightarrow \$1000$
c. $(6, 2500) \Rightarrow 6$ years

43. **Answers may vary.**   **45.** **Answers may vary.**   **47.** **Answers may vary.**

Exercise 3.2 (page 170)

1. $\dfrac{x}{8} = -12$

$8 \cdot \dfrac{x}{8} = 8(-12)$

$x = -96$

3. expression

5. $rb = a$

$0.005(250) = a$

$1.25 = a$

7. $-2.5 - (-2.6) = -2.5 + 2.6 = 0.1$

9. two

11. independent, dependent

13. linear

15. y-intercept

17. $x - 2y = -4$

$4 - 2(4) \overset{?}{=} -4$

$4 - 8 \overset{?}{=} -4$

$-4 = -4 \Rightarrow (4, 4)$ is a solution.

19. $y = \dfrac{2}{3}x + 5$

$12 \overset{?}{=} \dfrac{2}{3}(6) + 5$

$12 \overset{?}{=} 4 + 5$

$12 \neq 9 \Rightarrow (6, 12)$ is not a solution.

21.

$$y = x - 3$$

$x = 0$	$x = 1$	$x = -2$	$x = -4$
$y = 0 - 3$	$y = 1 - 3$	$y = -2 - 3$	$y = -4 - 3$
$y = -3$	$y = -2$	$y = -5$	$y = -7$

x	y
0	-3
1	-2
-2	-5
-4	-7

23.

$$y = -2x$$

$x = 0$	$x = 1$	$x = 3$	$x = -2$
$y = -2(0)$	$y = -2(1)$	$y = -2(3)$	$y = -2(-2)$
$y = 0$	$y = -2$	$y = -6$	$y = 4$

input	output
0	0
1	-2
3	-6
-2	4

25.

$$y = 2x$$

$x = 0$	$x = 1$	$x = -1$
$y = 2(0)$	$y = 2(1)$	$y = 2(-1)$
$y = 0$	$y = 2$	$y = -2$

x	y
0	0
1	2
-1	-2

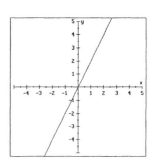

27.

$$y = 2x - 1$$

$x = 0$	$x = 1$	$x = -1$
$y = 2(0) - 1$	$y = 2(1) - 1$	$y = 2(-1) - 1$
$y = 0 - 1$	$y = 2 - 1$	$y = -2 - 1$
$y = -1$	$y = 1$	$y = -3$

x	y
0	-1
1	1
-1	-3

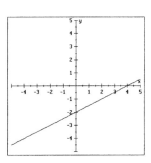

29.

$$y = \frac{x}{2} - 2$$

$x = 0$	$x = 2$	$x = -2$
$y = \dfrac{0}{2} - 2$	$y = \dfrac{2}{2} - 2$	$y = \dfrac{-2}{2} - 2$
$y = 0 - 2$	$y = 1 - 2$	$y = -1 - 2$
$y = -2$	$y = -1$	$y = -3$

x	y
0	-2
2	-1
-2	-3

31.

$$x + y = 7$$

$x = 0$	$y = 0$
$0 + y = 7$	$x + 0 = 7$
$y = 7$	$x = 7$
$(0, 7)$	$(7, 0)$

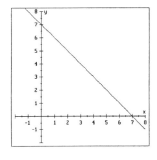

33.

$$x - y = 7$$

$x = 0$	$y = 0$
$0 - y = 7$	$x - 0 = 7$
$-y = 7$	$x = 7$
$y = -7$	$(7, 0)$
$(0, -7)$	

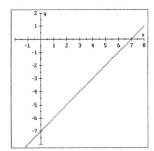

35.
$$y = -2x + 5$$

$x = 0$	$y = 0$
$y = -2(0) + 5$	$0 = -2x + 5$
$y = 0 + 5$	$2x = 5$
$y = 5$	$x = \frac{5}{2}$
$(0, 5)$	$\left(\frac{5}{2}, 0\right)$

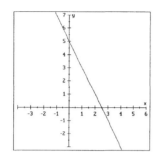

37.
$$2x + 3y = 12$$

$x = 0$	$y = 0$
$2(0) + 3y = 12$	$2x + 3(0) = 12$
$0 + 3y = 12$	$2x + 0 = 12$
$3y = 12$	$2x = 12$
$y = 4$	$x = 6$
$(0, 4)$	$(6, 0)$

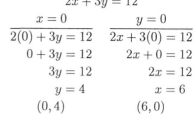

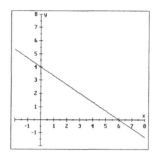

39.
$$y = -5$$
horizontal, y-coordinate $= -5$

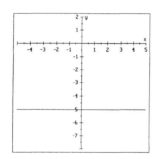

41.
$$x = 5$$
vertical, x-coordinate $= 5$

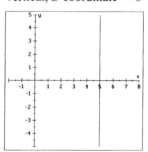

43.
$$y = 0$$
horizontal, y-coordinate $= 0$

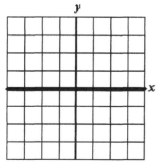

45.
$$2x = 5$$
$$x = \frac{5}{2} = 2\frac{1}{2}$$
vertical, x-coordinate $= \frac{5}{2}$

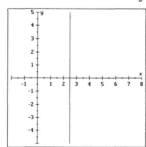

47. **a.** $c = 50 + 25u$

b.

u	c
4	$50 + 25(4) = 150$
8	$50 + 25(8) = 250$
14	$50 + 25(14) = 400$

c. The service fee is $50.

d. cost for 18 units $= \$500$
cost for 12 units $= \$350$
Total cost $= \$850$

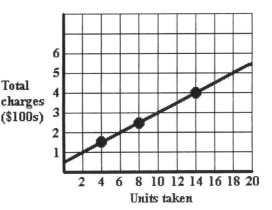

Total charges ($100s) vs. Units taken

49. **a.**

r	h
7	$3.9(7) + 28.9 = 56.2$
8.5	$3.9(8.5) + 28.9 = 62.1$
9	$3.9(9) + 28.9 = 64.0$

b. ...taller the woman is.

c. 58 inches tall

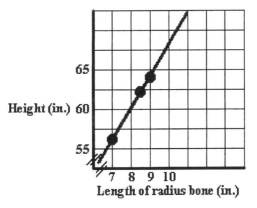

Height (in.) vs. Length of radius bone (in.)

51. Answers may vary. **53.** Answers may vary. **55.** Answers may vary.

57. $x_M = \dfrac{a+c}{2} = \dfrac{5+7}{2} = \dfrac{12}{2} = 6$

$y_M = \dfrac{b+d}{2} = \dfrac{3+9}{2} = \dfrac{12}{2} = 6$

$M(6,6)$

59. $x_M = \dfrac{a+c}{2} = \dfrac{2+(-3)}{2} = \dfrac{-1}{2} = -\dfrac{1}{2}$

$y_M = \dfrac{b+d}{2} = \dfrac{-7+12}{2} = \dfrac{5}{2}$

$M\left(-\dfrac{1}{2}, \dfrac{5}{2}\right)$

61. $x_M = \dfrac{a+c}{2} = \dfrac{4+10}{2} = \dfrac{14}{2} = 7$

$y_M = \dfrac{b+d}{2} = \dfrac{6+6}{2} = \dfrac{12}{2} = 6$

$M(7,6)$

Exercise 3.3 (page 182)

1. $4(a - 3) + 2a = 4a - 12 + 2a = 6a - 12$

3. $4z - 6(z + w) + 2w = 4z - 6z - 6w + 2w$
$$= -2z - 4w$$

5. $3(a - b) - 2(a + b) = 3a - 3b - 2a - 2b = a - 5b$

7. $y; x$

9. run

11. undefined

13. perpendicular

15. $m = \dfrac{y_2 - y_1}{x_2 - x_1} = \dfrac{9 - 0}{3 - 0} = \dfrac{9}{3} = 3$

17. $m = \dfrac{y_2 - y_1}{x_2 - x_1} = \dfrac{1 - 8}{6 - (-1)} = \dfrac{-7}{7} = -1$

19. $m = \dfrac{y_2 - y_1}{x_2 - x_1} = \dfrac{2 - (-1)}{-6 - 3} = \dfrac{3}{-9} = -\dfrac{1}{3}$

21. $m = \dfrac{y_2 - y_1}{x_2 - x_1} = \dfrac{5 - 5}{-9 - 7} = \dfrac{0}{-16} = 0$

23. $m = \dfrac{y_2 - y_1}{x_2 - x_1} = \dfrac{-2 - (-5)}{-7 - (-7)} = \dfrac{3}{0}$
The slope is undefined.

25. $m = \dfrac{y_2 - y_1}{x_2 - x_1} = \dfrac{2 - (-3)}{-3 - 2} = \dfrac{5}{-5} = -1$

27. Pick two x-coordinates at random and find the corresponding y-coordinates. For example, use $x = 0$ and $x = 4$.

x	0	4
y	6	0

$m = \dfrac{y_2 - y_1}{x_2 - x_1} = \dfrac{0 - 6}{4 - 0} = \dfrac{-6}{4} = -\dfrac{3}{2}$

29. Pick two x-coordinates at random and find the corresponding y-coordinates. For example, use $x = 2$ and $x = 6$.

x	2	6
y	2	5

$m = \dfrac{y_2 - y_1}{x_2 - x_1} = \dfrac{5 - 2}{6 - 2} = \dfrac{3}{4}$

31. Pick two x-coordinates at random and find the corresponding y-coordinates. For example, use $x = 4$ and $x = 8$.

x	4	8
y	0	2

$m = \dfrac{y_2 - y_1}{x_2 - x_1} = \dfrac{2 - 0}{8 - 4} = \dfrac{2}{4} = \dfrac{1}{2}$

33. $4y = 3(y + 2)$ horizontal line
$4y = 3y + 6$ $m = 0$
$y = 6$

35. negative

37. positive

39. undefined

41. $3 \neq -\dfrac{1}{3}$
$3\left(-\dfrac{1}{3}\right) = -1$
perpendicular

43. $4 \neq 0.25$
$4(0.25) \neq -1$
neither

45. $\dfrac{3}{4} = \dfrac{6}{8}$
parallel

47. $m = \dfrac{y_2 - y_1}{x_2 - x_1} = \dfrac{2 - 4}{4 - 3} = \dfrac{-2}{1} = -2$

parallel

49. $m = \dfrac{y_2 - y_1}{x_2 - x_1} = \dfrac{5 - 1}{6 - (-2)} = \dfrac{4}{8} = \dfrac{1}{2}$

perpendicular

51. $m = \dfrac{y_2 - y_1}{x_2 - x_1} = \dfrac{6 - 4}{6 - 5} = \dfrac{2}{1} = 2 \Rightarrow$ neither

53. $PQ: \; m = \dfrac{y_2 - y_1}{x_2 - x_1} = \dfrac{8 - 4}{4 - (-2)} = \dfrac{4}{6} = \dfrac{2}{3}$ $\quad PR: \; m = \dfrac{y_2 - y_1}{x_2 - x_1} = \dfrac{12 - 4}{8 - (-2)} = \dfrac{8}{10} = \dfrac{4}{5}$

not on same line

55. $PQ: \; m = \dfrac{y_2 - y_1}{x_2 - x_1} = \dfrac{0 - 10}{-6 - (-4)} = \dfrac{-10}{-2} = 5$ $\quad PR: \; m = \dfrac{y_2 - y_1}{x_2 - x_1} = \dfrac{5 - 10}{-1 - (-4)} = \dfrac{-5}{3}$

not on same line

57. $PQ: \; m = \dfrac{y_2 - y_1}{x_2 - x_1} = \dfrac{8 - 4}{0 - (-2)} = \dfrac{4}{2} = 2$ $\quad PR: \; m = \dfrac{y_2 - y_1}{x_2 - x_1} = \dfrac{12 - 4}{2 - (-2)} = \dfrac{8}{4} = 2$

on same line

59. x-axis: $y = 0$, $m = 0$

61. $m = \dfrac{\text{rise}}{\text{run}} = \dfrac{24}{5280} = \dfrac{1}{220}$

63. $m = \dfrac{\text{rise}}{\text{run}} = \dfrac{4}{12} = \dfrac{1}{3}$

65. Let $x =$ the number of years the program has been offered, and let $y =$ the enrollment. Then two points are $(1, 12)$ and $(5, 26)$. Find the slope:

$m = \dfrac{y_2 - y_1}{x_2 - x_1} = \dfrac{26 - 12}{5 - 1} = \dfrac{14}{4} = 3.5$

The enrollment is growing by 3.5 students per year.

67. Let $x =$ the year, and let $y =$ the price. Then two points are $(-10, 6700)$ and $(-3, 2200)$. Find the slope:

$m = \dfrac{y_2 - y_1}{x_2 - x_1} = \dfrac{6700 - 2200}{-3 - (-10)} = \dfrac{4500}{7}$

≈ 642.86

The price has decreased by about \$642.86 per year.

69. **Answers may vary.**

71. Slope from $(3, a)$ to $(5, 7)$: $m = \dfrac{y_2 - y_1}{x_2 - x_1} = \dfrac{7 - a}{5 - 3} = \dfrac{7 - a}{2}$

Slope from $(5, 7)$ to $(7, 10)$: $m = \dfrac{y_2 - y_1}{x_2 - x_1} = \dfrac{10 - 7}{7 - 5} = \dfrac{3}{2}$

Set slopes equal: $\dfrac{7 - a}{2} = \dfrac{3}{2}$

$2 \cdot \dfrac{7 - a}{2} = 2 \cdot \dfrac{3}{2}$

$7 - a = 3$

$4 = a$

Exercise 3.4 (page 189)

1. $3(x + 2) + x = 5x$
$3x + 6 + x = 5x$
$4x + 6 = 5x$
$6 = x$

3. $\dfrac{5(2 - x)}{3} - 1 = x + 5$
$3\left[\dfrac{5(2 - x)}{3} - 1\right] = 3(x + 5)$
$5(2 - x) - 3 = 3x + 15$
$10 - 5x - 3 = 3x + 15$
$-5x + 7 = 3x + 15$
$-8x = 8$
$x = -1$

5. $rb = a$
$0.67(15000) = a$
$10050 = a \Rightarrow$ About 10,050 of the ads are likely to be read.

7. $y - y_1 = m(x - x_1)$

9. $(1, 2); 2; 3$

11. $y - y_1 = m(x - x_1)$
$y - 0 = 4(x - 0)$

13. $y - y_1 = m(x - x_1)$
$y - 3 = 2(x - (-5))$

15. $y - y_1 = m(x - x_1)$
$y - 5 = -\dfrac{6}{7}(x - 6)$

17. $y - y_1 = m(x - x_1)$
$y - (-8) = 0.5(x - (-1))$

19. $y - y_1 = m(x - x_1)$
$y - 0 = -5(x - (-7))$

21. $y - y_1 = m(x - x_1)$
$y - 7 = 5(x - 0)$
$y - 7 = 5x$
$y = 5x + 7$

23. $y - y_1 = m(x - x_1)$
$y - 0 = -3(x - 2)$
$y = -3x + 6$

25. $y - y_1 = m(x - x_1)$
$y - 2 = \dfrac{6}{7}(x - 7)$
$y - 2 = \dfrac{6}{7}x - 6$
$y = \dfrac{6}{7}x - 4$

27. $y - y_1 = m(x - x_1)$
$y - (-1) = -\dfrac{3}{4}(x - 4)$
$y + 1 = -\dfrac{3}{4}x + 3$
$y = -\dfrac{3}{4}x + 2$

29. $y - y_1 = m(x - x_1)$
$y - 8 = 0.5(x - 10)$
$y - 8 = 0.5x - 5$
$y = 0.5x + 3$

31. Use the points $(-1, 3)$ and $(2, 5)$:

$$m = \frac{y_2 - y_1}{x_2 - x_1} = \frac{5 - 3}{2 - (-1)} = \frac{2}{3}$$

$$y - y_1 = m(x - x_1)$$

$$y - 3 = \frac{2}{3}(x - (-1))$$

$$y - 3 = \frac{2}{3}(x + 1)$$

$$y - 3 = \frac{2}{3}x + \frac{2}{3}$$

$$y = \frac{2}{3}x + \frac{2}{3} + \frac{3}{1}$$

$$y = \frac{2}{3}x + \frac{2}{3} + \frac{9}{3}$$

$$y = \frac{2}{3}x + \frac{11}{3}$$

33. Use the points $(-2, 0)$ and $(0, -1)$:

$$m = \frac{y_2 - y_1}{x_2 - x_1} = \frac{-1 - 0}{0 - (-2)} = \frac{-1}{2} = -\frac{1}{2}$$

$$y - y_1 = m(x - x_1)$$

$$y - 0 = -\frac{1}{2}(x - (-2))$$

$$y = -\frac{1}{2}(x + 2)$$

$$y = -\frac{1}{2}x - 1$$

35. $m = \dfrac{y_2 - y_1}{x_2 - x_1} = \dfrac{4 - 0}{4 - 0} = \dfrac{4}{4} = 1$

$$y - y_1 = m(x - x_1)$$

$$y - 0 = 1(x - 0)$$

$$y = x$$

37. $m = \dfrac{y_2 - y_1}{x_2 - x_1} = \dfrac{-3 - 4}{0 - 3} = \dfrac{-7}{-3} = \dfrac{7}{3}$

$$y - y_1 = m(x - x_1)$$

$$y - 4 = \frac{7}{3}(x - 3)$$

$$y - 4 = \frac{7}{3}x - 7$$

$$y = \frac{7}{3}x - 3$$

39. $m = \dfrac{y_2 - y_1}{x_2 - x_1} = \dfrac{-4 - 2}{-3 - 1} = \dfrac{-6}{-4} = \dfrac{3}{2}$

$$y - y_1 = m(x - x_1)$$

$$y - 2 = \frac{3}{2}(x - 1)$$

$$y - 2 = \frac{3}{2}x - \frac{3}{2}$$

$$y = \frac{3}{2}x - \frac{3}{2} + \frac{2}{1}$$

$$y = \frac{3}{2}x - \frac{3}{2} + \frac{4}{2}$$

$$y = \frac{3}{2}x + \frac{1}{2}$$

41. $(x_1, y_1) = (1, 3); m = 2 = \dfrac{+2}{+1} = \dfrac{\Delta y}{\Delta x}$

Start at $(1, 3)$ and go up 2 and right 1.

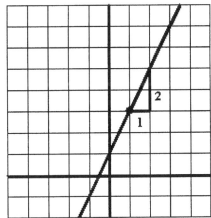

43. $(x_1, y_1) = (2, 4)$; $m = -\dfrac{2}{3} = \dfrac{-2}{+3} = \dfrac{\Delta y}{\Delta x}$

Start at $(2, 4)$ and go down 2 and right 3.

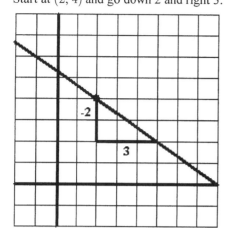

45. $(x_1, y_1) = (-2, 1)$; $m = -\dfrac{1}{2} = \dfrac{-1}{+2} = \dfrac{\Delta y}{\Delta x}$

Start at $(-2, 1)$ and go down 1 and right 2.

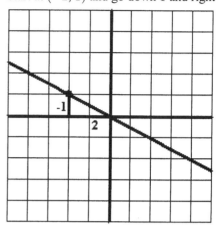

47. $(x_1, y_1) = (2, -2)$; $m = 3 = \dfrac{+3}{+1} = \dfrac{\Delta y}{\Delta x}$

Start at $(2, -2)$ and go up 3 and right 1.

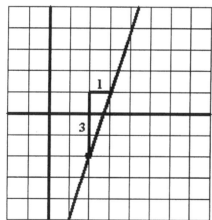

49. Let x = the # of kilowatt-hours, and let y = the bill. Then two points are known: $(200, 42)$ and $(250, 47)$. Find the slope:

$$m = \frac{y_2 - y_1}{x_2 - x_1} = \frac{47 - 42}{250 - 200} = \frac{5}{50} = \frac{1}{10}$$

Find the equation of the line:

$$y - y_1 = m(x - x_1)$$
$$y - 42 = \tfrac{1}{10}(x - 200)$$
$$y - 42 = \tfrac{1}{10}x - 20$$
$$y = \tfrac{1}{10}x + 22$$

Let $x = 300$ and find y:

$$y = \tfrac{1}{10}x + 22$$
$$= \tfrac{1}{10}(300) + 22 = 30 + 22 = 52$$

Her bill will be $52.

51. Let $x =$ the number of years owned, and let $y =$ the value. Then two points are known: $(0, 370)$ and $(2, 450)$. Find the slope:

$$m = \frac{y_2 - y_1}{x_2 - x_1} = \frac{450 - 370}{2 - 0} = \frac{80}{2} = 40$$

Find the equation of the line:

$$y - y_1 = m(x - x_1)$$
$$y - 370 = 40(x - 0)$$
$$y - 370 = 40x$$
$$y = 40x + 370$$

Let $x = 13$: $y = 40x + 370$
$$= 40(13) + 370$$
$$= 520 + 370 = \$890$$

The table will be worth \$890 after 13 years.

53. Let $x =$ the number of hours, and let $y =$ the total charge.

$$m = \frac{y_2 - y_1}{x_2 - x_1} = \frac{105 - 70}{4 - 2} = \frac{35}{2} = \$17.50 \text{ per hr}$$

55. Let $x =$ the year after purchase, and let $y =$ the value. Then two points are known: $(3, 147700)$ and $(10, 172200)$.

$$m = \frac{y_2 - y_1}{x_2 - x_1} = \frac{172200 - 147700}{10 - 3}$$
$$= \frac{24500}{7} = 3500$$

Find the equation of the line:

$$y - y_1 = m(x - x_1)$$
$$y - 147700 = 3500(x - 3)$$
$$y - 147700 = 3500x - 10500$$
$$y = 3500x + 137200$$

Let $x = 0$:
$$y = 3500(0) + 137200 = 137200$$
The original price was \$137,200.

57. **Answers may vary.**

59. The slope of a vertical line is undefined, so the equation cannot be written in point-slope form.

Exercise 3.5 (page 198)

1.
$$2x + 3 = 7$$
$$2x + 3 - 3 = 7 - 3$$
$$2x = 4$$
$$\frac{2x}{2} = \frac{4}{2}$$
$$x = 2$$

3.
$$3(y - 2) = y + 1$$
$$3y - 6 = y + 1$$
$$3y - y - 6 = y - y + 1$$
$$2y - 6 = 1$$
$$2y - 6 + 6 = 1 + 6$$
$$2y = 7$$
$$\frac{2y}{2} = \frac{7}{2}$$
$$y = \frac{7}{2}$$

5. $y = mx + b$

7. reciprocals

9.
$$y = mx + b$$
$$y = 3x + b$$
$$17 = 3(0) + b$$
$$17 = 0 + b$$
$$17 = b$$
$$y = 3x + 17$$

11.
$$y = mx + b$$
$$y = -7x + b$$
$$5 = -7(7) + b$$
$$5 = -49 + b$$
$$54 = b$$
$$y = -7x + 54$$

13.
$$y = mx + b$$
$$y = 0x + b$$
$$-4 = 0(2) + b$$
$$-4 = 0 + b$$
$$-4 = b$$
$$y = 0x + (-4)$$
$$y = -4$$

15.
$$y = mx + b$$
$$y = \frac{2}{3}x + b$$
$$4 = \frac{2}{3}(-3) + b$$
$$4 = -2 + b$$
$$6 = b$$
$$y = \frac{2}{3}x + 6$$

17.
$$y = mx + b$$
$$y = -\frac{4}{3}x + b$$
$$-2 = -\frac{4}{3}(6) + b$$
$$-2 = -8 + b$$
$$6 = b$$
$$y = -\frac{4}{3}x + 6$$

19.
$$m = \frac{y_2 - y_1}{x_2 - x_1}$$
$$= \frac{-11 - (-5)}{3 - 1}$$
$$= \frac{-6}{2} = -3$$
$$y = mx + b$$
$$-5 = -3(1) + b$$
$$-5 = -3 + b$$
$$-2 = b$$
$$y = -3x - 2$$

21.
$$m = \frac{y_2 - y_1}{x_2 - x_1} = \frac{10 - 8}{2 - 6} = \frac{2}{-4} = -\frac{1}{2}$$
$$y = mx + b$$
$$8 = -\frac{1}{2}(6) + b$$
$$8 = -3 + b$$
$$11 = b \Rightarrow y = -\frac{1}{2}x + 11$$

23.
$$m = \frac{y_2 - y_1}{x_2 - x_1} = \frac{-9 - (-1)}{-3 - 3} = \frac{-8}{-6} = \frac{4}{3}$$
$$y = mx + b$$
$$-1 = \frac{4}{3}(3) + b$$
$$-1 = 4 + b$$
$$-5 = b \Rightarrow y = \frac{4}{3}x - 5$$

25.
$$x - y = 1$$
$$-y = -x + 1$$
$$y = x - 1$$
$$m = 1, (0, -1)$$

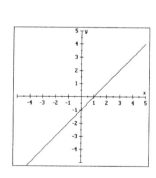

27.
$$2x = 3y - 6$$
$$2x + 6 = 3y$$
$$\frac{2x + 6}{3} = y$$
$$\frac{2}{3}x + \frac{6}{3} = y$$
$$\frac{2}{3}x + 2 = y$$
$$m = \frac{2}{3}, (0, 2)$$

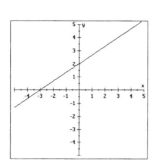

29.
$$3y = -2x + 18$$
$$y = \frac{-2x + 18}{3}$$
$$y = -\frac{2}{3}x + \frac{18}{3}$$
$$y = -\frac{2}{3}x + 6$$
$$m = -\frac{2}{3}, (0, 6)$$

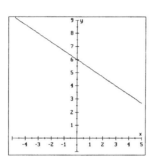

31.
$$3x - 2y = 8$$
$$-2y = -3x + 8$$
$$y = \frac{-3x + 8}{-2}$$
$$y = \frac{3}{2}x - \frac{8}{2}$$
$$y = \frac{3}{2}x - 4$$
$$m = \frac{3}{2}, (0, -4)$$

33.
$$-2x - 6y = 5$$
$$-6y = 2x + 5$$
$$y = \frac{2x + 5}{-6}$$
$$y = \frac{2}{-6}x + \frac{5}{-6}$$
$$y = -\frac{1}{3}x - \frac{5}{6}$$
$$m = -\frac{1}{3}, \left(0, -\frac{5}{6}\right)$$

35.
$$7x = 2y - 4$$
$$-2y = -7x - 4$$
$$y = \frac{-7x - 4}{-2}$$
$$y = \frac{-7}{-2}x - \frac{4}{-2}$$
$$y = \frac{7}{2}x + 2$$
$$m = \frac{7}{2}, (0, 2)$$

37.
$$y = 8x + 4 \quad y = 8x - 7$$
$$m = 8 \qquad m = 8$$
$$\text{parallel}$$

39.
$$x + y = 6 \qquad y = x + 8$$
$$y = -x + 6 \qquad m = 1$$
$$m = -1$$
$$\text{perpendicular}$$

41.
$$y = 3x + 9 \quad 2y = 6x - 10$$
$$m = 3 \qquad y = 3x - 5$$
$$m = 3$$
$$\text{parallel}$$

43.
$$x = 3y + 9 \quad y = -3x + 2$$
$$x - 9 = 3y \qquad m = -3$$
$$\frac{1}{3}x - 3 = y$$
$$m = \frac{1}{3}$$
$$\text{perpendicular}$$

45. $y = 8$ $x = 4$
horizontal   vertical
perpendicular

47. $3x = y - 2$ $3(y - 3) + x = 0$
$3x + 2 = y$ $3y - 9 + x = 0$
$m = 3$ $3y = -x + 9$
$$y = -\frac{1}{3}x + 3$$
$$m = -\frac{1}{3}$$
perpendicular

49. Find the slope of the given line:
$y = 4x - 9 \Rightarrow m = 4$
Use the parallel slope.
$y - y_1 = m(x - x_1)$
$y - 0 = 4(x - 0)$
$y = 4x$

51. Find the slope of the given line:
$4x - y = 7$
$4x - 7 = y \Rightarrow m = 4$
Use the parallel slope.
$y - y_1 = m(x - x_1)$
$y - 5 = 4(x - 2)$
$y - 5 = 4x - 8$
$y = 4x - 3$

53. Find the slope of the given line:
$$x = \frac{5}{4}y - 2$$
$4x = 5y - 8$
$4x + 8 = 5y$
$$\frac{4}{5}x + \frac{8}{5} = y \Rightarrow m = \frac{4}{5}$$
Use the parallel slope.
$y - y_1 = m(x - x_1)$
$$y - (-2) = \frac{4}{5}(x - 4)$$
$$y + 2 = \frac{4}{5}x - \frac{16}{5}$$
$$y = \frac{4}{5}x - \frac{16}{5} - \frac{10}{5}$$
$$y = \frac{4}{5}x - \frac{26}{5}$$

55. Find the slope of the given line:
$y = 4x - 9 \Rightarrow m = 4$
Use the perpendicular slope.
$y - y_1 = m(x - x_1)$
$$y - 0 = -\frac{1}{4}(x - 0)$$
$$y = -\frac{1}{4}x$$

57. Find the slope of the given line:
$$4x - y = 7$$
$$4x - 7 = y \Rightarrow m = 4$$
Use the perpendicular slope.
$$y - y_1 = m(x - x_1)$$
$$y - 5 = -\frac{1}{4}(x - 2)$$
$$y - 5 = -\frac{1}{4}x + \frac{1}{2}$$
$$y = -\frac{1}{4}x + \frac{11}{2}$$

59. Find the slope of the given line:
$$x = \frac{5}{4}y - 2$$
$$4x = 5y - 8$$
$$4x + 8 = 5y$$
$$\frac{4}{5}x + \frac{8}{5} = y \Rightarrow m = \frac{4}{5}$$
Use the perpendicular slope.
$$y - y_1 = m(x - x_1)$$
$$y - (-2) = -\frac{5}{4}(x - 4)$$
$$y + 2 = -\frac{5}{4}x + 5$$
$$y = -\frac{5}{4}x + 3$$

61.
$$4x + 5y = 20 \qquad 5x - 4y = 20$$
$$A = 4, B = 5 \qquad A = 5, B = -4$$
$$m_1 = -\frac{A}{B} = -\frac{4}{5} \qquad m_2 = -\frac{A}{B} = -\frac{5}{-4}$$
$$= \frac{5}{4}$$

perpendicular

63.
$$2x + 3y = 12 \qquad 6x + 9y = 32$$
$$A = 2, B = 3 \qquad A = 6, B = 9$$
$$m_1 = -\frac{A}{B} = -\frac{2}{3} \qquad m_2 = -\frac{A}{B} = -\frac{6}{9}$$
$$= -\frac{2}{3}$$

parallel

65. The line $y = 5$ is horizontal. A perpendicular line will be vertical. Find the vertical line through $(-2, 7)$. $\boxed{x = -2}$

67. The line $x = 8$ is vertical. A parallel line will also be vertical. Find the vertical line through $(5, 2)$. $\boxed{x = 5}$

69. $Ax + By = C$
$$By = -Ax + C$$
$$y = \frac{-Ax + C}{B}$$
$$y = -\frac{A}{B}x + \frac{C}{B}$$

71. Let x = the year of operation, and let y = the value. Then two points are known: $(0, 24300)$ and $(7, 1900)$. Find the slope:
$$m = \frac{y_2 - y_1}{x_2 - x_1} = \frac{1900 - 24300}{7 - 0}$$
$$= \frac{-22400}{7} = -3200$$
Find the equation of the line:
$$y - y_1 = m(x - x_1)$$
$$y - 24300 = -3200(x - 0)$$
$$y - 24300 = -3200x$$
$$y = -3200x + 24300$$

73. Let x = the year of operation, and let y = the value. Then two points are known: $(0, 475000)$ and $(10, 950000)$.

$$m = \frac{y_2 - y_1}{x_2 - x_1} = \frac{950000 - 475000}{10 - 0}$$
$$= \frac{475000}{10} = 47500$$

Find the equation of the line:
$$y - y_1 = m(x - x_1)$$
$$y - 475000 = 47500(x - 0)$$
$$y - 475000 = 47500x$$
$$y = 47500x + 475000$$

75. Let x = the year of operation, and let y = the value. Then two points are known: $(0, 1900)$ and $(3, 1190)$. Find the slope:

$$m = \frac{y_2 - y_1}{x_2 - x_1} = \frac{1190 - 1900}{3 - 0} = \frac{-710}{3}$$

Find the equation of the line:
$$y - y_1 = m(x - x_1)$$
$$y - 1900 = -\tfrac{710}{3}(x - 0)$$
$$y - 1900 = -\tfrac{710}{3}x$$
$$y = -\tfrac{710}{3}x + 1900$$

77. Let x = the age, and let y = the value. Then two points are known: $(0, 1050)$ and $(8, 90)$. Find the slope:

$$m = \frac{y_2 - y_1}{x_2 - x_1} = \frac{90 - 1050}{8 - 0} = \frac{-960}{8}$$
$$= -120$$

The depreciation rate is $120 per year.

79. Let x = the number of copies (in 100's), and let y = total charge. Then two points are known: $(7, 375)$ and $(10, 525)$

$$m = \frac{y_2 - y_1}{x_2 - x_1} = \frac{525 - 375}{10 - 7} = \frac{150}{3} = 50$$
$$y = mx + b$$
$$375 = 50(7) + b$$
$$375 = 350 + b$$
$$25 = b$$
$$y = 50x + 25 \Rightarrow \text{The setup cost is } \$25.$$

81. Let x = the year ($0 = 1993$), and let y = the avg. pension. Then 2 pts are known: $(0, 22176)$ and $(9, 42144)$. Find the slope:

$$m = \frac{y_2 - y_1}{x_2 - x_1} = \frac{42144 - 22176}{9 - 0}$$
$$= \frac{19968}{9} = 2218.6667$$

Find the equation of the line:
$$y - y_1 = m(x - x_1)$$
$$y - 22176 = 2218.6667(x - 0)$$
$$y - 22176 = 2218.6667x$$
$$y = 2218.6667x + 22176$$
Let $x = 14$:
$$y = 2218.6667x + 22176$$
$$= 2218.6667(14) + 22176$$
$$= 31061.33 + 22176 = \$53237.33$$

83. $0.80(465000) = \boxed{372000}$
Let x = the # of yrs, and let y = the value. Then 2 points are known: $(0, 465000)$ and $(40, 372000)$. Find the slope:

$$m = \frac{y_2 - y_1}{x_2 - x_1} = \frac{372000 - 465000}{40 - 0}$$
$$= \frac{-93000}{40} = \boxed{-2325}$$

Find the equation of the line:
$$y - y_1 = m(x - x_1)$$
$$y - 465000 = -2325(x - 0)$$
$$y - 465000 = -2325x$$
$$\boxed{y = -2325x + 465000}$$

85. Answers may vary.

87. The slope of a vertical line is undefined, so the equation cannot be written in slope-intercept form.

89. $a < 0, b > 0$

91. **Answers may vary.**

93. **Answers may vary.**

95. **Answers may vary.**

Exercise 3.6 (page 206)

1. $4x = 3(x + 2)$
$4x = 3x + 6$
$x = 6$

3. $5(2 - a) = 3(a + 6)$
$10 - 5a = 3a + 18$
$-8a = 8$
$a = -1$

5. input, function

7. range

9. independent

11. cannot

13. $$y = x + 2$$
Pick some values for x and find y for each:

x	-2	0	2
y	0	2	4

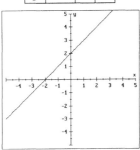

y is a function of x.
domain = the set of all real numbers
range = the set of all real numbers

15. $$y = -\tfrac{1}{2}x + 2$$
Pick some values for x and find y for each:

x	-2	0	2
y	3	2	1

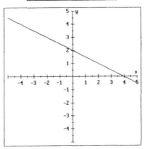

y is a function of x.
domain = the set of all real numbers
range = the set of all real numbers

17. $f(3) = 3(3) = 9$
$f(0) = 3(0) = 0$
$f(-1) = 3(-1) = -3$

19. $f(3) = 2(3) - 3 = 6 - 3 = 3$
$f(0) = 2(0) - 3 = 0 - 3 = -3$
$f(-1) = 2(-1) - 3 = -2 - 3 = -5$

21. $f(3) = 7 + 5(3) = 7 + 15 = 22$
$f(0) = 7 + 5(0) = 7 + 0 = 7$
$f(-1) = 7 + 5(-1) = 7 - 5 = 2$

23. $f(3) = 9 - 2(3) = 9 - 6 = 3$
$f(0) = 9 - 2(0) = 9 - 0 = 9$
$f(-1) = 9 - 2(-1) = 9 + 2 = 11$

25. $f(3) = \tfrac{1}{2}(3) + \tfrac{3}{2} = \tfrac{3}{2} + \tfrac{3}{2} = \tfrac{6}{2} = 3$
$f(0) = \tfrac{1}{2}(0) + \tfrac{3}{2} = 0 + \tfrac{3}{2} = \tfrac{3}{2}$
$f(-1) = \tfrac{1}{2}(-1) + \tfrac{3}{2} = -\tfrac{1}{2} + \tfrac{3}{2} = \tfrac{2}{2} = 1$

27. $f(1) = 1^2 = 1$
$f(-2) = (-2)^2 = 4$
$f(3) = 3^2 = 9$

29. $f(1) = 1^3 - 1 = 1 - 1 = 0$
$f(-2) = (-2)^3 - 1 = -8 - 1 = -9$
$f(3) = 3^3 - 1 = 27 - 1 = 26$

31. $f(1) = (1 + 1)^2 = 2^2 = 4$
$f(-2) = (-2 + 1)^2 = (-1)^2 = 1$
$f(3) = (3 + 1)^2 = 4^2 = 16$

33. $f(1) = 2(1)^2 - 1 = 2(1) - 1 = 2 - 1 = 1$
$f(-2) = 2(-2)^2 - (-2) = 2(4) + 2 = 8 + 2 = 10$
$f(3) = 2(3)^2 - 3 = 2(9) - 3 = 18 - 3 = 15$

35.
$$y = f(x) = \tfrac{1}{2}|x|$$

x	-2	-1	0	1	2
y	1	$\tfrac{1}{2}$	0	$\tfrac{1}{2}$	1

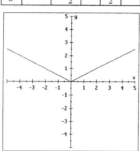

domain = the set of all real numbers
range = the set of real numbers greater
than or equal to 0

37.
$$y = f(x) = |x| - 1$$

x	-2	-1	0	1	2
y	1	0	-1	0	1

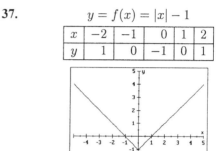

domain = the set of all real numbers
range = the set of real numbers greater
than or equal to -1

39. function

41. not a function

43. $C = 0.10n + 12 = 0.10(20) + 12$
$\qquad = 2 + 12 = \$14$

45. $C = 0.10n + 12 = 0.10(100) + 12$
$\qquad = 10 + 12 = \$22$

47. $C = 0.08n + 17 = 0.08(500) + 17$
$\qquad = 40 + 17 = \$57$

49. $C = 0.08n + 17 = 0.08(700) + 17$
$\qquad = 56 + 17 = \$73$

51. **Answers may vary.**

53. $f(x) + g(x) = (2x + 1) + (x) = 3x + 1; g(x) + f(x) = (x) + (2x + 1) = 3x + 1$
Thus, $f(x) + g(x) = g(x) + f(x)$.

55. $f(x) \cdot g(x) = (2x + 1)(x) = 2x^2 + x; g(x) \cdot f(x) = (x)(2x + 1) = 2x^2 + x$
Thus, $f(x) \cdot g(x) = g(x) \cdot f(x)$.

Exercise 3.7 (page 212)

1. $x - 2 > 5$
$\qquad x > 7$

3. $-4x + 2 > 18$
$\qquad -4x > 16$
$\qquad \dfrac{-4x}{-4} < \dfrac{16}{-4}$
$\qquad x < -4$

5. direct

7. constant

9. joint

11. $d = kn$

13. $l = \dfrac{k}{w}$

15. $A = kr^2$

17. $d = kst$

19. $I = \dfrac{kV}{R}$

21. $y = kx \qquad y = 5x$
$10 = k(2) \quad y = 5(7)$
$5 = k \qquad y = 35$

23. $r = ks \qquad r = \dfrac{7}{2}s$
$21 = k(6) \qquad r = \dfrac{7}{2}(12)$
$\dfrac{21}{6} = k \qquad r = 42$
$\dfrac{7}{2} = k$

25. $s = kt^2 \qquad s = \dfrac{3}{4}t^2$
$12 = k(4)^2 \quad s = \dfrac{3}{4}(30)^2$
$12 = 16k \qquad s = \dfrac{3}{4}(900)$
$\dfrac{3}{4} = k \qquad s = 675$

27. $y = \dfrac{k}{x} \qquad y = \dfrac{8}{x}$
$8 = \dfrac{k}{1} \qquad y = \dfrac{8}{8}$
$8 = k \qquad y = 1$

29. $r = \dfrac{k}{s} \qquad r = \dfrac{400}{s}$
$40 = \dfrac{k}{10} \qquad r = \dfrac{400}{15}$
$400 = k \qquad r = \dfrac{80}{3}$

31. $y = \dfrac{k}{x^2} \qquad y = \dfrac{96}{x^2}$
$6 = \dfrac{k}{4^2} \qquad y = \dfrac{96}{2^2}$
$6 = \dfrac{k}{16} \qquad y = \dfrac{96}{4}$
$96 = k \qquad y = 24$

33. $y = krs \qquad y = \dfrac{1}{3}rs$
$4 = k(2)(6)$
$4 = 12k \qquad y = \dfrac{1}{3}(3)(4)$
$\dfrac{1}{3} = k \qquad y = 4$

35. $D = kpq \qquad D = \dfrac{4}{5}pq$
$20 = k(5)(5) \qquad D = \dfrac{4}{5}(10)(10)$
$20 = 25k \qquad D = 80$
$\dfrac{4}{5} = k$

37. $y = \dfrac{ka}{b} \qquad y = \dfrac{5a}{b}$
$1 = \dfrac{k(2)}{10} \qquad y = \dfrac{5(7)}{14}$
$10 = 2k \qquad y = \dfrac{35}{14} = \dfrac{5}{2}$
$5 = k$

39. $d = kt^2 \qquad d = 16t^2$
$256 = k(4)^2 \quad d = 16(6)^2$
$256 = 16k \qquad d = 16(36)$
$16 = k \qquad d = 576$
The object will fall 576 ft.

41. $I = kt \qquad I = 350t$
$700 = k(2) \quad I = 350(7)$
$350 = k \qquad I = 2450$
The interest will be \$2,450.

43.
$$t = \frac{k}{s} \qquad t = \frac{150}{s}$$
$$3 = \frac{k}{50} \qquad t = \frac{150}{60}$$
$$150 = k \qquad t = \frac{5}{2}$$
The trip will take 2.5 hours.

45.
$$V = \frac{k}{P} \qquad V = \frac{320}{P}$$
$$40 = \frac{k}{8} \qquad V = \frac{320}{6}$$
$$320 = k \qquad V = \frac{160}{3}$$
The volume is $53\frac{1}{3}$ m³.

47.
$$I = krt \qquad I = 750rt$$
$$120 = k(0.08)(2) \qquad I = 750(0.12)(3)$$
$$120 = k(0.16) \qquad I = 270$$
$$750 = k$$
The interest would be $270.

49.
$$I = \frac{kV}{R} \qquad I = \frac{1V}{R}$$
$$4 = \frac{k(36)}{9} \qquad I = \frac{1(42)}{11}$$
$$36 = 36k \qquad I = \frac{42}{11}$$
$$1 = k$$
The current is $3\frac{9}{11}$ amperes.

51. Answers may vary.

53. $\frac{y}{x} = k \Rightarrow y = kx$; yes

Chapter 3 Summary (page 215)

1-6.

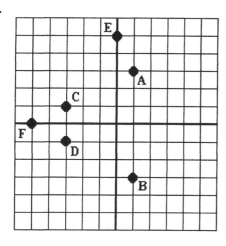

7. $(3, 1)$
8. $(-4, 5)$
9. $(-3, -4)$
10. $(2, -3)$
11. $(0, 0)$
12. $(0, 4)$
13. $(-5, 0)$
14. $(0, -3)$

15.
$$3x - 4y = 12$$
$$3(2) - 4(1) \overset{?}{=} 12$$
$$6 - 4 \overset{?}{=} 12$$
$$2 \neq 12$$
$(2, 1)$ is not a solution.

16.
$$3x - 4y = 12$$
$$3(3) - 4\left(-\frac{3}{4}\right) \overset{?}{=} 12$$
$$9 + 3 \overset{?}{=} 12$$
$$12 = 12$$
$\left(3, -\frac{3}{4}\right)$ is a solution.

17. $y = x - 5$

x	y
0	-5
2	-3

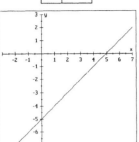

18. $y = 2x + 1$

x	y
0	1
2	5

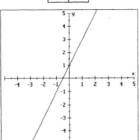

19. $y = \dfrac{x}{2} + 2$

x	y
0	2
2	3

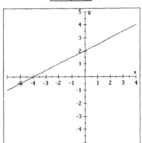

20. $y = 3$

x	y
0	3
2	3

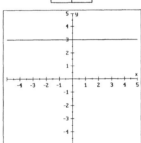

21. $x + y = 4$

x	y
0	4
4	0

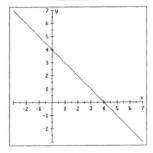

22. $x - y = -3$

x	y
0	3
-3	0

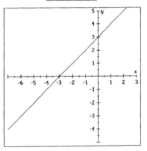

23. $3x + 5y = 15$

x	y
0	3
5	0

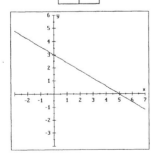

24. $7x - 4y = 28$

x	y
0	-7
4	0

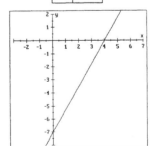

25. $m = \dfrac{y_2 - y_1}{x_2 - x_1} = \dfrac{3 - 4}{2 - 1} = \dfrac{-1}{1} = -1$

26. $m = \dfrac{y_2 - y_1}{x_2 - x_1} = \dfrac{-2 - 3}{3 - (-1)} = \dfrac{-5}{4} = -\dfrac{5}{4}$

27. $m = \dfrac{y_2 - y_1}{x_2 - x_1} = \dfrac{0 - (-1)}{-3 - (-1)} = \dfrac{1}{-2} = -\dfrac{1}{2}$

28. $m = \dfrac{y_2 - y_1}{x_2 - x_1} = \dfrac{2 - 2}{3 - (-8)} = \dfrac{0}{11} = 0$

29. $4x - 3y = 12$
$-3y = -4x + 12$
$y = \dfrac{4}{3}x - 4 \Rightarrow m = \dfrac{4}{3}$

30. $x = 4y + 2$
$x - 2 = 4y$
$\dfrac{1}{4}x - \dfrac{1}{2} = y \Rightarrow m = \dfrac{1}{4}$

31. positive **32.** 0 **33.** undefined **34.** negative

35. perpendicular **36.** parallel **37.** neither **38.** parallel

39. perpendicular **40.** parallel

41. $m = \dfrac{\text{rise}}{\text{run}} = \dfrac{2}{8} = \dfrac{1}{4}$

42. $m = \dfrac{y_2 - y_1}{x_2 - x_1} = \dfrac{66000 - 25000}{3 - 1}$
$= \dfrac{41000}{2} = \$20,500 \text{ per year}$

43. $y - y_1 = m(x - x_1)$
$y - 0 = 3(x - 0)$
$y = 3x$
$0 = 3x - y$
$3x - y = 0$

44. $y - y_1 = m(x - x_1)$
$y - \dfrac{2}{3} = -\dfrac{1}{3}(x - 1)$
$y - \dfrac{2}{3} = -\dfrac{1}{3}x + \dfrac{1}{3}$
$\dfrac{1}{3}x + y = 1$
$x + 3y = 3$

45. $y - y_1 = m(x - x_1)$
$y - (-2) = \dfrac{1}{9}(x - (-27))$
$y + 2 = \dfrac{1}{9}x + 3$
$-1 = \dfrac{1}{9}x - y$
$-9 = x - 9y$
$x - 9y = -9$

46. $y - y_1 = m(x - x_1)$
$y - \left(-\dfrac{1}{5}\right) = -\dfrac{3}{5}(x - 1)$
$y + \dfrac{1}{5} = -\dfrac{3}{5}x + \dfrac{3}{5}$
$\dfrac{3}{5}x + y = \dfrac{2}{5}$
$3x + 5y = 2$

47. $y = 5x + 2$
$m = 5, (0, 2)$

48. $y = -\dfrac{x}{2} + 4$
$m = -\dfrac{1}{2}, (0, 4)$

49. $y + 3 = 0$
$\quad\quad y = -3$
horizontal
$m = 0, (0, -3)$

50. $x + 3y = 1$
$\quad\quad 3y = -x + 1$
$\quad\quad y = \dfrac{-x + 1}{3}$
$\quad\quad y = -\frac{1}{3}x + \frac{1}{3}$
$\quad\quad m = -\frac{1}{3}, \left(0, \frac{1}{3}\right)$

51. $\quad\quad y = mx + b$
$\quad\quad y = -3x + 2$
$\quad\quad 3x + y = 2$

52. $y = mx + b$
$\quad\quad y = 0x + (-7)$
$\quad\quad y = -7$

53. $\quad\quad y = mx + b$
$\quad\quad y = 7x + 0$
$\quad\quad 0 = 7x - y$
$\quad\quad 7x - y = 0$

54. $\quad\quad y = mx + b$
$\quad\quad y = \frac{1}{2}x + \left(-\frac{3}{2}\right)$
$\quad\quad 2y = x - 3$
$\quad\quad 3 = x - 2y$
$\quad\quad x - 2y = 3$

55. $(-1, 3), m = -2$

56. $(1, -2), m = 2$

57. $\left(0, \frac{1}{2}\right), m = \frac{3}{2}$

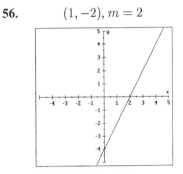

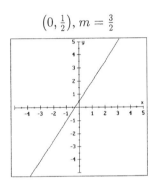

58. $(-3, 0), m = -\frac{5}{2}$

59. $y = 3x \quad\quad x = 3y$
$\quad m = 3 \quad\quad \frac{1}{3}x = y$
$\quad\quad\quad\quad\quad m = \frac{1}{3}$
$\quad\quad\quad$ neither

60. $3x = y \quad\quad x = -3y$
$\quad m = 3 \quad\quad -\frac{1}{3}x = y$
$\quad\quad\quad\quad\quad m = -\frac{1}{3}$
$\quad\quad\quad$ perpendicular

61. $x + 2y = y - x \quad\quad 2x + y = 3$
$\quad\quad\quad y = -2x \quad\quad\quad y = -2x + 3$
$\quad\quad\quad m = -2 \quad\quad\quad m = -2$
$\quad\quad\quad\quad\quad$ parallel

62. $3x + 2y = 7 \quad\quad 2x - 3y = 8$
$\quad\quad 2y = -3x + 7 \quad\quad -3y = -2x + 8$
$\quad\quad y = -\frac{3}{2}x + \frac{7}{2} \quad\quad y = \frac{2}{3}x - \frac{8}{3}$
$\quad\quad m = -\frac{3}{2} \quad\quad\quad m = \frac{2}{3}$
$\quad\quad\quad\quad$ perpendicular

63. Find the slope of the given line:
$$y = 7x - 18 \Rightarrow m = 7$$
Use the parallel slope.
$$y - y_1 = m(x - x_1)$$
$$y - 5 = 7(x - 2)$$
$$y - 5 = 7x - 14$$
$$9 = 7x - y, \text{ or } 7x - y = 9$$

64. Find the slope of the given line:
$$3x + 2y = 7$$
$$2y = -3x + 7$$
$$y = -\frac{3}{2}x + \frac{7}{2} \Rightarrow m = -\frac{3}{2}$$
Use the parallel slope.
$$y - y_1 = m(x - x_1)$$
$$y - 5 = -\frac{3}{2}(x - (-3))$$
$$2(y - 5) = 2\left(-\frac{3}{2}\right)(x + 3)$$
$$2y - 10 = -3(x + 3)$$
$$2y - 10 = -3x - 9$$
$$3x + 2y = 1$$

65. Find the slope of the given line:
$$2x - 5y = 12$$
$$2x - 12 = 5y$$
$$\frac{2}{5}x - \frac{12}{5} = y \Rightarrow m = \frac{2}{5}$$
Use the perpendicular slope.
$$y - y_1 = m(x - x_1)$$
$$y - 0 = -\frac{5}{2}(x - 0)$$
$$2y = -5x$$
$$5x + 2y = 0$$

66. Find the slope of the given line:
$$y = \frac{x}{3} + 17 \Rightarrow m = \frac{1}{3}$$
Use the perpendicular slope.
$$y - y_1 = m(x - x_1)$$
$$y - (-4) = -3(x - 0)$$
$$y + 4 = -3x$$
$$3x + y = -4$$

67.
$$4x + 5y = 10$$
$$A = 4, B = 5$$
$$m_1 = -\frac{A}{B} = -\frac{4}{5}$$

68.
$$3x - 5y = 17$$
$$A = 3, B = -5$$
$$m_1 = -\frac{A}{B} = -\frac{3}{-5} = \frac{3}{5}$$

69. Let x = the year of operation, and let y = the value. Then two points are known:
$(0, 2700)$ and $(5, 200)$. Find the slope: $m = \dfrac{y_2 - y_1}{x_2 - x_1} = \dfrac{200 - 2700}{5 - 0} = \dfrac{-2500}{5} = -500$
Find the equation of the line: $\quad y - y_1 = m(x - x_1)$
$$y - 2700 = -500(x - 0)$$
$$y - 2700 = -500x$$
$$y = -500x + 2700$$
Let $x = 3$: $y = -500x + 2700 = -500(3) + 2700 = -1500 + 2700 = \$1200$

70. $y = \frac{1}{2}x - 1$

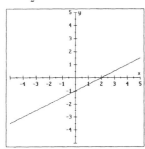

y is a function of x.

domain $=$ the set of all real numbers

range $=$ the set of all real numbers

71. $f(0) = -2(0) + 5 = 0 + 5 = 5$

72. $f(3) = -2(3) + 5 = -6 + 5 = -1$

73. $f\left(-\frac{1}{2}\right) = -2\left(-\frac{1}{2}\right) + 5 = 1 + 5 = 6$

74. $f(6) = -2(6) + 5 = -12 + 5 = -7$

75.
$$y = f(x) = |x| - 3$$

x	-2	-1	0	1	2
y	-1	-2	-3	-2	-1

domain $=$ the set of all real numbers

range $=$ the set of all real numbers greater than or equal to -3

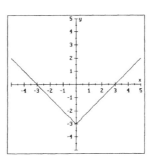

76. not a function

77. function

78.
$$\begin{array}{ll} s = kt^2 & s = 4t^2 \\ 64 = k(4)^2 & s = 4(10)^2 \\ 64 = 16k & s = 4(100) \\ 4 = k & s = 400 \end{array}$$

79.
$$l = \frac{k}{w}$$
$$30 = \frac{k}{20}$$
$$600 = k$$

80.
$$\begin{array}{ll} R = kbc & R = \frac{3}{4}bc \\ 72 = k(4)(24) & R = \frac{3}{4}(6)(18) \\ 72 = 96k & R = 81 \\ \frac{72}{96} = k & \\ \frac{3}{4} = k & \end{array}$$

81.
$$\begin{array}{ll} s = \frac{kw}{m^2} & s = \frac{7w}{m^2} \\ \frac{7}{4} = \frac{k(4)}{4^2} & s = \frac{7(5)}{7^2} \\ \frac{7}{4} = \frac{4k}{16} & s = \frac{35}{49} \\ \frac{7}{4} = \frac{k}{4} & s = \frac{5}{7} \\ 7 = k & \end{array}$$

Chapter 3 Test (page 220)

1.
$$y = \frac{x}{2} + 1$$

x	y
0	1
2	3

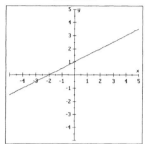

2.
$$2(x+1) - y = 4$$
$$2x + 2 - y = 4$$
$$2x - y = 2$$

x	y
0	-2
1	0

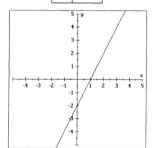

3.
$$x = 1$$

x	y
1	2
1	0

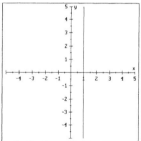

4.
$$2y = 8$$
$$y = 4$$

x	y
0	4
2	4

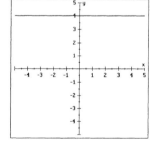

5. $m = \dfrac{y_2 - y_1}{x_2 - x_1} = \dfrac{8 - 0}{6 - 0} = \dfrac{8}{6} = \dfrac{4}{3}$

6. $m = \dfrac{y_2 - y_1}{x_2 - x_1} = \dfrac{-1 - 3}{3 - (-1)} = \dfrac{-4}{4} = -1$

7. $2x + y = 3$
$$y = -2x + 3$$
$$m = -2$$

8. $2y - 7(x+5) = 7$
$$2y - 7x - 35 = 7$$
$$2y = 7x + 42$$
$$y = \frac{7}{2}x + 21$$
y-intercept: $(0, 21)$

9. horizontal line $\Rightarrow m = 0$

10. vertical line $\Rightarrow m$ is undefined.

94

11. equal

12. -1

13. perpendicular

14. parallel

15. $m = \dfrac{\text{rise}}{\text{run}} = \dfrac{3}{12} = \dfrac{1}{4}$

16. $m = \dfrac{y_2 - y_1}{x_2 - x_1} = \dfrac{100000 - 50000}{5 - 2}$

$\qquad = \dfrac{50000}{3} \approx \$16,\!666.67$

17. 2

18. $-\frac{1}{2}$

19. $y - y_1 = m(x - x_1)$
$\quad y - 5 = 7(x - (-2))$
$\quad y - 5 = 7x + 14$
$\quad -19 = 7x - y, \text{ or } 7x - y = -19$

20. $y - y_1 = m(x - x_1)$
$\quad y - 3 = \dfrac{1}{2}(x - 0)$
$\quad y - 3 = \dfrac{1}{2}x$
$\quad 2y - 6 = x$
$\quad -6 = x - 2y, \text{ or } x - 2y = -6$

21. Parallel to y-axis $\Rightarrow$ vertical
$\quad x = -3$

22. Given line has $m = \frac{1}{3}$. Use $m = -3$.
$\quad y - y_1 = m(x - x_1)$
$\quad y - (-5) = -3(x - 3)$
$\quad y + 5 = -3x + 9$
$\quad 3x + y = 4$

23. $f(2) = 3(2) - 2 = 6 - 2 = 4$

24. $f(-3) = 3(-3) - 2 = -9 - 2 = -11$

25. $y = f(x) = -|x| + 4$

x	-2	-1	0	1	2
y	2	3	4	3	2

Domain = all real numbers
Range = all real numbers ≤ 4

26. not a function

27. function

28. $\quad y = kx \qquad y = 4x$
$\quad 32 = k(8) \qquad 4 = 4x$
$\quad 4 = k \qquad\quad 1 = x$

29. $\quad i = \dfrac{k}{d^2}$
$\quad 100 = \dfrac{k}{2^2}$
$\quad 100 = \dfrac{k}{4} \Rightarrow 400 = k$

Exercise 4.1 (page 229)

1.

$-3 \quad -1 \quad 1 \quad 3$

3. the product of 3 and the sum of x and y

5. $|2x| + 3$

7. $-5; 3$

9. $(3x)(3x)(3x)(3x)$

11. $y \cdot y \cdot y \cdot y \cdot y$

13. $x^n y^n$

15. a^{bc}

17. Answers may vary.

19. Base: 4
Exponent: 3

21. Base: x
Exponent: 5

23. Base: $2y$
Exponent: 3

25. Base: x
Exponent: 4

27. Base: x
Exponent: 1

29. Base: x
Exponent: 3

31. $5^3 = 5 \cdot 5 \cdot 5$

33. $x^7 = x \cdot x \cdot x \cdot x \cdot x \cdot x \cdot x$

35. $-4x^5 = -4 \cdot x \cdot x \cdot x \cdot x \cdot x$

37. $(3t)^5 = (3t)(3t)(3t)(3t)(3t)$

39. $2 \cdot 2 \cdot 2 = 2^3$

41. $x \cdot x \cdot x \cdot x = x^4$

43. $(2x)(2x)(2x) = (2x)^3$

45. $-4t \cdot t \cdot t \cdot t = -4t^4$

47. $5^4 = 5 \cdot 5 \cdot 5 \cdot 5 = 625$

49. $2^2 + 3^2 = 4 + 9 = 13$

51. $5^4 - 4^3 = 625 - 64 = 561$

53. $-5(3^4 + 4^3) = -5(81 + 64)$
$\qquad = -5(145) = -725$

55. $x^4 x^3 = x^{4+3} = x^7$

57. $x^5 x^5 = x^{5+5} = x^{10}$

59. $tt^2 = t^{1+2} = t^3$

61. $a^3 a^4 a^5 = a^{3+4+5} = a^{12}$

63. $y^3(y^2 y^4) = y^3 y^{2+4} = y^3 y^6 = y^{3+6} = y^9$

65. $4x^2(3x^5) = 4 \cdot 3x^2 x^5 = 12x^{2+5} = 12x^7$

67. $(-y^2)(4y^3) = -1 \cdot 4y^2 y^3 = -4y^{2+3}$
$\qquad = -4y^5$

69. $6x^3(-x^2)(-x^4) = 6(-1)(-1)x^3 x^2 x^4$
$\qquad = 6x^{3+2+4} = 6x^9$

71. $(3^2)^4 = 3^{2 \cdot 4} = 3^8$

73. $(y^5)^3 = y^{5 \cdot 3} = y^{15}$

75. $(a^3)^7 = a^{3 \cdot 7} = a^{21}$

77. $(x^2 x^3)^5 = (x^{2+3})^5 = (x^5)^5 = x^{5 \cdot 5} = x^{25}$

79. $(3zz^2 z^3)^5 = (3z^{1+2+3})^5 = (3z^6)^5 = 3^5(z^6)^5 = 243z^{6 \cdot 5} = 243z^{30}$

81. $(x^5)^2(x^7)^3 = x^{5 \cdot 2} x^{7 \cdot 3} = x^{10} x^{21} = x^{10+21} = x^{31}$

83. $(r^3 r^2)^4(r^3 r^5)^2 = (r^{3+2})^4(r^{3+5})^2 = (r^5)^4(r^8)^2 = r^{5 \cdot 4} r^{8 \cdot 2} = r^{20} r^{16} = r^{20+16} = r^{36}$

85. $(s^3)^3(s^2)^2(s^5)^4 = s^{3 \cdot 3} s^{2 \cdot 2} s^{5 \cdot 4} = s^9 s^4 s^{20} = s^{9+4+20} = s^{33}$

87. $(xy)^3 = x^3y^3$

89. $(r^3s^2)^2 = (r^3)^2(s^2)^2 = r^{3\cdot 2}s^{2\cdot 2} = r^6s^4$

91. $(4ab^2)^2 = 4^2a^2(b^2)^2 = 16a^2b^{2\cdot 2} = 16a^2b^4$

93. $(-2r^2s^3t)^3 = (-2)^3(r^2)^3(s^3)^3t^3$
$$= -8r^{2\cdot 3}s^{3\cdot 3}t^3 = -8r^6s^9t^3$$

95. $\left(\dfrac{a}{b}\right)^3 = \dfrac{a^3}{b^3}$

97. $\left(\dfrac{x^2}{y^3}\right)^5 = \dfrac{(x^2)^5}{(y^3)^5} = \dfrac{x^{2\cdot 5}}{y^{3\cdot 5}} = \dfrac{x^{10}}{y^{15}}$

99. $\left(\dfrac{-2a}{b}\right)^5 = \dfrac{(-2a)^5}{b^5} = \dfrac{(-2)^5a^5}{b^5} = \dfrac{-32a^5}{b^5}$

101. $\left(\dfrac{b^2}{3a}\right)^3 = \dfrac{(b^2)^3}{(3a)^3} = \dfrac{b^{2\cdot 3}}{3^3a^3} = \dfrac{b^6}{27a^3}$

103. $\dfrac{x^5}{x^3} = x^{5-3} = x^2$

105. $\dfrac{y^3y^4}{yy^2} = \dfrac{y^{3+4}}{y^{1+2}} = \dfrac{y^7}{y^3} = y^{7-3} = y^4$

107. $\dfrac{12a^2a^3a^4}{4(a^4)^2} = \dfrac{12a^{2+3+4}}{4a^{4\cdot 2}} = \dfrac{12a^9}{4a^8} = \dfrac{12}{4}a^{9-8} = 3a^1 = 3a$

109. $\dfrac{(ab^2)^3}{(ab)^2} = \dfrac{a^3(b^2)^3}{a^2b^2} = \dfrac{a^3b^{2\cdot 3}}{a^2b^2} = \dfrac{a^3b^6}{a^2b^2} = a^{3-2}b^{6-2} = a^1b^4 = ab^4$

111. $\dfrac{20(r^4s^3)^4}{6(rs^3)^3} = \dfrac{20(r^4)^4(s^3)^4}{6r^3(s^3)^3} = \dfrac{20r^{4\cdot 4}s^{3\cdot 4}}{6r^3s^{3\cdot 3}} = \dfrac{20r^{16}s^{12}}{6r^3s^9} = \dfrac{20}{6}r^{16-3}s^{12-9} = \dfrac{10}{3}r^{13}s^3 = \dfrac{10r^{13}s^3}{3}$

113. $\dfrac{17(x^4y^3)^8}{34(x^5y^2)^4} = \dfrac{17(x^4)^8(y^3)^8}{34(x^5)^4(y^2)^4} = \dfrac{17x^{4\cdot 8}y^{3\cdot 8}}{34x^{5\cdot 4}y^{2\cdot 4}} = \dfrac{17x^{32}y^{24}}{34x^{20}y^8} = \dfrac{17}{34}x^{32-20}y^{24-8} = \dfrac{1}{2}x^{12}y^{16} = \dfrac{x^{12}y^{16}}{2}$

115. $\left(\dfrac{y^3y}{2yy^2}\right)^3 = \left(\dfrac{y^{3+1}}{2y^{1+2}}\right)^3 = \left(\dfrac{y^4}{2y^3}\right)^3 = \left(\dfrac{1}{2}y^{4-3}\right)^3 = \left(\dfrac{1}{2}y^1\right)^3 = \left(\dfrac{1}{2}\right)^3(y^1)^3 = \dfrac{1}{8}y^{1\cdot 3} = \dfrac{y^3}{8}$

117. $\left(\dfrac{-2r^3r^3}{3r^4r}\right)^3 = \left(\dfrac{-2r^{3+3}}{3r^{4+1}}\right)^3 = \left(\dfrac{-2r^6}{3r^5}\right)^3 = \left(\dfrac{-2}{3}r^{6-5}\right)^3 = \dfrac{(-2)^3}{3^3}(r^1)^3 = \dfrac{-8}{27}r^{1\cdot 3} = -\dfrac{8r^3}{27}$

119.

Bounce	1	2	3	4
Height	$\frac{1}{2}(32)$	$\frac{1}{2}\left[\frac{1}{2}(32)\right] = 32\left(\frac{1}{2}\right)^2$	$\frac{1}{2}\left[32\left(\frac{1}{2}\right)^2\right] = 32\left(\frac{1}{2}\right)^3$	$\frac{1}{2}\left[32\left(\frac{1}{2}\right)^3\right] = 32\left(\frac{1}{2}\right)^4$

$32\left(\frac{1}{2}\right)^4 = 32\left(\frac{1^4}{2^4}\right) = 32\left(\frac{1}{16}\right) = \frac{32}{16} = 2$ ft

121.

Years	7	14	21	28
Value	2,000	4,000	8,000	16,000

123. $A = P(1+r)^t = 8000(1+0.06)^{30} = 8000(1.06)^{30} \approx 8000(5.743491) \approx \$45,947.93$

125. Answers may vary.

127. No. $2^3 = 8, 3^2 = 9 \Rightarrow 2^3 \neq 3^2$

Exercise 4.2 (page 234)

1. $\dfrac{3a^2 + 4b + 8}{a + 2b^2} = \dfrac{3(-2)^2 + 4(3) + 8}{-2 + 2(3)^2} = \dfrac{3(4) + 12 + 8}{-2 + 2(9)} = \dfrac{12 + 12 + 8}{-2 + 18} = \dfrac{32}{16} = 2$

3.
$$5\left(x - \frac{1}{2}\right) = \frac{7}{2}$$
$$5x - \frac{5}{2} = \frac{7}{2}$$
$$5x - \frac{5}{2} + \frac{5}{2} = \frac{7}{2} + \frac{5}{2}$$
$$5x = \frac{12}{2}$$
$$5x = 6$$
$$\frac{5x}{5} = \frac{6}{5}$$
$$x = \frac{6}{5}$$

5.
$$P = L + \frac{s}{f}i$$
$$P - L = \frac{s}{f}i$$
$$f(P - L) = f \cdot \frac{s}{f}i$$
$$f(P - L) = si$$
$$\frac{f(P - L)}{i} = \frac{si}{i}$$
$$\frac{f(P - L)}{i} = s, \text{ or } s = \frac{f(P - L)}{i}$$

7. 1

9. 1

11. $2^5 \cdot 2^{-2} = 2^{5+(-2)} = 2^3 = 8$

13. $4^{-3} \cdot 4^{-2} \cdot 4^5 = 4^{-3+(-2)+5} = 4^0 = 1$

15. $\dfrac{3^5 \cdot 3^{-2}}{3^3} = \dfrac{3^{5+(-2)}}{3^3} = \dfrac{3^3}{3^3} = 3^{3-3} = 3^0 = 1$

17. $\dfrac{2^5 \cdot 2^7}{2^6 \cdot 2^{-3}} = \dfrac{2^{5+7}}{2^{6+(-3)}} = \dfrac{2^{12}}{2^3} = 2^{12-3}$
$$= 2^9 = 512$$

19. $2x^0 = 2 \cdot x^0 = 2 \cdot 1 = 2$

21. $(-x)^0 = 1$

23. $\left(\dfrac{a^2 b^3}{ab^4}\right)^0 = 1$

25. $\dfrac{x^0 - 5x^0}{2x^0} = \dfrac{1 - 5(1)}{2(1)} = \dfrac{1 - 5}{2} = \dfrac{-4}{2} = -2$

27. $x^{-2} = \dfrac{1}{x^2}$

29. $b^{-5} = \dfrac{1}{b^5}$

31. $(2y)^{-4} = \dfrac{1}{(2y)^4} = \dfrac{1}{2^4 y^4} = \dfrac{1}{16y^4}$

33. $\left(ab^2\right)^{-3} = \dfrac{1}{(ab^2)^3} = \dfrac{1}{a^3 (b^2)^3} = \dfrac{1}{a^3 b^6}$

35. $\dfrac{y^4}{y^5} = y^{4-5} = y^{-1} = \dfrac{1}{y}$

37. $\dfrac{(r^2)^3}{(r^3)^4} = \dfrac{r^6}{r^{12}} = r^{6-12} = r^{-6} = \dfrac{1}{r^6}$

39. $\dfrac{y^4 y^3}{y^4 y^{-2}} = \dfrac{y^7}{y^2} = y^{7-2} = y^5$

41. $\dfrac{a^4 a^{-2}}{a^2 a^0} = \dfrac{a^2}{a^2} = a^{2-2} = a^0 = 1$

43. $\left(ab^2\right)^{-2} = \dfrac{1}{\left(ab^2\right)^2} = \dfrac{1}{a^2\left(b^2\right)^2} = \dfrac{1}{a^2 b^4}$

45. $\left(x^2 y\right)^{-3} = \dfrac{1}{\left(x^2 y\right)^3} = \dfrac{1}{\left(x^2\right)^3 y^3} = \dfrac{1}{x^6 y^3}$

47. $\left(x^{-4} x^3\right)^3 = \left(x^{-1}\right)^3 = x^{-3} = \dfrac{1}{x^3}$

49. $\left(y^3 y^{-2}\right)^{-2} = \left(y^1\right)^{-2} = y^{-2} = \dfrac{1}{y^2}$

51. $\left(a^{-2} b^{-3}\right)^{-4} = \left(a^{-2}\right)^{-4}\left(b^{-3}\right)^{-4} = a^8 b^{12}$

53. $\left(-2x^3 y^{-2}\right)^{-5} = (-2)^{-5}\left(x^3\right)^{-5}\left(y^{-2}\right)^{-5} = \dfrac{1}{(-2)^5} x^{-15} y^{10} = -\dfrac{y^{10}}{32 x^{15}}$

55. $\left(\dfrac{a^3}{a^{-4}}\right)^2 = \left(a^{3-(-4)}\right)^2 = \left(a^7\right)^2 = a^{14}$

57. $\left(\dfrac{b^5}{b^{-2}}\right)^{-2} = \left(b^{5-(-2)}\right)^{-2} = \left(b^7\right)^{-2} = b^{-14}$
$= \dfrac{1}{b^{14}}$

59. $\left(\dfrac{4x^2}{3x^{-5}}\right)^4 = \left(\dfrac{4}{3} x^{2-(-5)}\right)^4 = \left(\dfrac{4}{3} x^7\right)^4 = \left(\dfrac{4}{3}\right)^4 \left(x^7\right)^4 = \dfrac{256}{81} x^{28} = \dfrac{256 x^{28}}{81}$

61. $\left(\dfrac{12y^3 z^{-2}}{3y^{-4} z^3}\right)^2 = \left(\dfrac{12}{3} y^{3-(-4)} z^{-2-3}\right)^2 = \left(4 y^7 z^{-5}\right)^2 = 4^2 \left(y^7\right)^2 \left(z^{-5}\right)^2 = 16 y^{14} z^{-10} = \dfrac{16 y^{14}}{z^{10}}$

63. $\left(\dfrac{2x^3 y^{-2}}{4xy^2}\right)^7 = \left(\dfrac{2}{4} x^{3-1} y^{-2-2}\right)^7 = \left(\dfrac{1}{2} x^2 y^{-4}\right)^7 = \left(\dfrac{1}{2}\right)^7 \left(x^2\right)^7 \left(y^{-4}\right)^7 = \dfrac{1}{128} x^{14} y^{-28} = \dfrac{x^{14}}{128 y^{28}}$

65. $\left(\dfrac{14u^{-2} v^3}{21u^{-3} v}\right)^4 = \left(\dfrac{14}{21} u^{-2-(-3)} v^{3-1}\right)^4 = \left(\dfrac{2}{3} u^1 v^2\right)^4 = \left(\dfrac{2}{3}\right)^4 \left(u^1\right)^4 \left(v^2\right)^4 = \dfrac{16}{81} u^4 v^8 = \dfrac{16 u^4 v^8}{81}$

67. $\left(\dfrac{6a^2 b^3}{2ab^2}\right)^{-2} = \left(\dfrac{6}{2} a^{2-1} b^{3-2}\right)^{-2} = \left(3a^1 b^1\right)^{-2} = \dfrac{1}{(3ab)^2} = \dfrac{1}{3^2 a^2 b^2} = \dfrac{1}{9a^2 b^2}$

69. $\left(\dfrac{18a^2 b^3 c^{-4}}{3a^{-1} b^2 c}\right)^{-3} = \left(\dfrac{18}{3} a^{2-(-1)} b^{3-2} c^{-4-1}\right)^{-3} = \left(6a^3 b^1 c^{-5}\right)^{-3} = 6^{-3}\left(a^3\right)^{-3}\left(b^1\right)^{-3}\left(c^{-5}\right)^{-3}$
$= \dfrac{1}{6^3} a^{-9} b^{-3} c^{15} = \dfrac{c^{15}}{216 a^9 b^3}$

71. $\dfrac{\left(2x^{-2} y\right)^{-3}}{\left(4x^2 y^{-1}\right)^3} = \dfrac{2^{-3}\left(x^{-2}\right)^{-3} y^{-3}}{4^3\left(x^2\right)^3\left(y^{-1}\right)^3} = \dfrac{x^6 y^{-3}}{2^3 \cdot 4^3 x^6 y^{-3}} = \dfrac{1}{8 \cdot 64} x^{6-6} y^{-3-(-3)} = \dfrac{1}{512} x^0 y^0 = \dfrac{1}{512}$

73. $\dfrac{(17x^5y^{-5}z)^{-3}}{(17x^{-5}y^3z^2)^{-4}} = \dfrac{17^{-3}(x^5)^{-3}(y^{-5})^{-3}z^{-3}}{17^{-4}(x^{-5})^{-4}(y^3)^{-4}(z^2)^{-4}} = \dfrac{17^{-3}x^{-15}y^{15}z^{-3}}{17^{-4}x^{20}y^{-12}z^{-8}}$

$$= 17^{-3-(-4)}x^{-15-20}y^{15-(-12)}z^{-3-(-8)}$$

$$= 17^1x^{-35}y^{27}z^5 = \dfrac{17y^{27}z^5}{x^{35}}$$

75. $x^{2m}x^m = x^{2m+m} = x^{3m}$

77. $u^{2m}v^{3n}u^{3m}v^{-3n} = u^{2m+3m}v^{3n+(-3n)}$
$$= u^{5m}v^0 = u^{5m}$$

79. $y^{3m+2}y^{-m} = y^{3m+2+(-m)} = y^{2m+2}$

81. $\dfrac{y^{3m}}{y^{2m}} = y^{3m-2m} = y^m$

83. $\dfrac{x^{3n}}{x^{6n}} = x^{3n-6n} = x^{-3n} = \dfrac{1}{x^{3n}}$

85. $\left(x^{m+1}\right)^2 = x^{(m+1)\cdot 2} = x^{2m+2}$

87. $\left(x^{3-2n}\right)^{-4} = x^{(3-2n)(-4)} = x^{-12+8n} = x^{8n-12}$

89. $\left(y^{2-n}\right)^{-4} = y^{(2-n)(-4)} = y^{-8+4n} = y^{4n-8}$

Problems 91-95 are to be solved using a calculator. The keystrokes needed to solve each problem using a TI-83 graphing calculator appear in each solution. There may be other solutions. Keystrokes for other calculators may be slightly different.

91. $P = A(1+i)^{-n} = 100,000(1+0.07)^{-40}$

| 1 | 0 | 0 | 0 | 0 | 0 | (| 1 | + | . | 0 | 7 |) | ^ | (−) | 4 | 0 | = |

The original deposit should be $6,678.04.

93. $P = A(1+i)^{-n} = 100,000(1+0.09)^{-40}$

| 1 | 0 | 0 | 0 | 0 | 0 | (| 1 | + | . | 0 | 9 |) | ^ | (−) | 4 | 0 | = |

The original deposit should be $3,183.76.

95. $P = A(1+i)^{-n} = 1,000,000(1+0.08)^{-60}$

| 1 | 0 | 0 | 0 | 0 | 0 | 0 | (| 1 | + | . | 0 | 8 |) | ^ | (−) | 6 | 0 | = |

The original deposit should be $9,875.85.

97. **Answers may vary.**

99. If $x > 1$, then x raised to a negative power is less than x. If $x = 1$, then x raised to a negative power is equal to x. If $0 < x < 1$, then x raised to a negative power is greater than x.

Exercise 4.3 (page 241)

1. $-5y^{55} = -5(-1)^{55} = -5(-1) = 5$

3. commutative property of addition

5. $3(x - 4) - 6 = 0$
$3x - 12 - 6 = 0$
$3x - 18 = 0$
$3x = 18$
$x = 6$

7. scientific notation

9. $23,000 = 2.3 \times 10^4$

11. $1,700,000 = 1.7 \times 10^6$

13. $0.062 = 6.2 \times 10^{-2}$

15. $0.0000051 = 5.1 \times 10^{-6}$

17. $42.5 \times 10^2 = 4.25 \times 10^1 \times 10^2$
$= 4.25 \times 10^3$

19. $0.25 \times 10^{-2} = 2.5 \times 10^{-1} \times 10^{-2}$
$= 2.5 \times 10^{-3}$

21. $2.3 \times 10^2 = 230$

23. $8.12 \times 10^5 = 812,000$

25. $1.15 \times 10^{-3} = 0.00115$

27. $9.76 \times 10^{-4} = 0.000976$

29. $25 \times 10^6 = 25,000,000$

31. $0.51 \times 10^{-3} = 0.00051$

33. $(3.4 \times 10^2)(2.1 \times 10^3) = (3.4)(2.1) \times 10^2 \times 10^3 = 7.14 \times 10^5 = 714,000$

35. $\dfrac{9.3 \times 10^2}{3.1 \times 10^{-2}} = \dfrac{9.3}{3.1} \times \dfrac{10^2}{10^{-2}} = 3 \times 10^4 = 30,000$

37. $\dfrac{96,000}{(12,000)(0.00004)} = \dfrac{9.6 \times 10^4}{(1.2 \times 10^4)(4 \times 10^{-5})} = \dfrac{9.6 \times 10^4}{4.8 \times 10^{-1}} = \dfrac{9.6}{4.8} \times \dfrac{10^4}{10^{-1}} = 2 \times 10^5 = 200,000$

39. $25,700,000,000,000 = 2.57 \times 10^{13}$ mi

41. $1.14 \times 10^8 = 114,000,000$ mi

43. $0.00622 = 6.22 \times 10^{-3}$ mi

45. 3.6×10^7 mi $\times 5280$ ft/mi $= 19008 \times 10^7$ ft
$= 1.9008 \times 10^{11}$ ft

47. $\dfrac{3.3 \times 10^4 \text{ cm}}{\text{sec}} = \dfrac{3.3 \times 10^4 \text{ cm}}{\text{sec}} \cdot \dfrac{1 \text{ m}}{100 \text{ cm}} \cdot \dfrac{1 \text{ km}}{1,000 \text{ m}} = \dfrac{3.3 \times 10^4}{10^5}$ km/sec $= 3.3 \times 10^{-1}$ km/sec

49. 1.5×10^{-4} in.; $25,000,000,000,000 = 2.5 \times 10^{13}$

51. Answers may vary.

53. Answers may vary.

Exercise 4.4 (page 250)

1. $5(u - 5) + 9 = 2(u + 4)$
$5u - 25 + 9 = 2u + 8$
$5u - 16 = 2u + 8$
$3u - 16 = 8$
$3u = 24$
$u = 8$

3. $-4(3y + 2) \leq 28$
$-12y - 8 \leq 28$
$-12y \leq 36$
$\dfrac{-12y}{-12} \geq \dfrac{36}{-12}$
$y \geq -3$

-3

5. $\left(x^2 x^4\right)^3 = \left(x^6\right)^3 = x^{18}$

7. $\left(\dfrac{y^2 y^5}{y^4}\right)^3 = \left(\dfrac{y^7}{y^4}\right)^3 = \left(y^3\right)^3 = y^9$

9. algebraic

11. polynomial

13. trinomial

15. degree

17. function

19. domain

21. yes

23. yes

25. binomial

27. trinomial

29. monomial

31. binomial

33. trinomial

35. none of these (4 terms)

37. 4th

39. 3rd

41. 8th $(3 + 5 = 8)$

43. 6th $(3 + 1 + 2 = 6)$

45. 12th

47. 0th

49. $5x - 3 = 5(2) - 3 = 10 - 3 = 7$

51. $5x - 3 = 5(-1) - 3 = -5 - 3 = -8$

53. $-x^2 - 4 = -(0)^2 - 4 = -0 - 4 = -4$

55. $-x^2 - 4 = -(-1)^2 - 4 = -1 - 4 = -5$

57. $f(0) = 5(0) + 1 = 0 + 1 = 1$

59. $f(-2) = 5(-2) + 1 = -10 + 1 = -9$

61. $f(0) = (0)^2 - 2(0) + 3 = 0 - 0 + 3 = 3$

63. $f(-2) = (-2)^2 - 2(-2) + 3$
$= 4 + 4 + 3 = 11$

65.

x	$x^2 - 3$
-2	$(-2)^2 - 3 = 4 - 3 = 1$
-1	$(-1)^2 - 3 = 1 - 3 = -2$
0	$(0)^2 - 3 = 0 - 3 = -3$
1	$(1)^2 - 3 = 1 - 3 = -2$
2	$(2)^2 - 3 = 4 - 3 = 1$

67.

x	$x^3 + 2$
-2	$(-2)^3 + 2 = -8 + 2 = -6$
-1	$(-1)^3 + 2 = -1 + 2 = 1$
0	$(0)^3 + 2 = 0 + 2 = 2$
1	$(1)^3 + 2 = 1 + 2 = 3$
2	$(2)^3 + 2 = 8 + 2 = 10$

69. $f(x) = x^2 - 1$

x	$f(x)$
-2	3
-1	0
0	-1
1	0
2	3

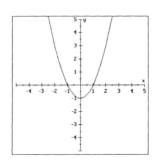

71. $f(x) = x^3 + 2$

x	$f(x)$
-2	-6
-1	1
0	2
1	3
2	10

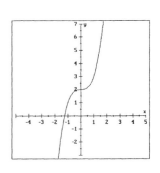

73. $h = -16t^2 + 64t$
$h = -16(2)^2 + 64(2)$
$h = -16(4) + 128$
$h = -64 + 128 = 64$ ft

75. $f(d) = -0.08d^2 + 100d$
$f(815) = -0.08(815)^2 + 100(815)$
$= -0.08(664225) + 81500$
$= -53138 + 81500$
$= \$28,362$

77. $d = 0.04v^2 + 0.9v$
$d = 0.04(30)^2 + 0.9(30)$
$d = 0.04(900) + 27$
$d = 36 + 27 = 63$ ft

79. **Answers may vary.**

81. There are many possible polynomials. One is $2x - 2$.

Exercise 4.5 (page 256)

1. $ab + cd = (3)(-2) + (-1)(2)$
$= -6 + (-2) = -8$

3. $a(b + c) = 3[-2 + (-1)] = 3(-3) = -9$

5. $-4(2x - 9) \geq 12$
$-8x + 36 \geq 12$
$-8x \geq -24$
$\dfrac{-8x}{-8} \leq \dfrac{-24}{-8}$
$x \leq 3$

7. monomial

9. coefficients, variables

11. like terms

13. like terms; $3y + 4y = 7y$

15. unlike terms

17. like terms; $3x^3 + 4x^3 + 6x^3 = 13x^3$

19. like terms; $-5x^3y^2 + 13x^3y^2 = 8x^3y^2$

21. like terms; $-23t^6 + 32t^6 + 56t^6 = 65t^6$

23. unlike terms

25. $4y + 5y = 9y$

27. $-8t^2 - 4t^2 = -12t^2$

29. $32u^3 - 16u^3 = 16u^3$

31. $18x^5y^2 - 11x^5y^2 = 7x^5y^2$

33. $3rst + 4rst + 7rst = 14rst$

35. $-4a^2bc + 5a^2bc - 7a^2bc = -6a^2bc$

37. $(3x)^2 - 4x^2 + 10x^2 = 9x^2 - 4x^2 + 10x^2$
$= 15x^2$

39. $5x^2y^2 + 2(xy)^2 - (3x^2)y^2 = 5x^2y^2 + 2x^2y^2 - 3x^2y^2 = 4x^2y^2$

41. $(-3x^2y)^4 + (4x^4y^2)^2 - 2x^8y^4 = 81x^8y^4 + 16x^8y^4 - 2x^8y^4 = 95x^8y^4$

43. $(3x + 7) + (4x - 3) = 3x + 7 + 4x - 3$
$= 7x + 4$

45. $(4a + 3) - (2a - 4) = 4a + 3 - 2a + 4$
$= 2a + 7$

47. $(2x + 3y) + (5x - 10y) = 2x + 3y + 5x - 10y = 7x - 7y$

49. $(-8x - 3y) - (11x + y) = -8x - 3y - 11x - y = -19x - 4y$

51. $(3x^2 - 3x - 2) + (3x^2 + 4x - 3) = 3x^2 - 3x - 2 + 3x^2 + 4x - 3 = 6x^2 + x - 5$

53. $(2b^2 + 3b - 5) - (2b^2 - 4b - 9) = 2b^2 + 3b - 5 - 2b^2 + 4b + 9 = 7b + 4$

55. $\left(2x^2 - 3x + 1\right) - \left(4x^2 - 3x + 2\right) + \left(2x^2 + 3x + 2\right)$
$= 2x^2 - 3x + 1 - 4x^2 + 3x - 2 + 2x^2 + 3x + 2 = 3x + 1$

57. $2(x + 3) + 3(x + 3) = 2x + 6 + 3x + 9 = 5x + 15$

59. $-8(x - y) + 11(x - y) = -8x + 8y + 11x - 11y = 3x - 3y$

61. $2\left(x^2 - 5x - 4\right) - 3\left(x^2 - 5x - 4\right) + 6\left(x^2 - 5x - 4\right)$
$= 2x^2 - 10x - 8 - 3x^2 + 15x + 12 + 6x^2 - 30x - 24 = 5x^2 - 25x - 20$

63.
$$\begin{array}{r} 3x^2 + 4x + 5 \\ + \quad 2x^2 - 3x + 6 \\ \hline 5x^2 + x + 11 \end{array}$$

65.
$$\begin{array}{r} 2x^3 - 3x^2 + 4x - 7 \\ + \quad -9x^3 - 4x^2 - 5x + 6 \\ \hline -7x^3 - 7x^2 - x - 1 \end{array}$$

67.
$$\begin{array}{r} -3x^2y + 4xy + 25y^2 \\ + \quad 5x^2y - 3xy - 12y^2 \\ \hline 2x^2y + xy + 13y^2 \end{array}$$

69.
$$\begin{array}{r} 3x^2 + 4x - 5 \\ - \quad 2x^2 - 2x + 3 \\ \hline 3x^2 + 4x - 5 \\ + \quad 2x^2 + 2x - 3 \\ \hline 5x^2 + 6x - 8 \end{array}$$

71.
$$\begin{array}{r} 4x^3 + 4x^2 - 3x + 10 \\ - \quad 5x^3 - 2x^2 - 4x - 4 \\ \hline 4x^3 + 4x^2 - 3x + 10 \\ + \quad -5x^3 + 2x^2 + 4x + 4 \\ \hline -x^3 + 6x^2 + x + 14 \end{array}$$

73.
$$\begin{array}{r} -2x^2y^2 - 4xy + 12y^2 \\ - \quad 10x^2y^2 + 9xy - 24y^2 \\ \hline -2x^2y^2 - 4xy + 12y^2 \\ + \quad -10x^2y^2 - 9xy + 24y^2 \\ \hline -12x^2y^2 - 13xy + 36y^2 \end{array}$$

75. $\left[\left(2x^2 - 3x + 4\right) + \left(3x^2 - 2\right)\right] + \left(x^2 + x - 3\right)$
$= \left[2x^2 - 3x + 4 + 3x^2 - 2\right] + x^2 + x - 3$
$= 5x^2 - 3x + 2 + x^2 + x - 3 = 6x^2 - 2x - 1$

77. $\left[\left(3t^3 + t^2\right) + \left(-t^3 + 6t - 3\right)\right] - \left(t^3 - 2t^2 + 2\right)$
$= \left[3t^3 + t^2 - t^3 + 6t - 3\right] - t^3 + 2t^2 - 2$
$= 2t^3 + t^2 + 6t - 3 - t^3 + 2t^2 - 2 = t^3 + 3t^2 + 6t - 5$

79. $\left[\left(-2x^2 - 7x + 1\right) + \left(-4x^2 + 8x - 1\right)\right] + \left(3x^2 + 4x - 7\right)$
$$= \left[-2x^2 - 7x + 1 - 4x^2 + 8x - 1\right] + 3x^2 + 4x - 7$$
$$= -6x^2 + x + 3x^2 + 4x - 7 = -3x^2 + 5x - 7$$

81. $2(x + 3) + 4(x - 2) = 2x + 6 + 4x - 8 = 6x - 2$

83. $-2(x^2 + 7x - 1) - 3(x^2 - 2x + 7) = -2x^2 - 14x + 2 - 3x^2 + 6x - 21 = -5x^2 - 8x - 19$

85. $2\left(2y^2 - 2y + 2\right) - 4\left(3y^2 - 4y - 1\right) + 4\left(y^3 - y^2 - y\right)$
$$= 4y^2 - 4y + 4 - 12y^2 + 16y + 4 + 4y^3 - 4y^2 - 4y = 4y^3 - 12y^2 + 8y + 8$$

87. $2\left(a^2b^2 - ab\right) - 3\left(ab + 2ab^2\right) + \left(b^2 - ab + a^2b^2\right) = 2a^2b^2 - 2ab - 3ab - 6ab^2 + b^2 - ab + a^2b^2$
$$= 3a^2b^2 - 6ab + b^2 - 6ab^2$$

89. $-4\left(x^2y^2 + xy^3 + xy^2z\right) - 2\left(x^2y^2 - 4xy^2z\right) - 2\left(8xy^3 - y\right)$
$$= -4x^2y^2 - 4xy^3 - 4xy^2z - 2x^2y^2 + 8xy^2z - 16xy^3 + 2y$$
$$= -6x^2y^2 + 4xy^2z - 20xy^3 + 2y$$

91. $y = 900x + 105{,}000$
$y = 900(10) + 105{,}000$
$y = 9{,}000 + 105{,}000 = \$114{,}000$

93. $y = 1{,}000x + 120{,}000$
$y = 1{,}000(12) + 120{,}000$
$y = 12{,}000 + 120{,}000 = \$132{,}000$

95. **a.** $y = 900x + 105{,}000$
$y = 900(20) + 105{,}000$
$y = 18{,}000 + 105{,}000 = \$123{,}000$

$y = 1{,}000x + 120{,}000$
$y = 1{,}000(20) + 120{,}000$
$y = 20{,}000 + 120{,}000 = \$140{,}000$

$$\$123{,}000 + \$140{,}000 = \$263{,}000$$

b. $y = 1{,}900x + 225{,}000 = 1{,}900(20) + 225{,}000 = 38{,}000 + 225{,}000 = \$263{,}000$

97. $y = -1{,}100x + 6{,}600$

99. $y = (-1{,}100x + 6{,}600) + (-1{,}700x + 9{,}200) = -2{,}800x + 15{,}800$

101. **Answers may vary.**

103. $P(x + h) + P(x) = [3(x + h) - 5] + [3x - 5] = 3x + 3h - 5 + 3x - 5 = 6x + 3h - 10$

105. $P(x) - Q(x) = \left(x^{23} + 5x^2 + 73\right) - \left(x^{23} + 4x^2 + 73\right) = x^{23} + 5x^2 + 73 - x^{23} - 4x^2 - 73 = x^2$
$P(7) - Q(7) = 7^2 = 49$

Exercise 4.6 (page 266)

1. distributive property

3. commutative property of multiplication

5.
$$\frac{5}{3}(5y + 6) - 10 = 0$$

$$3\left[\frac{5}{3}(5y + 6) - 10\right] = 3(0)$$

$$3 \cdot \frac{5}{3}(5y + 6) - 3 \cdot 10 = 0$$

$$5(5y + 6) - 30 = 0$$

$$25y + 30 - 30 = 0$$

$$25y = 0$$

$$y = 0$$

7. monomial

9. trinomial

11. $(2x)(3x) = 6x^2$

13. $(5)(3x) = 15x$

15. $(3x^2)(4x^3) = (3)(4)x^2x^3 = 12x^5$

17. $(3b^2)(-2b)(4b^3) = (3)(-2)(4)b^2bb^3$
$$= -24b^6$$

19. $(2x^2y^3)(3x^3y^2) = (2)(3)x^2x^3y^3y^2 = 6x^5y^5$

21. $(x^2y^5)(x^2z^5)(-3y^2z^3) = (1)(1)(-3)x^2x^2y^5y^2z^5x^3 = -3x^4y^7z^8$

23. $(x^2y^3)^5 = (x^2)^5(y^3)^5 = x^{10}y^{15}$

25. $(a^3b^2c)(abc^3)^2 = (a^3b^2c)\left[a^2b^2(c^3)^2\right] = (a^3b^2c)(a^2b^2c^6) = a^3a^2b^2b^2cc^6 = a^5b^4c^7$

27. $3(x + 4) = 3(x) + 3(4) = 3x + 12$

29. $-4(t + 7) = -4(t) + (-4)(7) = -4t - 28$

31. $3x(x - 2) = 3x(x) - 3x(2) = 3x^2 - 6x$

33. $-2x^2(3x^2 - x) = -2x^2(3x^2) - (-2x^2)(x)$
$$= -6x^4 + 2x^3$$

35. $3xy(x + y) = 3xy(x) + 3xy(y) = 3x^2y + 3xy^2$

37. $2x^2(3x^2 + 4x - 7) = 2x^2(3x^2) + 2x^2(4x) - 2x^2(7) = 6x^4 + 8x^3 - 14x^2$

39. $\frac{1}{4}x^2(8x^5 - 4) = \frac{1}{4}x^2(8x^5) - \frac{1}{4}x^2(4) = 2x^7 - x^2$

41. $-\frac{2}{3}r^2t^2(9r - 3t) = -\frac{2}{3}r^2t^2(9r) - \left(-\frac{2}{3}r^2t^2\right)(3t) = -6r^3t^2 + 2r^2t^3$

43. $(3xy)(-2x^2y^3)(x + y) = (-6x^3y^4)(x + y) = -6x^3y^4(x) + (-6x^3y^4)(y) = -6x^4y^4 - 6x^3y^5$

45. $(a + 4)(a + 5) = a(a) + a(5) + 4(a) + 4(5) = a^2 + 5a + 4a + 20 = a^2 + 9a + 20$

47. $(3x - 2)(x + 4) = 3x(x) + 3x(4) + (-2)(x) + (-2)(4) = 3x^2 + 12x - 2x - 8 = 3x^2 + 10x - 8$

49. $(2a + 4)(3a - 5) = 2a(3a) - 2a(5) + 4(3a) - 4(5) = 6a^2 - 10a + 12a - 20 = 6a^2 + 2a - 20$

51. $(3x - 5)(2x + 1) = 3x(2x) + 3x(1) + (-5)(2x) + (-5)(1) = 6x^2 + 3x - 10x - 5$
$$= 6x^2 - 7x - 5$$

53. $(x+3)(2x-3) = x(2x) - x(3) + 3(2x) - 3(3) = 2x^2 - 3x + 6x - 9 = 2x^2 + 3x - 9$

55. $(2s+3t)(3s-t) = 2s(3s) - 2s(t) + 3t(3s) - 3t(t) = 6s^2 - 2st + 9st - 3t^2 = 6s^2 + 7st - 3t^2$

57. $(x+y)(x+z) = x(x) + x(z) + y(x) + y(z) = x^2 + xz + xy + yz$

59. $(u+v)(u+2t) = u(u) + u(2t) + v(u) + v(2t) = u^2 + 2tu + uv + 2tv$

61. $(-2r-3s)(2r+7s) = (-2r)(2r) + (-2r)(7s) + (-3s)(2r) + (-3s)(7s)$
$$= -4r^2 - 14rs - 6rs - 21s^2 = -4r^2 - 20rs - 21s^2$$

63.
$$\begin{array}{r} 4x+\ \ 3 \\ x+\ \ 2 \\ \hline 4x^2+\ \ 3x \\ 8x+6 \\ \hline 4x^2+11x+6 \end{array}$$

65.
$$\begin{array}{r} 4x-\ \ 2y \\ 3x+\ \ 5y \\ \hline 12x^2-\ \ 6xy \\ 20xy-10y^2 \\ \hline 12x^2+14xy-10y^2 \end{array}$$

67.
$$\begin{array}{r} x^2+\ x+1 \\ x-1 \\ \hline x^3+x^2+x \\ -\,x^2-x-1 \\ \hline x^3\qquad\quad\ -1 \end{array}$$

69. $(2x+1)(x^2+3x-1) = (2x+1)(x^2) + (2x+1)(3x) - (2x+1)(1)$
$$= 2x(x^2) + 1(x^2) + 2x(3x) + 1(3x) - 2x(1) - 1(1)$$
$$= 2x^3 + x^2 + 6x^2 + 3x - 2x - 1 = 2x^3 + 7x^2 + x - 1$$

71. $(x+4)(x+4) = x(x) + x(4) + 4(x) + 4(4) = x^2 + 4x + 4x + 16 = x^2 + 8x + 16$

73. $(t-3)(t-3) = t(t) - t(3) + (-3)(t) - (-3)(3) = t^2 - 3t - 3t + 9 = t^2 - 6t + 9$

75. $(r+4)(r-4) = r(r) - r(4) + 4(r) - 4(4) = r^2 - 4r + 4r - 16 = r^2 - 16$

77. $(x+5)^2 = (x+5)(x+5) = x(x) + x(5) + 5(x) + 5(5) = x^2 + 5x + 5x + 25 = x^2 + 10x + 25$

79. $(2s+1)(2s+1) = 2s(2s) + 2s(1) + 1(2s) + 1(1) = 4s^2 + 2s + 2s + 1 = 4s^2 + 4s + 1$

81. $(4x+5)(4x-5) = 4x(4x) - 4x(5) + 5(4x) - 5(5) = 16x^2 - 20x + 20x - 25 = 16x^2 - 25$

83. $(x-2y)^2 = (x-2y)(x-2y) = x(x) - x(2y) + (-2y)(x) - (-2y)(2y)$
$$= x^2 - 2xy - 2xy + 4y^2 = x^2 - 4xy + 4y^2$$

85. $(2a-3b)^2 = (2a-3b)(2a-3b) = 2a(2a) - 2a(3b) + (-3b)(2a) - (-3b)(3b)$
$$= 4a^2 - 6ab - 6ab + 9b^2 = 4a^2 - 12ab + 9b^2$$

87. $(4x+5y)(4x-5y) = 4x(4x) - 4x(5y) + 5y(4x) - 5y(5y)$
$$= 16x^2 - 20xy + 20xy - 25y^2 = 16x^2 - 25y^2$$

89. $2(x-4)(x+1) = 2[x(x) + x(1) + (-4)(x) + (-4)(1)] = 2[x^2 + x - 4x - 4]$
$$= 2[x^2 - 3x - 4] = 2x^2 - 6x - 8$$

91. $3a(a+b)(a-b) = 3a[a(a) - a(b) + b(a) - b(b)] = 3a[a^2 - ab + ab - b^2]$
$$= 3a[a^2 - b^2] = 3a^3 - 3ab^2$$

93. $(4t+3)(t^2 + 2t + 3) = 4t(t^2) + 4t(2t) + 4t(3) + 3(t^2) + 3(2t) + 3(3)$
$$= 4t^3 + 8t^2 + 12t + 3t^2 + 6t + 9 = 4t^3 + 11t^2 + 18t + 9$$

95. $(-3x + y)(x^2 - 8xy + 16y^2)$
$$= -3x(x^2) - (-3x)(8xy) + (-3x)(16y^2) + y(x^2) - y(8xy) + y(16y^2)$$
$$= -3x^3 + 24x^2y - 48xy^2 + x^2y - 8xy^2 + 16y^3 = -3x^3 + 25x^2y - 56xy^2 + 16y^3$$

97. $(x - 2y)(x^2 + 2xy + 4y^2)$
$$= x(x^2) + x(2xy) + x(4y^2) + (-2y)(x^2) + (-2y)(2xy) + (-2y)(4y^2)$$
$$= x^3 + 2x^2y + 4xy^2 - 2x^2y - 4xy^2 - 8y^3 = x^3 - 8y^3$$

99. $2t(t+2) + 3t(t-5) = [2t(t) + 2t(2)] + [3t(t) - 3t(5)] = [2t^2 + 4t] + [3t^2 - 15t]$
$$= 2t^2 + 4t + 3t^2 - 15t = 5t^2 - 11t$$

101. $3xy(x+y) - 2x(xy - x) = [3xy(x) + 3xy(y)] - [(2x)(xy) - (2x)(x)]$
$$= [3x^2y + 3xy^2] - [2x^2y - 2x^2]$$
$$= 3x^2y + 3xy^2 - 2x^2y + 2x^2 = x^2y + 3xy^2 + 2x^2$$

103. $(x+y)(x-y) + x(x+y) = [x(x) - x(y) + y(x) - y(y)] + [x(x) + x(y)]$
$$= [x^2 - xy + xy - y^2] + [x^2 + xy]$$
$$= [x^2 - y^2] + [x^2 + xy] = x^2 - y^2 + x^2 + xy = 2x^2 + xy - y^2$$

105. $(x+2)^2 - (x-2)^2 = [(x+2)(x+2)] - [(x-2)(x-2)]$
$$= [x(x) + x(2) + 2(x) + 2(2)] - [x(x) - x(2) + (-2)(x) - (-2)(2)]$$
$$= [x^2 + 2x + 2x + 4] - [x^2 - 2x - 2x + 4]$$
$$= [x^2 + 4x + 4] - [x^2 - 4x + 4] = x^2 + 4x + 4 - x^2 + 4x - 4 = 8x$$

107. $(2s - 3)(s + 2) + (3s + 1)(s - 3)$
$$= [2s(s) + 2s(2) + (-3)(s) + (-3)(2)] + [3s(s) - 3s(3) + 1(s) - 1(3)]$$
$$= [2s^2 + 4s - 3s - 6] + [3s^2 - 9s + s - 3]$$
$$= [2s^2 + s - 6] + [3s^2 - 8s - 3] = 2s^2 + s - 6 + 3s^2 - 8s - 3 = 5s^2 - 7s - 9$$

109. $(s - 4)(s + 1) = s^2 + 5$
$s^2 + s - 4s - 4 = s^2 + 5$
$s^2 - 3s - 4 = s^2 + 5$
$-3s - 4 = 5$
$-3s = 9$
$s = -3$

111. $z(z + 2) = (z + 4)(z - 4)$
$z^2 + 2z = z^2 - 4z + 4z - 16$
$z^2 + 2z = z^2 - 16$
$2z = -16$
$z = -8$

113.
$$(x+4)(x-4) = (x-2)(x+6)$$
$$x^2 - 4x + 4x - 16 = x^2 + 6x - 2x - 12$$
$$x^2 - 16 = x^2 + 4x - 12$$
$$-16 = 4x - 12$$
$$-4 = 4x$$
$$-1 = x$$

115.
$$(a-3)^2 = (a+3)^2$$
$$(a-3)(a-3) = (a+3)(a+3)$$
$$a^2 - 3a - 3a + 9 = a^2 + 3a + 3a + 9$$
$$a^2 - 6a + 9 = a^2 + 6a + 9$$
$$-6a + 9 = 6a + 9$$
$$-6a = 6a$$
$$12a = 0$$
$$a = 0$$

117.
$$4 + (2y-3)^2 = (2y-1)(2y+3)$$
$$4 + (2y-3)(2y-3) = (2y-1)(2y+3)$$
$$4 + 4y^2 - 6y - 6y + 9 = 4y^2 + 6y - 2y - 3$$
$$4y^2 - 12y + 13 = 4y^2 + 4y - 3$$
$$-12y + 13 = 4y - 3$$
$$-12y = 4y - 16$$
$$-16y = -16$$
$$y = 1$$

119. Let $r =$ the smaller radius.
Then $r + 3 =$ the larger radius.

$$\boxed{\text{Larger area}} - \boxed{\text{Smaller area}} = 15\pi$$
$$\pi(r+3)^2 - \pi r^2 = 15\pi$$
$$\pi(r+3)(r+3) - \pi r^2 = 15\pi$$
$$\pi(r^2 + 3r + 3r + 9) - \pi r^2 = 15\pi$$
$$\pi(r^2 + 6r + 9) - \pi r^2 = 15\pi$$
$$\pi r^2 + 6\pi r + 9\pi - \pi r^2 = 15\pi$$
$$6\pi r + 9\pi = 15\pi$$
$$6\pi r = 6\pi$$
$$\frac{6\pi r}{6\pi} = \frac{6\pi}{6\pi}$$
$$r = 1$$

The larger radius $= 1 + 3 = 4$ m.

121. Let $s =$ the side of the softball field. Then $s + 30 =$ the side of the baseball field.

$$\boxed{\text{Larger area}} - \boxed{\text{Smaller area}} = 4500$$
$$(s+30)^2 - s^2 = 4500$$
$$(s+30)(s+30) - s^2 = 4500$$
$$s^2 + 30s + 30s + 900 - s^2 = 4500$$
$$60s + 900 = 4500$$
$$60s = 3600$$
$$s = 60$$

The baseball field has a side of $60 + 30 = 90$ feet.

123. Answers may vary.

125. Answers may vary.

Exercise 4.7 (page 273)

1. binomial

3. none of these

5. 2

7. polynomial

9. two

11. $\dfrac{a}{b}$

13. $\dfrac{5}{15} = \dfrac{5(1)}{5(3)} = \dfrac{1}{3}$

15. $\dfrac{-125}{75} = -\dfrac{25(5)}{25(3)} = -\dfrac{5}{3}$

17. $\dfrac{120}{160} = \dfrac{40(3)}{40(4)} = \dfrac{3}{4}$

19. $\dfrac{-3612}{-3612} = 1$

21. $\dfrac{-90}{360} = -\dfrac{90(1)}{90(4)} = -\dfrac{1}{4}$

23. $\dfrac{5880}{2660} = \dfrac{140(42)}{140(19)} = \dfrac{42}{19}$

25. $\dfrac{xy}{yz} = \dfrac{x}{z}$

27. $\dfrac{r^3 s^2}{rs^3} = r^{3-1} s^{2-3} = r^2 s^{-1} = \dfrac{r^2}{s}$

29. $\dfrac{8x^3 y^2}{4xy^3} = \dfrac{8}{4} x^{3-1} y^{2-3} = 2x^2 y^{-1} = \dfrac{2x^2}{y}$

31. $\dfrac{12u^5 v}{-4u^2 v^3} = \dfrac{12}{-4} u^{5-2} v^{1-3} = -3u^3 v^{-2}$
$$= \dfrac{-3u^3}{v^2}$$

33. $\dfrac{-16r^3 y^2}{-4r^2 y^4} = \dfrac{-16}{-4} r^{3-2} y^{2-4} = 4ry^{-2}$
$$= \dfrac{4r}{y^2}$$

35. $\dfrac{-65rs^2 t}{15r^2 s^3 t} = \dfrac{-65}{15} r^{1-2} s^{2-3} t^{1-1}$
$$= -\dfrac{13}{3} r^{-1} s^{-1} t^0 = -\dfrac{13}{3rs}$$

37. $\dfrac{x^2 x^3}{xy^6} = \dfrac{x^5}{xy^6} = \dfrac{x^{5-1}}{y^6} = \dfrac{x^4}{y^6}$

39. $\dfrac{\left(a^3 b^4\right)^3}{ab^4} = \dfrac{a^9 b^{12}}{ab^4} = a^{9-1} b^{12-4} = a^8 b^8$

41. $\dfrac{15\left(r^2 s^3\right)^2}{-5\left(rs^5\right)^3} = \dfrac{15r^4 s^6}{-5r^3 s^{15}} = \dfrac{15}{-5} r^{4-3} s^{6-15} = -3r^1 s^{-9} = -\dfrac{3r}{s^9}$

43. $\dfrac{-32\left(x^3 y\right)^3}{128\left(x^2 y^2\right)^3} = \dfrac{-32x^9 y^3}{128x^6 y^6} = \dfrac{-32}{128} x^{9-6} y^{3-6} = -\dfrac{1}{4} x^3 y^{-3} = -\dfrac{x^3}{4y^3}$

45. $\dfrac{\left(5a^2 b\right)^3}{\left(2a^2 b^2\right)^3} = \dfrac{125a^6 b^3}{8a^6 b^6} = \dfrac{125}{8} a^{6-6} b^{3-6} = \dfrac{125}{8} a^0 b^{-3} = \dfrac{125}{8b^3}$

47. $\dfrac{-\left(3x^3 y^4\right)^3}{-\left(9x^4 y^5\right)^2} = \dfrac{-27x^9 y^{12}}{-81x^8 y^{10}} = \dfrac{-27}{-81} x^{9-8} y^{12-10} = \dfrac{1}{3} xy^2 = \dfrac{xy^2}{3}$

49. $\dfrac{\left(a^2 a^3\right)^4}{\left(a^4\right)^3} = \dfrac{\left(a^5\right)^4}{a^{12}} = \dfrac{a^{20}}{a^{12}} = a^{20-12} = a^8$

51. $\dfrac{\left(z^3 z^{-4}\right)^3}{\left(z^{-3}\right)^2} = \dfrac{\left(z^{-1}\right)^3}{z^{-6}} = \dfrac{z^{-3}}{z^{-6}} = z^{-3-(-6)} = z^3$

53. $\dfrac{6x + 9y}{3xy} = \dfrac{6x}{3xy} + \dfrac{9y}{3xy} = \dfrac{2}{y} + \dfrac{3}{x}$

55. $\dfrac{5x - 10y}{25xy} = \dfrac{5x}{25xy} - \dfrac{10y}{25xy} = \dfrac{1}{5y} - \dfrac{2}{5x}$

57. $\dfrac{3x^2 + 6y^3}{3x^2 y^2} = \dfrac{3x^2}{3x^2 y^2} + \dfrac{6y^3}{3x^2 y^2} = \dfrac{1}{y^2} + \dfrac{2y}{x^2}$

59. $\dfrac{15a^3 b^2 - 10a^2 b^3}{5a^2 b^2} = \dfrac{15a^3 b^2}{5a^2 b^2} - \dfrac{10a^2 b^3}{5a^2 b^2}$
$$= 3a - 2b$$

61. $\dfrac{4x - 2y + 8z}{4xy} = \dfrac{4x}{4xy} - \dfrac{2y}{4xy} + \dfrac{8z}{4xy} = \dfrac{1}{y} - \dfrac{1}{2x} + \dfrac{2z}{xy}$

63. $\dfrac{12x^3y^2 - 8x^2y - 4x}{4xy} = \dfrac{12x^3y^2}{4xy} - \dfrac{8x^2y}{4xy} - \dfrac{4x}{4xy} = 3x^2y - 2x - \dfrac{1}{y}$

65. $\dfrac{-25x^2y + 30xy^2 - 5xy}{-5xy} = \dfrac{-25x^2y}{-5xy} + \dfrac{30xy^2}{-5xy} - \dfrac{5xy}{-5xy} = 5x - 6y + 1$

67. $\dfrac{5x(4x - 2y)}{2y} = \dfrac{20x^2 - 10xy}{2y} = \dfrac{20x^2}{2y} - \dfrac{10xy}{2y} = \dfrac{10x^2}{y} - 5x$

69. $\dfrac{(-2x)^3 + (3x^2)^2}{6x^2} = \dfrac{-8x^3 + 9x^4}{6x^2} = \dfrac{-8x^3}{6x^2} + \dfrac{9x^4}{6x^2} = -\dfrac{4x}{3} + \dfrac{3x^2}{2}$

71. $\dfrac{4x^2y^2 - 2(x^2y^2 + xy)}{2xy} = \dfrac{4x^2y^2 - 2x^2y^2 - 2xy}{2xy} = \dfrac{2x^2y^2 - 2xy}{2xy} = \dfrac{2x^2y^2}{2xy} - \dfrac{2xy}{2xy} = xy - 1$

73. $\dfrac{(3x - y)(2x - 3y)}{6xy} = \dfrac{6x^2 - 9xy - 2xy + 3y^2}{6xy} = \dfrac{6x^2 - 11xy + 3y^2}{6xy} = \dfrac{6x^2}{6xy} - \dfrac{11xy}{6xy} + \dfrac{3y^2}{6xy}$

$$= \dfrac{x}{y} - \dfrac{11}{6} + \dfrac{y}{2x}$$

75. $\dfrac{(a + b)^2 - (a - b)^2}{2ab} = \dfrac{(a + b)(a + b) - (a - b)(a - b)}{2ab}$

$$= \dfrac{[a^2 + ab + ab + b^2] - [a^2 - ab - ab + b^2]}{2ab}$$

$$= \dfrac{[a^2 + 2ab + b^2] - [a^2 - 2ab + b^2]}{2ab}$$

$$= \dfrac{a^2 + 2ab + b^2 - a^2 + 2ab - b^2}{2ab} = \dfrac{4ab}{2ab} = 2$$

77. $l = \dfrac{P - 2w}{2}$

$l = \dfrac{P}{2} - \dfrac{2w}{2}$

$l = \dfrac{P}{2} - w$

They are the same.

79. $C = \dfrac{0.15x + 12}{x}$

$C = \dfrac{0.15x}{x} + \dfrac{12}{x}$

$C = 0.15 + \dfrac{12}{x}$

They are the same.

81. Answers may vary.

83. $\dfrac{x^{500} - x^{499}}{x^{499}} = \dfrac{x^{500}}{x^{499}} - \dfrac{x^{499}}{x^{499}} = x - 1$

Let $x = 501$: $x - 1 = 501 - 1 = 500$

Exercise 4.8 (page 279)

1. $21, 22, 24, 25, 26, 27, 28$

3. $|a - b| = |-2 - 3| = |-5| = 5$

5. $-|a^2 - b^2| = -|(-2)^2 - 3^2| = -|4 - 9| = -|-5| = -(+5) = -5$

7. $3(2x^2 - 4x + 5) + 2(x^2 + 3x - 7) = 6x^2 - 12x + 15 + 2x^2 + 6x - 14 = 8x^2 - 6x + 1$

9. divisor; dividend

11. remainder

13. $4x^3 - 2x^2 + 7x + 6$

15. $6x^4 - x^3 + 2x^2 + 9x$

17. $0x^3$ and $0x$

19.
$$
\begin{array}{r}
x + 2 \\
x + 2\,\overline{\smash{\big)}\,x^2 + 4x + 4} \\
\underline{x^2 + 2x} \\
2x + 4 \\
\underline{2x + 4} \\
0
\end{array}
$$

21.
$$
\begin{array}{r}
y + 12 \\
y + 1\,\overline{\smash{\big)}\,y^2 + 13y + 12} \\
\underline{y^2 + y} \\
12y + 12 \\
\underline{12y + 12} \\
0
\end{array}
$$

23.
$$
\begin{array}{r}
a + b \\
a + b\,\overline{\smash{\big)}\,a^2 + 2ab + b^2} \\
\underline{a^2 + ab} \\
ab + b^2 \\
\underline{ab + b^2} \\
0
\end{array}
$$

25.
$$
\begin{array}{r}
3a - 2 \\
2a + 3\,\overline{\smash{\big)}\,6a^2 + 5a - 6} \\
\underline{6a^2 + 9a} \\
-4a - 6 \\
\underline{-4a - 6} \\
0
\end{array}
$$

27.
$$
\begin{array}{r}
b + 3 \\
3b + 2\,\overline{\smash{\big)}\,3b^2 + 11b + 6} \\
\underline{3b^2 + 2b} \\
9b + 6 \\
\underline{9b + 6} \\
0
\end{array}
$$

29.
$$
\begin{array}{r}
x - 3y \\
2x - y\,\overline{\smash{\big)}\,2x^2 - 7xy + 3y^2} \\
\underline{2x^2 - xy} \\
-6xy + 3y^2 \\
\underline{-6xy + 3y^2} \\
0
\end{array}
$$

31.
$$
\begin{array}{r}
2x + 1 \\
5x + 3\,\overline{\smash{\big)}\,10x^2 + 11x + 3} \\
\underline{10x^2 + 6x} \\
5x + 3 \\
\underline{5x + 3} \\
0
\end{array}
$$

33.
$$
\begin{array}{r}
x - 7 \\
2x + 4\,\overline{\smash{\big)}\,2x^2 - 10x - 28} \\
\underline{2x^2 + 4x} \\
-14x - 28 \\
\underline{-14x - 28} \\
0
\end{array}
$$

35.
$$
\begin{array}{r}
3x + 2y \\
2x - y\,\overline{\smash{\big)}\,6x^2 + xy - 2y^2} \\
\underline{6x^2 - 3xy} \\
4xy - 2y^2 \\
\underline{4xy - 2y^2} \\
0
\end{array}
$$

37.
$$
\begin{array}{r}
2x - y \\
x + 3y\,\overline{\smash{\big)}\,2x^2 + 5xy - 3y^2} \\
\underline{2x^2 + 6xy} \\
-xy - 3y^2 \\
\underline{-xy - 3y^2} \\
0
\end{array}
$$

39.
$$
\begin{array}{r}
x + 5y \\
3x - 2y\,\overline{\smash{\big)}\,3x^2 + 13xy - 10y^2} \\
\underline{3x^2 - 2xy} \\
15xy - 10y^2 \\
\underline{15xy - 10y^2} \\
0
\end{array}
$$

41.
$$
\begin{array}{r}
x - 5y \\
4x + y\,\overline{\smash{\big)}\,4x^2 - 19xy - 5y^2} \\
\underline{4x^2 + xy} \\
-20xy - 5y^2 \\
\underline{-20xy - 5y^2} \\
0
\end{array}
$$

43.
$$
\begin{array}{r}
x^2 + 2x - 1 \\
2x+3 \overline{\smash{\big)}\ 2x^3 + 7x^2 + 4x - 3} \\
\underline{2x^3 + 3x^2} \\
4x^2 + 4x \\
\underline{4x^2 + 6x} \\
-2x - 3 \\
\underline{-2x - 3} \\
0
\end{array}
$$

45.
$$
\begin{array}{r}
2x^2 + 2x + 1 \\
3x+2 \overline{\smash{\big)}\ 6x^3 + 10x^2 + 7x + 2} \\
\underline{6x^3 + 4x^2} \\
6x^2 + 7x \\
\underline{6x^2 + 4x} \\
3x + 2 \\
\underline{3x + 2} \\
0
\end{array}
$$

47.
$$
\begin{array}{r}
x^2 + xy + y^2 \\
2x+y \overline{\smash{\big)}\ 2x^3 + 3x^2y + 3xy^2 + y^3} \\
\underline{2x^3 + x^2y} \\
2x^2y + 3xy^2 \\
\underline{2x^2y + xy^2} \\
2xy^2 + y^3 \\
\underline{2xy^2 + y^3} \\
0
\end{array}
$$

49.
$$
\begin{array}{r}
x + 1 + \frac{-1}{2x+3} \\
2x+3 \overline{\smash{\big)}\ 2x^2 + 5x + 2} \\
\underline{2x^2 + 3x} \\
2x + 2 \\
\underline{2x + 3} \\
-1
\end{array}
$$

51.
$$
\begin{array}{r}
2x + 2 + \frac{-3}{2x+1} \\
2x+1 \overline{\smash{\big)}\ 4x^2 + 6x - 1} \\
\underline{4x^2 + 2x} \\
4x - 1 \\
\underline{4x + 2} \\
-3
\end{array}
$$

53.
$$
\begin{array}{r}
x^2 + 2x + 1 \\
x+1 \overline{\smash{\big)}\ x^3 + 3x^2 + 3x + 1} \\
\underline{x^3 + x^2} \\
2x^2 + 3x \\
\underline{2x^2 + 2x} \\
x + 1 \\
\underline{x + 1} \\
0
\end{array}
$$

55.
$$
\begin{array}{r}
x^2 + 2x - 1 + \frac{6}{2x+3} \\
2x+3 \overline{\smash{\big)}\ 2x^3 + 7x^2 + 4x + 3} \\
\underline{2x^3 + 3x^2} \\
4x^2 + 4x \\
\underline{4x^2 + 6x} \\
-2x + 3 \\
\underline{-2x - 3} \\
6
\end{array}
$$

57.
$$
\begin{array}{r}
2x^2 + 8x + 14 + \frac{31}{x-2} \\
x-2 \overline{\smash{\big)}\ 2x^3 + 4x^2 - 2x + 3} \\
\underline{2x^3 - 4x^2} \\
8x^2 - 2x \\
\underline{8x^2 - 16x} \\
14x + 3 \\
\underline{14x - 28} \\
31
\end{array}
$$

59.
$$
\begin{array}{r}
x + 1 \\
x-1 \overline{\smash{\big)}\ x^2 + 0x - 1} \\
\underline{x^2 - x} \\
x - 1 \\
\underline{x - 1} \\
0
\end{array}
$$

61.
$$
\begin{array}{r}
2x - 3 \\
2x+3 \overline{\smash{\big)}\ 4x^2 + 0x - 9} \\
\underline{4x^2 + 6x} \\
-6x - 9 \\
\underline{-6x - 9} \\
0
\end{array}
$$

63.
$$
\begin{array}{r}
x^2 - x + 1 \\
x+1 \overline{\smash{\big)}\ x^3 + 0x^2 + 0x + 1} \\
\underline{x^3 + x^2} \\
-x^2 + 0x \\
\underline{-x^2 - x} \\
x + 1 \\
\underline{x + 1} \\
0
\end{array}
$$

65.
$$\begin{array}{r}a^2 - 3a + 10 + \frac{-30}{a+3} \\ a+3 \,\overline{\smash{\big)}\, a^3 + 0a^2 + a + 0} \\ \underline{a^3 + 3a^2} \\ -3a^2 + a \\ \underline{-3a^2 - 9a} \\ 10a + 0 \\ \underline{10a + 30} \\ -30\end{array}$$

67.
$$\begin{array}{r}5x^2 - x + 4 + \frac{16}{3x-4} \\ 3x-4 \,\overline{\smash{\big)}\, 15x^3 - 23x^2 + 16x + 0} \\ \underline{15x^3 - 20x^2} \\ -3x^2 + 16x \\ \underline{-3x^2 + 4x} \\ 12x + 0 \\ \underline{12x - 16} \\ 16\end{array}$$

69. Answers may vary.

71. $x^2 - 2x$ is added to the dividend when it should be subtracted.

Chapter 4 Summary (page 281)

1. $(-3x)^4 = (-3x)(-3x)(-3x)(-3x)$

2. $\left(\frac{1}{2}pq\right)^3 = \left(\frac{1}{2}pq\right)\left(\frac{1}{2}pq\right)\left(\frac{1}{2}pq\right)$

3. $5^3 = 5 \cdot 5 \cdot 5 = 125$

4. $3^5 = 3 \cdot 3 \cdot 3 \cdot 3 \cdot 3 = 243$

5. $(-8)^2 = (-8)(-8) = 64$

6. $-8^2 = -1 \cdot 8 \cdot 8 = -64$

7. $3^2 + 2^2 = 9 + 4 = 13$

8. $(3+2)^2 = (5)^2 = 25$

9. $x^3 x^2 = x^{3+2} = x^5$

10. $x^2 x^7 = x^{2+7} = x^9$

11. $\left(y^7\right)^3 = y^{7 \cdot 3} = y^{21}$

12. $\left(x^{21}\right)^2 = x^{21 \cdot 2} = x^{42}$

13. $(ab)^3 = a^3 b^3$

14. $(3x)^4 = 3^4 x^4 = 81x^4$

15. $b^3 b^4 b^5 = b^{3+4+5} = b^{12}$

16. $-z^2(z^3 y^2) = -z^{2+3} y^2 = -y^2 z^5$

17. $(16s)^2 s = 16^2 s^2 s = 256 s^3$

18. $-3y(y^5) = -3y^6$

19. $\left(x^2 x^3\right)^3 = \left(x^5\right)^3 = x^{15}$

20. $\left(2x^2 y\right)^2 = 2^2\left(x^2\right)^2 y^2 = 4x^4 y^2$

21. $\dfrac{x^7}{x^3} = x^{7-3} = x^4$

22. $\left(\dfrac{x^2 y}{xy^2}\right)^2 = \dfrac{x^4 y^2}{x^2 y^4} = x^2 y^{-2} = \dfrac{x^2}{y^2}$

23. $\dfrac{8\left(y^2 x\right)^2}{4\left(yx^2\right)^2} = \dfrac{8y^4 x^2}{4y^2 x^4} = \dfrac{8}{4} y^2 x^{-2} = \dfrac{2y^2}{x^2}$

24. $\dfrac{\left(5y^2 z^3\right)^3}{25(yz)^5} = \dfrac{125 y^6 z^9}{25 y^5 z^5} = 5yz^4$

25. $x^0 = 1$

26. $\left(3x^2 y^2\right)^0 = 1$

27. $\left(3x^0\right)^2 = (3 \cdot 1)^2 = 3^2 = 9$

28. $\left(3x^2 y^0\right)^2 = \left(3x^2 \cdot 1\right)^2 = \left(3x^2\right)^2 = 9x^4$

29. $x^{-3} = \dfrac{1}{x^3}$

30. $x^{-2}x^3 = x^1 = x$

31. $y^4 y^{-3} = y^1 = y$

32. $\dfrac{x^3}{x^{-7}} = x^{3-(-7)} = x^{10}$

33. $\left(x^{-3}x^4\right)^{-2} = \left(x^1\right)^{-2} = x^{-2} = \dfrac{1}{x^2}$

34. $\left(a^{-2}b\right)^{-3} = a^6 b^{-3} = \dfrac{a^6}{b^3}$

35. $\left(\dfrac{x^2}{x}\right)^{-5} = (x)^{-5} = \dfrac{1}{x^5}$

36. $\left(\dfrac{15z^4}{5z^3}\right)^{-2} = (3z)^{-2} = \dfrac{1}{(3z)^2} = \dfrac{1}{9z^2}$

37. $728 = 7.28 \times 10^2$

38. $9{,}370 = 9.37 \times 10^3$

39. $0.0136 = 1.36 \times 10^{-2}$

40. $0.00942 = 9.42 \times 10^{-3}$

41. $7.73 = 7.73 \times 10^0$

42. $753 \times 10^3 = 7.53 \times 10^2 \times 10^3$
$= 7.53 \times 10^5$

43. $0.018 \times 10^{-2} = 1.8 \times 10^{-2} \times 10^{-2}$
$= 1.8 \times 10^{-4}$

44. $600 \times 10^2 = 6.00 \times 10^2 \times 10^2$
$= 6.00 \times 10^4$

45. $7.26 \times 10^5 = 726{,}000$

46. $3.91 \times 10^{-4} = 0.000391$

47. $2.68 \times 10^0 = 2.68$

48. $5.76 \times 10^1 = 57.6$

49. $739 \times 10^{-2} = 7.39$

50. $0.437 \times 10^{-3} = 0.000437$

51. $\dfrac{(0.00012)(0.00004)}{0.00000016} = \dfrac{(1.2 \times 10^{-4})(4 \times 10^{-5})}{1.6 \times 10^{-7}} = \dfrac{4.8 \times 10^{-9}}{1.6 \times 10^{-7}} = 3 \times 10^{-2} = 0.03$

52. $\dfrac{(4{,}800)(20{,}000)}{600{,}000} = \dfrac{(4.8 \times 10^3)(2 \times 10^4)}{6 \times 10^5} = \dfrac{9.6 \times 10^7}{6 \times 10^5} = 1.6 \times 10^2 = 160$

53. $13x^7$: monomial, degree $= 7$

54. $5^3 x + x^2$: binomial, degree $= 2$

55. $-3x^5 + x - 1$: trinomial, degree $= 5$

56. $9x + 21x^3 y^2$: binomial,
degree $= 3 + 2 = 5$

57. $3x + 2 = 3(3) + 2 = 9 + 2 = 11$

58. $3x + 2 = 3(0) + 2 = 0 + 2 = 2$

59. $3x + 2 = 3(-2) + 2 = -6 + 2 = -4$

60. $3x + 2 = 3\left(\frac{2}{3}\right) + 2 = 2 + 2 = 4$

61. $5x^4 - x = 5(3)^4 - 3 = 5(81) - 3$
$= 405 - 3 = 402$

62. $5x^4 - x = 5(0)^4 - 0 = 5(0) - 0$
$= 0 - 0 = 0$

63. $5x^4 - x = 5(-2)^4 - (-2) = 5(16) + 2$
$\qquad = 80 + 2 = 82$

64. $5x^4 - x = 5(-0.3)^4 - (-0.3)$
$\qquad = 5(0.0081) + 0.3$
$\qquad = 0.0405 + 3 = 0.3405$

65. $f(0) = 0^2 - 4 = 0 - 4 = -4$

66. $f(5) = 5^2 - 4 = 25 - 4 = 21$

67. $f(-2) = (-2)^2 - 4 = 4 - 4 = 0$

68. $f\left(\frac{1}{2}\right) = \left(\frac{1}{2}\right)^2 - 4 = \frac{1}{4} - 4 = -\frac{15}{4}$

69. $f(x) = x^2 - 5$

x	-2	-1	0	1	2
$f(x)$	-1	-4	-5	-4	-1

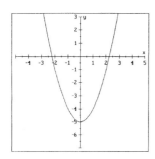

70. $f(x) = x^3 - 2$

x	-2	-1	0	1	2
$f(x)$	-10	-3	-2	-1	6

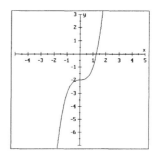

71. $3x + 5x - x = 7x$

72. $3x + 2y$: not like terms

73. $(xy)^2 + 3x^2y^2 = x^2y^2 + 3x^2y^2 = 4x^2y^2$

74. $-2x^2yz + 3x^2yz = x^2yz$

75. $(3x^2 + 2x) + (5x^2 - 8x) = 3x^2 + 2x + 5x^2 - 8x = 8x^2 - 6x$

76. $(7a^2 + 2a - 5) - (3a^2 - 2a + 1) = 7a^2 + 2a - 5 - 3a^2 + 2a - 1 = 4a^2 + 4a - 6$

77. $3(9x^2 + 3x + 7) - 2(11x^2 - 5x + 9) = 27x^2 + 9x + 21 - 22x^2 + 10x - 18 = 5x^2 + 19x + 3$

78. $4(4x^3 + 2x^2 - 3x - 8) - 5(2x^3 - 3x + 8) = 16x^3 + 8x^2 - 12x - 32 - 10x^3 + 15x - 40$
$\qquad = 6x^3 + 8x^2 + 3x - 72$

79. $(2x^2y^3)(5xy^2) = 10x^3y^5$

80. $(xyz^3)(x^3z)^2 = xyz^3x^6z^2 = x^7yz^5$

81. $5(x + 3) = 5x + 15$

82. $3(2x + 4) = 6x + 12$

83. $x^2(3x^2 - 5) = 3x^4 - 5x^2$

84. $2y^2(y^2 + 5y) = 2y^4 + 10y^3$

85. $-x^2y(y^2 - xy) = -x^2y^3 + x^3y^2$

86. $-3xy(xy - x) = -3x^2y^2 + 3x^2y$

87. $(x + 3)(x + 2) = x^2 + 2x + 3x + 6$
$\qquad = x^2 + 5x + 6$

88. $(2x + 1)(x - 1) = 2x^2 - 2x + x - 1$
$\qquad = 2x^2 - x - 1$

89. $(3a - 3)(2a + 2) = 6a^2 + 6a - 6a - 6$
$$= 6a^2 - 6$$

90. $6(a - 1)(a + 1) = 6(a^2 + a - a - 1)$
$$= 6(a^2 - 1) = 6a^2 - 6$$

91. $(a - b)(2a + b) = 2a^2 + ab - 2ab - b^2$
$$= 2a^2 - ab - b^2$$

92. $(3x - y)(2x + y) = 6x^2 + 3xy - 2xy - y^2$
$$= 6x^2 + xy - y^2$$

93. $(x + 3)(x + 3) = x^2 + 3x + 3x + 9$
$$= x^2 + 6x + 9$$

94. $(x + 5)(x - 5) = x^2 - 5x + 5x - 25$
$$= x^2 - 25$$

95. $(y - 2)(y + 2) = y^2 + 2y - 2y - 4$
$$= y^2 - 4$$

96. $(x + 4)^2 = (x + 4)(x + 4)$
$$= x^2 + 4x + 4x + 16$$
$$= x^2 + 8x + 16$$

97. $(x - 3)^2 = (x - 3)(x - 3) = x^2 - 3x - 3x + 9 = x^2 - 6x + 9$

98. $(y - 1)^2 = (y - 1)(y - 1) = y^2 - y - y + 1 = y^2 - 2y + 1$

99. $(2y + 1)^2 = (2y + 1)(2y + 1) = 4y^2 + 2y + 2y + 1 = 4y^2 + 4y + 1$

100. $(y^2 + 1)(y^2 - 1) = y^4 - y^2 + y^2 - 1 = y^4 - 1$

101. $(3x + 1)(x^2 + 2x + 1) = 3x^3 + 6x^2 + 3x + x^2 + 2x + 1 = 3x^3 + 7x^2 + 5x + 1$

102. $(2a - 3)(4a^2 + 6a + 9) = 8a^3 + 12a^2 + 18a - 12a^2 - 18a - 27 = 8a^3 - 27$

103. $x^2 + 3 = x(x + 3)$
$$x^2 + 3 = x^2 + 3x$$
$$3 = 3x$$
$$1 = x$$

104. $x^2 + x = (x + 1)(x + 2)$
$$x^2 + x = x^2 + 2x + x + 2$$
$$x^2 + x = x^2 + 3x + 2$$
$$x = 3x + 2$$
$$-2x = 2$$
$$x = -1$$

105. $(x + 2)(x - 5) = (x - 4)(x - 1)$
$$x^2 - 5x + 2x - 10 = x^2 - x - 4x + 4$$
$$x^2 - 3x - 10 = x^2 - 5x + 4$$
$$-3x - 10 = -5x + 4$$
$$-3x = -5x + 14$$
$$2x = 14$$
$$x = 7$$

106. $(x - 1)(x - 2) = (x - 3)(x + 1)$
$$x^2 - 2x - x + 2 = x^2 + x - 3x - 3$$
$$x^2 - 3x + 2 = x^2 - 2x - 3$$
$$-3x + 2 = -2x - 3$$
$$-3x = -2x - 5$$
$$-x = -5$$
$$x = 5$$

107.
$$x^2 + x(x+2) = x(2x+1) + 1$$
$$x^2 + x^2 + 2x = 2x^2 + x + 1$$
$$2x^2 + 2x = 2x^2 + x + 1$$
$$2x = x + 1$$
$$x = 1$$

108.
$$(x+5)(3x+1) = x^2 + (2x-1)(x-5)$$
$$3x^2 + 16x + 5 = 3x^2 - 11x + 5$$
$$16x + 5 = -11x + 5$$
$$16x = -11x$$
$$27x = 0$$
$$x = 0$$

109. $\dfrac{3x+6y}{2xy} = \dfrac{3x}{2xy} + \dfrac{6y}{2xy} = \dfrac{3}{2y} + \dfrac{3}{x}$

110. $\dfrac{14xy - 21x}{7xy} = \dfrac{14xy}{7xy} - \dfrac{21x}{7xy} = 2 - \dfrac{3}{y}$

111. $\dfrac{15a^2bc + 20ab^2c - 25abc^2}{-5abc} = \dfrac{15a^2bc}{-5abc} + \dfrac{20ab^2c}{-5abc} - \dfrac{25abc^2}{-5abc} = -3a - 4b + 5c$

112.
$$\frac{(x+y)^2 + (x-y)^2}{-2xy} = \frac{(x+y)(x+y) + (x-y)(x-y)}{-2xy}$$
$$= \frac{x^2 + xy + xy + y^2 + x^2 - xy - xy + y^2}{-2xy}$$
$$= \frac{2x^2 + 2y^2}{-2xy} = \frac{2x^2}{-2xy} + \frac{2y^2}{-2xy} = -\frac{x}{y} - \frac{y}{x}$$

113.
$$\begin{array}{r} x + 1 + \frac{3}{x+2} \\ x+2 \overline{\smash{)}\; x^2 + 3x + 5} \\ \underline{x^2 + 2x} \\ x + 5 \\ \underline{x + 2} \\ 3 \end{array}$$

114.
$$\begin{array}{r} x - 5 \\ x-1 \overline{\smash{)}\; x^2 - 6x + 5} \\ \underline{x^2 - x} \\ -5x + 5 \\ \underline{-5x + 5} \\ 0 \end{array}$$

115.
$$\begin{array}{r} 2x + 1 \\ x+3 \overline{\smash{)}\; 2x^2 + 7x + 3} \\ \underline{2x^2 + 6x} \\ x + 3 \\ \underline{x + 3} \\ 0 \end{array}$$

116.
$$\begin{array}{r} x + 5 + \frac{3}{3x-1} \\ 3x-1 \overline{\smash{)}\; 3x^2 + 14x - 2} \\ \underline{3x^2 - x} \\ 15x - 2 \\ \underline{15x - 5} \\ 3 \end{array}$$

117.
$$\begin{array}{r} 3x^2 + 2x + 1 + \frac{2}{2x-1} \\ 2x-1 \overline{\smash{)}\; 6x^3 + x^2 + 0x + 1} \\ \underline{6x^3 - 3x^2} \\ 4x^2 + 0x \\ \underline{4x^2 - 2x} \\ 2x + 1 \\ \underline{2x - 1} \\ 2 \end{array}$$

118.
$$\begin{array}{r} 3x^2 - x - 4 \\ 3x+1 \overline{\smash{)}\; 9x^3 + 0x^2 - 13x - 4} \\ \underline{9x^3 + 3x^2} \\ -3x^2 - 13x \\ \underline{-3x^2 - x} \\ -12x - 4 \\ \underline{-12x - 4} \\ 0 \end{array}$$

Chapter 4 Test (page 285)

1. $2xxxyyyy = 2x^3y^4$

2. $3^2 + 5^3 = 9 + 125 = 134$

3. $y^2(yy^3) = y^2y^4 = y^6$

4. $\left(-3b^2\right)\left(2b^3\right)\left(-b^2\right) = (-3)(2)(-1)b^2b^3b^2 = 6b^7$

5. $(2x^3)^5(x^2)^3 = 32x^{15}x^6 = 32x^{21}$ **6.** $(2rr^2r^3)^3 = (2r^6)^3 = 8r^{18}$

7. $3x^0 = 3(1) = 3$ **8.** $2y^{-5}y^2 = 2y^{-3} = \dfrac{2}{y^3}$ **9.** $\dfrac{y^2}{yy^{-2}} = \dfrac{y^2}{y^{-1}} = y^3$

10. $\left(\dfrac{a^2b^{-1}}{4a^3b^{-2}}\right)^{-3} = \left(\dfrac{1}{4}a^{-1}b^1\right)^{-3} = \left(\dfrac{1}{4}\right)^{-3}a^3b^{-3} = \dfrac{64a^3}{b^3}$

11. $28{,}000 = 2.8 \times 10^4$ **12.** $0.0025 = 2.5 \times 10^{-3}$ **13.** $7.4 \times 10^3 = 7{,}400$

14. $9.3 \times 10^{-5} = 0.000093$ **15.** $3x^2 + 2$: binomial

16. $2 + 3 + 5 = 10$ **17.** $x^2 + x + 2 = (-2)^2 + (-2) - 2$
$$= 4 - 2 - 2 = 0$$

18. $f(x) = x^2 + 2$

x	$f(x)$
-2	6
-1	3
0	2
1	3
2	6

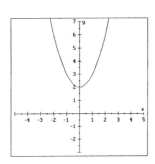

19. $-6(x - y) + 2(x + y) - 3(x + 2y) = -6x + 6y + 2x + 2y - 3x - 6y = -7x + 2y$

20. $-2(x^2 + 3x - 1) - 3(x^2 - x + 2) + 5(x^2 + 2) = -2x^2 - 6x + 2 - 3x^2 + 3x - 6 + 5x^2 + 10$
$$= -3x + 6$$

21. $\begin{array}{r} 3x^3 + 4x^2 - x - 7 \\ + 2x^3 - 2x^2 + 3x + 2 \\ \hline 5x^3 + 2x^2 + 2x - 5 \end{array}$ **22.** $\begin{array}{r} 2x^2 - 7x + 3 \\ - 3x^2 - 2x - 1 \\ \hline \end{array} \Rightarrow \begin{array}{r} 2x^2 - 7x + 3 \\ + -3x^2 + 2x + 1 \\ \hline -x^2 - 5x + 4 \end{array}$

23. $(-2x^3)(2x^2y) = -4x^5y$ **24.** $3y^2(y^2 - 2y + 3) = 3y^4 - 6y^3 + 9y^2$

25. $(2x - 5)(3x + 4) = 6x^2 + 8x - 15x - 20 = 6x^2 - 7x - 20$

26. $(2x - 3)(x^2 - 2x + 4) = 2x^3 - 4x^2 + 8x - 3x^2 + 6x - 12 = 2x^3 - 7x^2 + 14x - 12$

27.
$$(a+2)^2 = (a-3)^2$$
$$(a+2)(a+2) = (a-3)(a-3)$$
$$a^2 + 4a + 4 = a^2 - 6a + 9$$
$$4a + 4 = -6a + 9$$
$$10a = 5$$
$$a = \frac{5}{10} = \frac{1}{2}$$

28. $\dfrac{8x^2y^3z^4}{16x^3y^2z^4} = \dfrac{1}{2}x^{-1}y^1z^0 = \dfrac{y}{2x}$

29.
$$\frac{6a^2 - 12b^2}{24ab} = \frac{6a^2}{24ab} - \frac{12b^2}{24ab}$$
$$= \frac{a}{4b} - \frac{b}{2a}$$

30.
$$\begin{array}{r} x - 2 \\ 2x+3\,\overline{\smash{\big)}\,2x^2 - x - 6} \\ \underline{2x^2 + 3x} \\ -4x - 6 \\ \underline{-4x - 6} \\ 0 \end{array}$$

Cumulative Review Exercises (page 286)

1. $5 + 3 \cdot 2 = 5 + 6 = 11$

2. $3 \cdot 5^2 - 4 = 3 \cdot 25 - 4 = 75 - 4 = 71$

3. $\dfrac{3x - y}{xy} = \dfrac{3(2) - (-5)}{2(-5)} = \dfrac{6+5}{-10} = -\dfrac{11}{10}$

4. $\dfrac{x^2 - y^2}{x + y} = \dfrac{2^2 - (-5)^2}{2 + (-5)} = \dfrac{4 - 25}{-3} = \dfrac{-21}{-3} = 7$

5.
$$\frac{4}{5}x + 6 = 18$$
$$\frac{4}{5}x = 12$$
$$5 \cdot \frac{4}{5}x = 5(12)$$
$$4x = 60$$
$$x = 15$$

6.
$$x - 2 = \frac{x+2}{3}$$
$$3(x-2) = 3 \cdot \frac{x+2}{3}$$
$$3x - 6 = x + 2$$
$$2x - 6 = 2$$
$$2x = 8$$
$$x = 4$$

7.
$$2(5x + 2) = 3(3x - 2)$$
$$10x + 4 = 9x - 6$$
$$x + 4 = -6$$
$$x = -10$$

8.
$$4(y + 1) = -2(4 - y)$$
$$4y + 4 = -8 + 2y$$
$$2y + 4 = -8$$
$$2y = -12$$
$$y = -6$$

9.
$$5x - 3 > 7$$
$$5x > 10$$
$$x > 2$$

10.
$$7x - 9 < 5$$
$$7x < 14$$
$$x < 2$$

11. $-2 < -x + 3 < 5$

$\quad -5 < \quad -x \quad < 2$

$\quad \dfrac{-5}{-1} > \quad \dfrac{-x}{-1} \quad > \dfrac{2}{-1}$

$\quad 5 > \quad x \quad > -2$

$\quad -2 < \quad x \quad < 5$

12. $0 \leq \dfrac{4-x}{3} \leq 2$

$\quad 3(0) \leq 3 \cdot \dfrac{4-x}{3} \leq 3(2)$

$\quad 0 \leq \quad 4 - x \quad \leq 6$

$\quad -4 \leq \quad -x \quad \leq 2$

$\quad \dfrac{-4}{-1} \geq \quad \dfrac{-x}{-1} \quad \geq \dfrac{2}{-1}$

$\quad 4 \geq \quad x \quad \geq -2$

$\quad -2 \leq \quad x \quad \leq 4$

13. $A = p + prt$

$A - p = p - p + prt$

$A - p = prt$

$\dfrac{A - p}{pt} = \dfrac{prt}{pt}$

$\dfrac{A - p}{pt} = r$, or $r = \dfrac{A - p}{pt}$

14. $A = \dfrac{1}{2}bh$

$2A = 2 \cdot \dfrac{1}{2}bh$

$2A = bh$

$\dfrac{2A}{b} = \dfrac{bh}{b}$

$\dfrac{2A}{b} = h$, or $h = \dfrac{2A}{b}$

15. $3x - 4y = 12$

$x = 0$	$y = 0$
$3(0) - 4y = 12$	$3x - 4(0) = 12$
$0 - 4y = 12$	$3x - 0 = 12$
$-4y = 12$	$3x = 12$
$y = -3$	$x = 4$

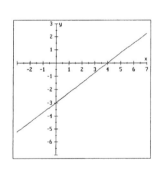

16. $y - 2 = \dfrac{1}{2}(x - 4)$

$x = 0$	$x = 4$
$y - 2 = \dfrac{1}{2}(0 - 4)$	$y - 2 = \dfrac{1}{2}(4 - 4)$
$y - 2 = \dfrac{1}{2}(-4)$	$y - 2 = \dfrac{1}{2}(0)$
$y - 2 = -2$	$y - 2 = 0$
$y = 0$	$y = 2$

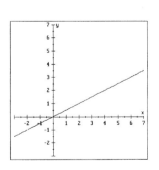

17. $f(0) = 5(0) - 2 = 0 - 2 = -2$

18. $f(3) = 5(3) - 2 = 15 - 2 = 13$

19. $f(-2) = 5(-2) - 2 = -10 - 2 = -12$

20. $f\left(\frac{1}{5}\right) = 5\left(\frac{1}{5}\right) - 2 = 1 - 2 = -1$

21. $(y^3 y^5)y^6 = y^8 y^6 = y^{14}$

22. $\dfrac{x^3 y^4}{x^2 y^3} = x^{3-2} y^{4-3} = x^1 y^1 = xy$

23. $\dfrac{a^4 b^{-3}}{a^{-3} b^3} = a^{4-(-3)} b^{-3-3} = a^7 b^{-6} = \dfrac{a^7}{b^6}$

24. $\left(\dfrac{-x^{-2} y^3}{x^{-3} y^2}\right)^2 = \left(-x^{-2-(-3)} y^{3-2}\right)^2$

$= \left(-x^1 y^1\right)^2 = x^2 y^2$

25. $(3x^2 + 2x - 7) - (2x^2 - 2x + 7) = 3x^2 + 2x - 7 - 2x^2 + 2x - 7 = x^2 + 4x - 14$

26. $(3x - 7)(2x + 8) = 6x^2 + 24x - 14x - 56 = 6x^2 + 10x - 56$

27. $(x - 2)(x^2 + 2x + 4) = x^3 + 2x^2 + 4x - 2x^2 - 4x - 8 = x^3 - 8$

28.

$$
\begin{array}{r}
2x + 1 \\
x - 3 \overline{\smash{\big)}\, 2x^2 - 5x - 3} \\
\underline{2x^2 - 6x} \\
x - 3 \\
\underline{x - 3} \\
0
\end{array}
$$

29. $(1.6 \times 10^2)(3 \times 10^{16}) = 4.8 \times 10^{18}$ m

30.
$A = 2lw + 2wd + 2ld$
$202 = 2(9)(5) + 2(5)d + 2(9)d$
$202 = 90 + 28d$
$112 = 28d \Rightarrow d = 4$ inches

31.
$A = \pi(R + r)(R - r)$
$= \pi(17 + 3)(17 - 3)$
$= \pi(20)(14)$
$= 280\pi \approx 879.6$ in.^{2}

32. Let $r =$ the regular price. Then an employee can purchase the TV for $0.75r$.

Purchase price + Sales tax = Total price

$0.75r + 0.08(0.75r) = 414.72$

$0.81r = 414.72$

$\dfrac{0.81r}{0.81} = \dfrac{414.72}{0.81}$

$r = 512 \Rightarrow$ The regular price is \$512.

Exercise 5.1 (page 295)

1. $3x - 2(x + 1) = 5$
$3x - 2x - 2 = 5$
$x = 7$

3. $\dfrac{2x - 7}{5} = 3$

$5 \cdot \dfrac{2x - 7}{5} = 5(3)$
$2x - 7 = 15$
$2x = 22$
$x = 11$

5. prime

7. largest (or greatest)

9. $a + b$

11. $12 = 2 \cdot 6 = 2 \cdot (2 \cdot 3) = 2^2 \cdot 3$

13. $15 = 3 \cdot 5$

15. $40 = 4 \cdot 10 = (2 \cdot 2) \cdot (2 \cdot 5) = 2^3 \cdot 5$

17. $98 = 2 \cdot 49 = 2 \cdot (7 \cdot 7) = 2 \cdot 7^2$

19. $225 = 15 \cdot 15 = (3 \cdot 5) \cdot (3 \cdot 5) = 3^2 \cdot 5^2$

21. $288 = 24 \cdot 12 = (4 \cdot 6) \cdot (4 \cdot 3)$
$= (2^2 \cdot 2 \cdot 3) \cdot (2^2 \cdot 3) = 2^5 \cdot 3^2$

23. $4a + 12 = \boxed{4}\,(a + 3)$

25. $r^4 + r^2 = r^2(\boxed{r^2} + 1)$

27. $4y^2 + 8y - 2xy = 2y(2y + \boxed{4} - \boxed{x})$

29. $a(x + y) + b(x + y) = (x + y)\boxed{(a + b)}$

31. $(r - s)p - (r - s)q = (r - s)\boxed{(p - q)}$

33. $3x + 6 = 3(x + 2)$

35. $xy - xz = x(y - z)$

37. $t^3 + 2t^2 = t^2(t + 2)$

39. $r^4 - r^2 = r^2(r^2 - 1)$

41. $a^3b^3z^3 - a^2b^3z^2 = a^2b^3z^2(az - 1)$

43. $24x^2y^3z^4 + 8xy^2z^3 = 8xy^2z^3(3xyz + 1)$

45. $12uvw^3 - 18uv^2w^2 = 6uvw^2(2w - 3v)$

47. $3x + 3y - 6z = 3(x + y - 2z)$

49. $ab + ac - ad = a(b + c - d)$

51. $4y^2 + 8y - 2xy = 2y(2y + 4 - x)$

53. $12r^2 - 3rs + 9r^2s^2 = 3r(4r - s + 3rs^2)$

55. $abx - ab^2x + abx^2 = abx(1 - b + x)$

57. $4x^2y^2z^2 - 6xy^2z^2 + 12xyz^2 = 2xyz^2(2xy - 3y + 6)$

59. $70a^3b^2c^2 + 49a^2b^3c^3 - 21a^2b^2c^2 = 7a^2b^2c^2(10a + 7bc - 3)$

61. $-a - b = -(a + b)$

63. $-2x + 5y = -(2x - 5y)$

65. $-2a + 3b = -(2a - 3b)$

67. $-3m - 4n + 1 = -(3m + 4n - 1)$

69. $-3xy + 2z + 5w = -(3xy - 2z - 5w)$

71. $-3ab - 5ac + 9bc = -(3ab + 5ac - 9bc)$

73. $-3x^2y - 6xy^2 = -3xy(x + 2y)$

75. $-4a^2b^3 + 12a^3b^2 = -4a^2b^2(b - 3a)$

77. $-4a^2b^2c^2 + 14a^2b^2c - 10ab^2c^2 = -2ab^2c(2ac - 7a + 5c)$

79. $-14a^6b^6 + 49a^2b^3 - 21ab = -7ab(2a^5b^5 - 7ab^2 + 3)$

81. $-5a^2b^3c + 15a^3b^4c^2 - 25a^4b^3c = -5a^2b^3c(1 - 3abc + 5a^2)$

83. $(x + y)2 + (x + y)b = (x + y)(2 + b)$ **85.** $3(x + y) - a(x + y) = (x + y)(3 - a)$

87. $3(r - 2s) - x(r - 2s) = (r - 2s)(3 - x)$

89. $(x - 3)^2 + (x - 3) = (x - 3)(x - 3) + 1(x - 3) = (x - 3)(x - 3 + 1) = (x - 3)(x - 2)$

91. $2x(a^2 + b) + 2y(a^2 + b) = (a^2 + b)(2x + 2y) = 2(a^2 + b)(x + y)$

93. $3x^2(r + 3s) - 6y^2(r + 3s) = (r + 3s)(3x^2 - 6y^2) = 3(r + 3s)(x^2 - 2y^2)$

95. $3x(a + b + c) - 2y(a + b + c) = (a + b + c)(3x - 2y)$

97. $14x^2y(r + 2s - t) - 21xy(r + 2s - t) = (r + 2s - t)(14x^2y - 21xy) = 7xy(r + 2s - t)(2x - 3)$

99. $(x + 3)(x + 1) - y(x + 1) = (x + 1)(x + 3 - y)$

101. $(3x - y)(x^2 - 2) + (x^2 - 2) = (3x - y)(x^2 - 2) + 1(x^2 - 2) = (x^2 - 2)(3x - y + 1)$

103. $2x + 2y + ax + ay = 2(x + y) + a(x + y) = (x + y)(2 + a)$

105. $7r + 7s - kr - ks = 7(r + s) - k(r + s) = (r + s)(7 - k)$

107. $xr + xs + yr + ys = x(r + s) + y(r + s) = (r + s)(x + y)$

109. $2ax + 2bx + 3a + 3b = 2x(a + b) + 3(a + b) = (a + b)(2x + 3)$

111. $2ab + 2ac + 3b + 3c = 2a(b + c) + 3(b + c) = (b + c)(2a + 3)$

113. $2x^2 + 2xy - 3x - 3y = 2x(x + y) - 3(x + y) = (x + y)(2x - 3)$

115. $3tv - 9tw + uv - 3uw = 3t(v - 3w) + u(v - 3w) = (v - 3w)(3t + u)$

117. $9mp + 3mq - 3np - nq = 3m(3p + q) - n(3p + q) = (3p + q)(3m - n)$

119. $mp - np - m + n = p(m - n) - 1(m - n) = (m - n)(p - 1)$

121. $x(a - b) + y(b - a) = x(a - b) - y(a - b) = (a - b)(x - y)$

123. $ax^3 + bx^3 + 2ax^2y + 2bx^2y = x^2[ax + bx + 2ay + 2by] = x^2[x(a+b) + 2y(a+b)]$
$$= x^2(a+b)(x+2y)$$

125. $4a^2b + 12a^2 - 8ab - 24a = 4a[ab + 3a - 2b - 6] = 4a[a(b+3) - 2(b+3)]$
$$= 4a(b+3)(a-2)$$

127. $x^3 + 2x^2 + x + 2 = x^2(x+2) + 1(x+2) = (x+2)(x^2+1)$

129. $x^3y - x^2y - xy^2 + y^2 = y[x^3 - x^2 - xy + y] = y[x^2(x-1) - y(x-1)]$
$$= y(x-1)(x^2-y)$$

131. $2r - bs - 2s + br = 2r - 2s + br - bs = 2(r-s) + b(r-s) = (r-s)(2+b)$

133. $ax + by + bx + ay = ax + ay + bx + by = a(x+y) + b(x+y) = (x+y)(a+b)$

135. $ac + bd - ad - bc = ac - bc - ad + bd = c(a-b) - d(a-b) = (a-b)(c-d)$

137. $ar^2 - brs + ars - br^2 = r[ar - bs + as - br] = r[ar - br + as - bs] = r[r(a-b) + s(a-b)]$
$$= r(a-b)(r+s)$$

139. $ba + 3 + a + 3b = ba + a + 3b + 3 = a(b+1) + 3(b+1) = (b+1)(a+3)$

141. $pr + qs - ps - qr = pr - qr - ps + qs = r(p-q) - s(p-q) = (p-q)(r-s)$

143. Answers may vary.   **145.** Answers may vary.

147. $ax + ay + bx + by = a(x+y) + b(x+y) = (x+y)(a+b)$
$ax + bx + ay + by = x(a+b) + y(a+b) = (a+b)(x+y)$

Exercise 5.2   (page 301)

1. $\dfrac{p}{w} + \dfrac{v^2}{2g} + h = k$

$$\dfrac{p}{w} = k - h - \dfrac{v^2}{2g}$$

$$w \cdot \dfrac{p}{w} = w\left(k - h - \dfrac{v^2}{2g}\right)$$

$$p = w\left(k - h - \dfrac{v^2}{2g}\right)$$

3. difference of two squares

5. $(p-q)$

7. $x^2 - 9 = (x+3)\boxed{(x-3)}$

9. $4m^2 - 9n^2 = (2m+3n)\boxed{(2m-3n)}$

11. $x^2 - 16 = x^2 - 4^2 = (x+4)(x-4)$

13. $y^2 - 49 = y^2 - 7^2 = (y+7)(y-7)$

15. $4y^2 - 49 = (2y)^2 - 7^2 = (2y+7)(2y-7)$

17. $9x^2 - y^2 = (3x)^2 - y^2 = (3x+y)(3x-y)$

19. $25t^2 - 36u^2 = (5t)^2 - (6u)^2$
$$= (5t + 6u)(5t - 6u)$$

21. $16a^2 - 25b^2 = (4a)^2 - (5b)^2$
$$= (4a + 5b)(4a - 5b)$$

23. $a^2 + b^2$: sum of squares $\Rightarrow$ prime

25. $a^4 - 4b^2 = (a^2)^2 - (2b)^2$
$$= (a^2 + 2b)(a^2 - 2b)$$

27. $49y^2 - 225z^4 = (7y)^2 - (15z^2)^2 = (7y + 15z^2)(7y - 15z^2)$

29. $196x^4 - 169y^2 = (14x^2)^2 - (13y)^2 = (14x^2 + 13y)(14x^2 - 13y)$

31. $8x^2 - 32y^2 = 8[x^2 - 4y^2] = 8[x^2 - (2y)^2] = 8(x + 2y)(x - 2y)$

33. $2a^2 - 8y^2 = 2[a^2 - 4y^2] = 2[a^2 - (2y)^2] = 2(a + 2y)(a - 2y)$

35. $3r^2 - 12s^2 = 3[r^2 - 4s^2] = 3[r^2 - (2s)^2] = 3(r + 2s)(r - 2s)$

37. $x^3 - xy^2 = x(x^2 - y^2) = x(x + y)(x - y)$

39. $4a^2x - 9b^2x = x[4a^2 - 9b^2] = x[(2a)^2 - (3b)^2] = x(2a + 3b)(2a - 3b)$

41. $3m^3 - 3mn^2 = 3m(m^2 - n^2) = 3m(m + n)(m - n)$

43. $4x^4 - x^2y^2 = x^2(4x^2 - y^2) = x^2[(2x)^2 - y^2] = x^2(2x + y)(2x - y)$

45. $2a^3b - 242ab^3 = 2ab[a^2 - 121b^2] = 2ab[a^2 - (11b)^2] = 2ab(a + 11b)(a - 11b)$

47. $x^4 - 81 = (x^2 + 9)(x^2 - 9)$
$$= (x^2 + 9)(x + 3)(x - 3)$$

49. $a^4 - 16 = (a^2 + 4)(a^2 - 4)$
$$= (a^2 + 4)(a + 2)(a - 2)$$

51. $a^4 - b^4 = (a^2 + b^2)(a^2 - b^2) = (a^2 + b^2)(a + b)(a - b)$

53. $81r^4 - 256s^4 = (9r^2 + 16s^2)(9r^2 - 16s^2) = (9r^2 + 16s^2)(3r + 4s)(3r - 4s)$

55. $a^4 - b^8 = (a^2 + b^4)(a^2 - b^4) = (a^2 + b^4)(a + b^2)(a - b^2)$

57. $x^8 - y^8 = (x^4 + y^4)(x^4 - y^4) = (x^4 + y^4)(x^2 + y^2)(x^2 - y^2) = (x^4 + y^4)(x^2 + y^2)(x + y)(x - y)$

59. $2x^4 - 2y^4 = 2(x^4 - y^4) = 2(x^2 + y^2)(x^2 - y^2) = 2(x^2 + y^2)(x + y)(x - y)$

61. $a^4b - b^5 = b(a^4 - b^4) = b(a^2 + b^2)(a^2 - b^2) = b(a^2 + b^2)(a + b)(a - b)$

63. $48m^4n - 243n^5 = 3n(16m^4 - 81n^4) = 3n(4m^2 + 9n^2)(4m^2 - 9n^2)$
$$= 3n(4m^2 + 9n^2)(2m + 3n)(2m - 3n)$$

65. $3a^5y + 6ay^5 = 3ay(a^4 + 2y^4)$

67. $3a^{10} - 3a^2b^4 = 3a^2(a^8 - b^4) = 3a^2(a^4 + b^2)(a^4 - b^2) = 3a^2(a^4 + b^2)(a^2 + b)(a^2 - b)$

69. $2x^8y^2 - 32y^6 = 2y^2(x^8 - 16y^4) = 2y^2(x^4 + 4y^2)(x^4 - 4y^2) = 2y^2(x^4 + 4y^2)(x^2 + 2y)(x^2 - 2y)$

71. $a^6b^2 - a^2b^6c^4 = a^2b^2(a^4 - b^4c^4) = a^2b^2(a^2 + b^2c^2)(a^2 - b^2c^2)$
$$= a^2b^2(a^2 + b^2c^2)(a + bc)(a - bc)$$

73. $a^2b^7 - 625a^2b^3 = a^2b^3(b^4 - 625) = a^2b^3(b^2 + 25)(b^2 - 25) = a^2b^3(b^2 + 25)(b + 5)(b - 5)$

75. $243r^5s - 48rs^5 = 3rs(81r^4 - 16s^4) = 3rs(9r^2 + 4s^2)(9r^2 - 4s^2)$
$$= 3rs(9r^2 + 4s^2)(3r + 2s)(3r - 2s)$$

77. $16(x - y)^2 - 9 = [4(x - y)]^2 - 3^2 = [4(x - y) + 3][4(x - y) - 3] = (4x - 4y + 3)(4x - 4y - 3)$

79. $a^3 - 9a + 3a^2 - 27 = a(a^2 - 9) + 3(a^2 - 9) = (a^2 - 9)(a + 3) = (a + 3)(a - 3)(a + 3)$

81. $y^3 - 16y - 3y^2 + 48 = y(y^2 - 16) - 3(y^2 - 16) = (y^2 - 16)(y - 3) = (y + 4)(y - 4)(y - 3)$

83. $3x^3 - 12x + 3x^2 - 12 = 3[x^3 - 4x + x^2 - 4] = 3[x(x^2 - 4) + 1(x^2 - 4)]$
$$= 3(x^2 - 4)(x + 1) = 3(x + 2)(x - 2)(x + 1)$$

85. $3m^3 - 3mn^2 + 3am^2 - 3an^2 = 3[m^3 - mn^2 + am^2 - an^2] = 3[m(m^2 - n^2) + a(m^2 - n^2)]$
$$= 3(m^2 - n^2)(m + a)$$
$$= 3(m + n)(m - n)(m + a)$$

87. $2m^3n^2 - 32mn^2 + 8m^2 - 128 = 2[m^3n^2 - 16mn^2 + 4m^2 - 64]$
$$= 2[mn^2(m^2 - 16) + 4(m^2 - 16)]$$
$$= 2(m^2 - 16)(mn^2 + 4) = 2(m + 4)(m - 4)(mn^2 + 4)$$

89. **Answers may vary.**

91. $399 \cdot 401 = (400 - 1)(400 + 1) = 400^2 - 1^2 = 160,000 - 1 = 159,999$

Exercise 5.3 (page 310)

1. $x - 3 > 5$
$$x > 8$$

3. $-3x - 5 \geq 4$
$$-3x \geq 9$$
$$\frac{-3x}{-3} \leq \frac{9}{-3}$$
$$x \leq -3$$

5.
$$\frac{3(x-1)}{4} < 12$$
$$4 \cdot \frac{3x-3}{4} < 4(12)$$
$$3x - 3 < 48$$
$$3x < 51$$
$$x < 17$$

7. $-2 < x \le 4$

9. $x^2 + 2xy + y^2 = (x+y)^2$

11. $y^2 + 6y + 8 = \left(y + \boxed{4}\,\right)\left(y + \boxed{2}\,\right)$

13. $x^2 - xy - 2y^2 = \left(x + \boxed{y}\,\right)\left(x - \boxed{2y}\,\right)$

15. $x^2 + 3x + 2 = (x+2)(x+1)$

17. $z^2 + 12z + 11 = (z+1)(z+11)$

19. $a^2 - 4a - 5 = (a-5)(a+1)$

21. $t^2 - 9t + 14 = (t-2)(t-7)$

23. $u^2 + 10u + 15$: prime

25. $y^2 - y - 30 = (y-6)(y+5)$

27. $a^2 + 6a - 16 = (a+8)(a-2)$

29. $t^2 - 5t - 50 = (t-10)(t+5)$

31. $r^2 - 9r - 12$: prime

33. $y^2 + 2yz + z^2 = (y+z)(y+z)$

35. $x^2 + 4xy + 4y^2 = (x+2y)(x+2y)$

37. $m^2 + 3mn - 10n^2 = (m+5n)(m-2n)$

39. $a^2 - 4ab - 12b^2 = (a-6b)(a+2b)$

41. $u^2 + 2uv - 15v^2 = (u+5v)(u-3v)$

43. $-x^2 - 7x - 10 = -\left(x^2 + 7x + 10\right)$
$$= -(x+5)(x+2)$$

45. $-y^2 - 2y + 15 = -\left(y^2 + 2y - 15\right)$
$$= -(y+5)(y-3)$$

47. $-t^2 - 15t + 34 = -\left(t^2 + 15t - 34\right)$
$$= -(t+17)(t-2)$$

49. $-r^2 + 14r - 40 = -\left(r^2 - 14r + 40\right)$
$$= -(r-4)(r-10)$$

51. $-a^2 - 4ab - 3b^2 = -\left(a^2 + 4ab + 3b^2\right)$
$$= -(a+3b)(a+b)$$

53. $-x^2 + 6xy + 7y^2 = -\left(x^2 - 6xy - 7y^2\right)$
$$= -(x-7y)(x+y)$$

55. $4 - 5x + x^2 = x^2 - 5x + 4$
$$= (x-4)(x-1)$$

57. $10y + 9 + y^2 = y^2 + 10y + 9$
$$= (y+9)(y+1)$$

59. $c^2 - 5 + 4c = c^2 + 4c - 5$
$$= (c+5)(c-1)$$

61. $-r^2 + 2s^2 + rs = -r^2 + rs + 2s^2$
$$= -\left(r^2 - rs - 2s^2\right)$$
$$= -(r-2s)(r+s)$$

63. $4rx + r^2 + 3x^2 = r^2 + 4rx + 3x^2$
$$= (r+3x)(r+x)$$

65. $-3ab + a^2 + 2b^2 = a^2 - 3ab + 2b^2$
$$= (a - 2b)(a - b)$$

67. $2x^2 + 10x + 12 = 2\left(x^2 + 5x + 6\right)$
$$= 2(x + 2)(x + 3)$$

69. $3y^3 + 6y^2 + 3y = 3y\left(y^2 + 2y + 1\right)$
$$= 3y(y + 1)(y + 1)$$

71. $-5a^2 + 25a - 30 = -5\left(a^2 - 5a + 6\right)$
$$= -5(a - 3)(a - 2)$$

73. $3z^2 - 15tz + 12t^2 = 3(z^2 - 5tz + 4t^2) = 3(z - 4t)(z - t)$

75. $12xy + 4x^2y - 72y = 4y(3x + x^2 - 18) = 4y(x^2 + 3x - 18) = 4y(x + 6)(x - 3)$

77. $-4x^2y - 4x^3 + 24xy^2 = -4x(xy + x^2 - 6y^2) = -4x(x^2 + xy - 6y^2) = -4x(x + 3y)(x - 2y)$

79. $x^2 + 4x + 4 - y^2 = (x + 2)(x + 2) - y^2 = (x + 2)^2 - y^2 = (x + 2 + y)(x + 2 - y)$

81. $b^2 - 6b + 9 - c^2 = (b - 3)(b - 3) - c^2 = (b - 3)^2 - c^2 = (b - 3 + c)(b - 3 - c)$

83. $a^2 + 2ab + b^2 - 4 = (a + b)(a + b) - 4 = (a + b)^2 - 2^2 = (a + b + 2)(a + b - 2)$

85. $b^2 - y^2 - 4y - 4 = b^2 - \left(y^2 + 4y + 4\right) = b^2 - (y + 2)(y + 2) = b^2 - (y + 2)^2$
$$= [b + (y + 2)][b - (y + 2)]$$
$$= (b + y + 2)(b - y - 2)$$

87. $x^2 + 6x + 9 = (x + 3)(x + 3)$

89. $y^2 - 8y + 16 = (y - 4)(y - 4)$

91. $t^2 + 20t + 100 = (t + 10)(t + 10)$

93. $u^2 - 18u + 81 = (u - 9)(u - 9)$

95. $x^2 + 4xy + 4y^2 = (x + 2y)(x + 2y)$

97. $r^2 - 10rs + 25s^2 = (r - 5s)(r - 5s)$

99. **Answers may vary.**

101. Both answers check. Both may be factored more completely:
$$(2x + 6)(x + 7) = 2(x + 3)(x + 7)$$
$$(x + 3)(2x + 14) = (x + 3) \cdot 2(x + 7)$$

Exercise 5.4 (page 318)

1. $l = f + (n - 1)d$
$$l = f + nd - d$$
$$l - f + d = nd$$
$$\frac{l - f + d}{d} = \frac{nd}{d}$$
$$\frac{l - f + d}{d} = n$$

3. descending

5. opposites

7. $6x^2 + x - 2 = \left(3x + \boxed{2}\right)\left(2x - \boxed{1}\right)$

9. $12x^2 - 7xy + y^2 = \left(3x - \boxed{y}\right)\left(4x - \boxed{y}\right)$

11. $2x^2 - 3x + 1 = (2x - 1)(x - 1)$

13. $3a^2 + 13a + 4 = (3a + 1)(a + 4)$

15. $4z^2 + 13z + 3 = (z + 3)(4z + 1)$

17. $6y^2 + 7y + 2 = (3y + 2)(2y + 1)$

19. $6x^2 - 7x + 2 = (2x - 1)(3x - 2)$

21. $3a^2 - 4a - 4 = (3a + 2)(a - 2)$

23. $2x^2 - 3x - 2 = (2x + 1)(x - 2)$

25. $2m^2 + 5m - 12 = (2m - 3)(m + 4)$

27. $10y^2 - 3y - 1 = (5y + 1)(2y - 1)$

29. $12y^2 - 5y - 2 = (4y + 1)(3y - 2)$

31. $5t^2 + 13t + 6 = (5t + 3)(t + 2)$

33. $16m^2 - 14m + 3 = (8m - 3)(2m - 1)$

35. $3x^2 - 4xy + y^2 = (3x - y)(x - y)$

37. $2u^2 + uv - 3v^2 = (2u + 3v)(u - v)$

39. $4a^2 - 4ab + b^2 = (2a - b)(2a - b)$

41. $6r^2 + rs - 2s^2 = (2r - s)(3r + 2s)$

43. $4x^2 + 8xy + 3y^2 = (2x + y)(2x + 3y)$

45. $4a^2 - 15ab + 9b^2 = (4a - 3b)(a - 3b)$

47. $-13x + 3x^2 - 10 = 3x^2 - 13x - 10$
$ = (3x + 2)(x - 5)$

49. $15 + 8a^2 - 26a = 8a^2 - 26a + 15$
$ = (4a - 3)(2a - 5)$

51. $12y^2 + 12 - 25y = 12y^2 - 25y + 12$
$ = (3y - 4)(4y - 3)$

53. $3x^2 + 6 + x = 3x^2 + x + 6 \Rightarrow$ prime

55. $2a^2 + 3b^2 + 5ab = 2a^2 + 5ab + 3b^2$
$ = (2a + 3b)(a + b)$

57. $pq + 6p^2 - q^2 = 6p^2 + pq - q^2$
$ = (2p + q)(3p - q)$

59. $b^2 + 4a^2 + 16ab = 4a^2 + 16ab + b^2$
PRIME

61. $12x^2 + 10y^2 - 23xy = 12x^2 - 23xy + 10y^2$
$ = (3x - 2y)(4x - 5y)$

63. $4x^2 + 10x - 6 = 2(2x^2 + 5x - 3)$
$ = 2(2x - 1)(x + 3)$

65. $y^3 + 13y^2 + 12y = y(y^2 + 13y + 12)$
$ = y(y + 12)(y + 1)$

67. $6x^3 - 15x^2 - 9x = 3x(2x^2 - 5x - 3)$
$ = 3x(2x + 1)(x - 3)$

69. $30r^5 + 63r^4 - 30r^3 = 3r^3(10r^2 + 21r - 10)$
$ = 3r^3(5r - 2)(2r + 5)$

71. $4a^2 - 4ab - 8b^2 = 4(a^2 - ab - 2b^2)$
$ = 4(a - 2b)(a + b)$

73. $8x^2 - 12xy - 8y^2 = 4(2x^2 - 3xy - 2y^2)$
$ = 4(2x + y)(x - 2y)$

75. $-16m^3n - 20m^2n^2 - 6mn^3 = -2mn(8m^2 + 10mn + 3n^2) = -2mn(4m + 3n)(2m + n)$

77. $-28u^3v^3 + 26u^2v^4 - 6uv^5 = -2uv^3(14u^2 - 13uv + 3v^2) = -2uv^3(7u - 3v)(2u - v)$

79. $4x^2 + 12x + 9 = (2x + 3)(2x + 3)$
$ = (2x + 3)^2$

81. $9x^2 + 12x + 4 = (3x + 2)(3x + 2)$
$ = (3x + 2)^2$

83. $16x^2 - 8xy + y^2 = (4x - y)(4x - y) = (4x - y)^2$

85. $4x^2 + 4xy + y^2 - 16 = \left(4x^2 + 4xy + y^2\right) - 16 = (2x + y)(2x + y) - 16$
$$= (2x + y)^2 - 4^2 = (2x + y + 4)(2x + y - 4)$$

87. $9 - a^2 - 4ab - 4b^2 = 9 - \left(a^2 + 4ab + 4b^2\right) = 9 - (a + 2b)(a + 2b)$
$$= 3^2 - (a + 2b)^2$$
$$= (3 + (a + 2b))(3 - (a + 2b))$$
$$= (3 + a + 2b)(3 - a - 2b)$$

89. $9p^2 + 1 + 6p - q^2 = 9p^2 + 6p + 1 - q^2 = (3p + 1)(3p + 1) - q^2 = (3p + 1)^2 - q^2$
$$= (3p + 1 + q)(3p + 1 - q)$$

91. **Answers may vary.**

93. $(6x + 1)(x + 6) = 6x^2 + 37x + 6;\ (6x - 1)(x - 6) = 6x^2 - 37x + 6 \Rightarrow b = \pm 37$
$(6x + 2)(x + 3) = 6x^2 + 20x + 6;\ (6x - 2)(x - 3) = 6x^2 - 20x + 6 \Rightarrow b = \pm 20$
$(6x + 3)(x + 2) = 6x^2 + 15x + 6;\ (6x - 3)(x - 2) = 6x^2 - 15x + 6 \Rightarrow b = \pm 15$
$(6x + 6)(x + 1) = 6x^2 + 12x + 6;\ (6x - 6)(x - 1) = 6x^2 - 12x + 6 \Rightarrow b = \pm 12$
$(2x + 1)(3x + 6) = 6x^2 + 15x + 6;\ (2x - 1)(3x - 6) = 6x^2 - 15x + 6 \Rightarrow b = \pm 15$
$(2x + 2)(3x + 3) = 6x^2 + 12x + 6;\ (2x - 2)(3x - 3) = 6x^2 - 12x + 6 \Rightarrow b = \pm 12$
$(2x + 3)(3x + 2) = 6x^2 + 13x + 6;\ (2x - 3)(3x - 2) = 6x^2 - 13x + 6 \Rightarrow b = \pm 13$
$(2x + 6)(3x + 1) = 6x^2 + 20x + 6;\ (2x - 6)(3x - 1) = 6x^2 - 20x + 6 \Rightarrow b = \pm 20$
$b = \pm 12,\ \pm 13,\ \pm 15,\ \pm 20,\ \pm 37$

Exercise 5.5   (page 323)

1. $1 \times 10^{-13} = 0.0000000000001$ cm

3. $x^3 + y^3 = (x + y)\boxed{\left(x^2 - xy + y^2\right)}$

5. $y^3 + 1 = y^3 + 1^3$
$$= (y + 1)\left(y^2 - 1y + 1^2\right)$$
$$= (y + 1)\left(y^2 - y + 1\right)$$

7. $a^3 - 27 = a^3 - 3^3$
$$= (a - 3)\left(a^2 + 3a + 3^2\right)$$
$$= (a - 3)\left(a^2 + 3a + 9\right)$$

9. $8 + x^3 = 2^3 + x^3 = (2 + x)\left(2^2 - 2x + x^2\right)$
$$= (2 + x)\left(4 - 2x + x^2\right)$$

11. $s^3 - t^3 = (s - t)(s^2 + st + t^2)$

13. $27x^3 + y^3 = (3x)^3 + y^3$
$$= (3x + y)\left((3x)^2 - 3xy + y^2\right)$$
$$= (3x + y)\left(9x^2 - 3xy + y^2\right)$$

15. $a^3 + 8b^3 = a^3 + (2b)^3$
$$= (a + 2b)\left(a^2 - 2ab + (2b)^2\right)$$
$$= (a + 2b)\left(a^2 - 2ab + 4b^2\right)$$

17. $64x^3 - 27 = (4x)^3 - 3^3 = (4x - 3)\left((4x)^2 + 3(4x) + 3^2\right) = (4x - 3)(16x^2 + 12x + 9)$

19. $27x^3 - 125y^3 = (3x)^3 - (5y)^3 = (3x - 5y)\big((3x)^2 + (3x)(5y) + (5y)^2\big)$
$$= (3x - 5y)\big(9x^2 + 15xy + 25y^2\big)$$

21. $a^6 - b^3 = (a^2)^3 - b^3 = (a^2 - b)\big((a^2)^2 + a^2 b + b^2\big) = (a^2 - b)(a^4 + a^2 b + b^2)$

23. $x^9 + y^6 = (x^3)^3 + (y^2)^3 = (x^3 + y^2)\big((x^3)^2 - x^3 y^2 + (y^2)^2\big) = (x^3 + y^2)(x^6 - x^3 y^2 + y^4)$

25. $2x^3 + 54 = 2(x^3 + 27)$
$$= 2(x^3 + 3^3)$$
$$= 2(x + 3)(x^2 - 3x + 3^2)$$
$$= 2(x + 3)(x^2 - 3x + 9)$$

27. $-x^3 + 216 = -(x^3 - 216)$
$$= -(x^3 - 6^3)$$
$$= -(x - 6)(x^2 + 6x + 6^2)$$
$$= -(x - 6)(x^2 + 6x + 36)$$

29. $64m^3 x - 8n^3 x = 8x(8m^3 - n^3) = 8x\big((2m)^3 - n^3\big) = 8x(2m - n)\big((2m)^2 + 2mn + n^2\big)$
$$= 8x(2m - n)\big(4m^2 + 2mn + n^2\big)$$

31. $x^4 y + 216xy^4 = xy(x^3 + 216y^3) = xy\big(x^3 + (6y)^3\big) = xy(x + 6y)\big(x^2 - 6xy + (6y)^2\big)$
$$= xy(x + 6y)\big(x^2 - 6xy + 36y^2\big)$$

33. $81r^4 s^2 - 24rs^5 = 3rs^2(27r^3 - 8s^3) = 3rs^2\big((3r)^3 - (2s)^3\big)$
$$= 3rs^2(3r - 2s)\big((3r)^2 + (3r)(2s) + (2s)^2\big)$$
$$= 3rs^2(3r - 2s)\big(9r^2 + 6rs + 4s^2\big)$$

35. $125a^6 b^2 + 64a^3 b^5 = a^3 b^2(125a^3 + 64b^3) = a^3 b^2\big((5a)^3 + (4b)^3\big)$
$$= a^3 b^2(5a + 4b)\big((5a)^2 - (5a)(4b) + (4b)^2\big)$$
$$= a^3 b^2(5a + 4b)\big(25a^2 - 20ab + 16b^2\big)$$

37. $y^7 z - yz^4 = yz(y^6 - z^3) = yz\big((y^2)^3 - z^3\big) = yz(y^2 - z)\big((y^2) + y^2 z + z^2\big)$
$$= yz(y^2 - z)\big(y^4 + y^2 z + z^2\big)$$

39. $2mp^4 + 16mpq^3 = 2mp(p^3 + 8q^3) = 2mp\big(p^3 + (2q)^3\big) = 2mp(p + 2q)\big(p^2 - 2pq + (2q)^2\big)$
$$= 2mp(p + 2q)\big(p^2 - 2pq + 4q^2\big)$$

41. $x^6 - 1 = (x^3)^2 - 1^2 = (x^3 + 1)(x^3 - 1) = (x^3 + 1^3)(x^3 - 1^3)$
$$= (x + 1)(x^2 - 1x + 1^2)(x - 1)(x^2 + 1x + 1^2)$$
$$= (x + 1)(x^2 - x + 1)(x - 1)(x^2 + x + 1)$$

43. $x^{12} - y^6 = \left(x^6\right)^2 - \left(y^3\right)^2 = \left(x^6 + y^3\right)\left(x^6 - y^3\right)$
$$= \left(\left(x^2\right)^3 + y^3\right)\left(\left(x^2\right)^3 - y^3\right)$$
$$= \left(x^2 + y\right)\left(\left(x^2\right)^2 - x^2 y + y^2\right)\left(x^2 - y\right)\left(\left(x^2\right)^2 + x^2 y + y^2\right)$$
$$= \left(x^2 + y\right)\left(x^4 - x^2 y + y^2\right)\left(x^2 - y\right)\left(x^4 + x^2 y + y^2\right)$$

45. $3\left(x^3 + y^3\right) - z\left(x^3 + y^3\right) = \left(x^3 + y^3\right)(3 - z) = (x + y)\left(x^2 - xy + y^2\right)(3 - z)$

47. $\left(m^3 + 8n^3\right) + \left(m^3 x + 8n^3 x\right) = 1\left(m^3 + 8n^3\right) + x\left(m^3 + 8n^3\right)$
$$= \left(m^3 + 8n^3\right)(1 + x)$$
$$= (m + 2n)\left(m^2 - 2mn + (2n)^2\right)(1 + x)$$
$$= (m + 2n)\left(m^2 - 2mn + 4n^2\right)(1 + x)$$

49. $\left(a^4 + 27a\right) - \left(a^3 b + 27b\right) = a\left(a^3 + 27\right) - b\left(a^3 + 27\right) = \left(a^3 + 27\right)(a - b)$
$$= \left(a^3 + 3^3\right)(a - b)$$
$$= (a + 3)\left(a^2 - 3a + 3^2\right)(a - b)$$
$$= (a + 3)\left(a^2 - 3a + 9\right)(a - b)$$

51. $y^3\left(y^2 - 1\right) - 27\left(y^2 - 1\right) = \left(y^2 - 1\right)\left(y^3 - 27\right) = \left(y^2 - 1^2\right)\left(y^3 - 3^3\right)$
$$= (y + 1)(y - 1)(y - 3)\left(y^2 + 3y + 9\right)$$

53. **Answers may vary.**

55. $a^3 - b^3 = 11^3 - 7^3 = 1331 - 343 = \boxed{988}$
$(a - b)\left(a^2 + ab + b^2\right) = (11 - 7)\left(11^2 + 11 \cdot 7 + 7^2\right) = 4(121 + 77 + 49) = 4(247) = \boxed{988}$

Exercise 5.6 (page 327)

1. $2(t - 5) + t = 3(2 - t)$
$2t - 10 + t = 6 - 3t$
$3t - 10 = 6 - 3t$
$6t = 16$
$t = \frac{16}{6} = \frac{8}{3}$

3. $5 - 3(t + 1) = t + 2$
$5 - 3t - 3 = t + 2$
$-3t + 2 = t + 2$
$-4t = 0$
$t = \frac{0}{-4} = 0$

5. factors

7. binomials

9. $6x + 3 = 3(2x + 1)$

11. $x^2 - 6x - 7 = (x - 7)(x + 1)$

13. $6t^2 + 7t - 3 = (2t + 3)(3t - 1)$

15. $4x^2 - 25 = (2x)^2 - 5^2 = (2x + 5)(2x - 5)$

17. $t^2 - 2t + 1 = (t - 1)(t - 1)$

19. $a^3 - 8 = a^3 - 2^3 = (a - 2)\left(a^2 + 2a + 2^2\right)$
$$= (a - 2)\left(a^2 + 2a + 4\right)$$

21. $x^2y^2 - 2x^2 - y^2 + 2 = x^2(y^2 - 2) - 1(y^2 - 2) = (y^2 - 2)(x^2 - 1) = (y^2 - 2)(x^2 - 1^2)$
$$= (y^2 - 2)(x + 1)(x - 1)$$

23. $70p^4q^3 - 35p^4q^2 + 49p^5q^2 = 7p^4q^2(10q - 5 + 7p)$

25. $\begin{aligned} 2ab^2 + 8ab - 24a &= 2a(b^2 + 4b - 12) \\ &= 2a(b + 6)(b - 2) \end{aligned}$
27. $-8p^3q^7 - 4p^2q^3 = -4p^2q^3(2pq^4 + 1)$

29. $4a^2 - 4ab + b^2 - 9 = (4a^2 - 4ab + b^2) - 9 = (2a - b)^2 - 3^2 = (2a - b + 3)(2a - b - 3)$

31. $x^2 + 7x + 1 \Rightarrow$ prime

33. $-2x^5 + 128x^2 = -2x^2(x^3 - 64) = -2x^2(x^3 - 4^3) = -2x^2(x - 4)(x^2 + 4x + 16)$

35. $14t^3 - 40t^2 + 6t^4 = 2t^2(7t - 20 + 3t^2) = 2t^2(3t^2 + 7t - 20) = 2t^2(3t - 5)(t + 4)$

37. $a^2(x - a) - b^2(x - a) = (x - a)(a^2 - b^2) = (x - a)(a + b)(a - b)$

39. $\begin{aligned} 8p^6 - 27q^6 &= (2p^2)^3 - (3q^2)^3 = (2p^2 - 3q^2)\left((2p^2)^2 + (2p^2)(3q^2) + (3q^2)^2\right) \\ &= (2p^2 - 3q^2)(4p^4 + 6p^2q^2 + 9q^4) \end{aligned}$

41. $\begin{aligned} 125p^3 - 64y^3 &= (5p)^3 - (4y)^3 = (5p - 4y)\left((5p)^2 + (5p)(4y) + (4y)^2\right) \\ &= (5p - 4y)(25p^2 + 20py + 16y^2) \end{aligned}$

43. $-16x^4y^2z + 24x^5y^3z^4 - 15x^2y^3z^7 = -x^2y^2z(16x^2 - 24x^3yz^3 + 15yz^6)$

45. $81p^4 - 16q^4 = (9p^2)^2 - (4q^2)^2 = (9p^2 + 4q^2)(9p^2 - 4q^2) = (9p^2 + 4q^2)(3p + 2q)(3p - 2q)$

47. $4x^2 + 9y^2 \Rightarrow$ prime

49. $\begin{aligned} 54x^3 + 250y^6 &= 2(27x^3 + 125y^6) = 2\left((3x)^3 + (5y^2)^3\right) \\ &= 2(3x + 5y^2)\left((3x)^2 - (3x)(5y^2) + (5y^2)^2\right) \\ &= 2(3x + 5y^2)(9x^2 - 15xy^2 + 25y^4) \end{aligned}$

51. $10r^2 - 13r - 4 \Rightarrow$ prime
53. $\begin{aligned} 21t^3 - 10t^2 + t &= t(21t^2 - 10t + 1) \\ &= t(3t - 1)(7t - 1) \end{aligned}$

55. $\begin{aligned} x^5 - x^3y^2 + x^2y^3 - y^5 &= x^3(x^2 - y^2) + y^3(x^2 - y^2) = (x^2 - y^2)(x^3 + y^3) \\ &= (x + y)(x - y)(x + y)(x^2 - xy + y^2) \end{aligned}$

57. $2a^2c - 2b^2c + 4a^2d - 4b^2d = 2(a^2c - b^2c + 2a^2d - 2b^2d)$
$$= 2(c(a^2 - b^2) + 2d(a^2 - b^2))$$
$$= 2(a^2 - b^2)(c + 2d) = 2(a + b)(a - b)(c + 2d)$$

59. Answers may vary.

61. Answers may vary.

Exercise 5.7 (page 332)

1. $u^3 u^2 u^4 = u^{3+2+4} = u^9$

3. $\frac{a^3 b^4}{a^2 b^5} = a^1 b^{-1} = \frac{a}{b}$

5. quadratic

7. second

9. $(x - 2)(x + 3) = 0$
$x - 2 = 0$ **or** $x + 3 = 0$
$x = 2 \qquad\qquad x = -3$

11. $(x - 4)(x + 1) = 0$
$x - 4 = 0$ **or** $x + 1 = 0$
$x = 4 \qquad\qquad x = -1$

13. $(2x - 5)(3x + 6) = 0$
$2x - 5 = 0$ **or** $3x + 6 = 0$
$2x = 5 \qquad\qquad 3x = -6$
$x = \frac{5}{2} \qquad\qquad x = -2$

15. $(x - 1)(x + 2)(x - 3) = 0$
$x - 1 = 0$ **or** $x + 2 = 0$ **or** $x - 3 = 0$
$x = 1 \qquad\quad x = -2 \qquad\quad x = 3$

17. $x^2 - 3x = 0$
$x(x - 3) = 0$
$x = 0$ **or** $x - 3 = 0$
$x = 0 \qquad\qquad x = 3$

19. $2x^2 - 5x = 0$
$x(2x - 5) = 0$
$x = 0$ **or** $2x - 5 = 0$
$x = 0 \qquad\qquad 2x = 5$
$x = 0 \qquad\qquad x = \frac{5}{2}$

21. $x^2 - 7x = 0$
$x(x - 7) = 0$
$x = 0$ **or** $x - 7 = 0$
$x = 0 \qquad\qquad x = 7$

23. $3x^2 + 8x = 0$
$x(3x + 8) = 0$
$x = 0$ **or** $3x + 8 = 0$
$x = 0 \qquad\qquad 3x = -8$
$x = 0 \qquad\qquad x = -\frac{8}{3}$

25. $8x^2 - 16x = 0$
$8x(x - 2) = 0$
$8x = 0$ **or** $x - 2 = 0$
$x = 0 \qquad\qquad x = 2$

27. $10x^2 + 2x = 0$
$2x(5x + 1) = 0$
$2x = 0$ **or** $5x + 1 = 0$
$x = 0 \qquad\qquad 5x = -1$
$x = 0 \qquad\qquad x = -\frac{1}{5}$

29. $x^2 - 25 = 0$
$(x + 5)(x - 5) = 0$
$x + 5 = 0$ **or** $x - 5 = 0$
$x = -5 \qquad\qquad x = 5$

31. $y^2 - 49 = 0$
$(y + 7)(y - 7) = 0$
$y + 7 = 0$ **or** $y - 7 = 0$
$y = -7 \qquad\qquad y = 7$

33.
$$4x^2 - 1 = 0$$
$$(2x + 1)(2x - 1) = 0$$
$$2x + 1 = 0 \quad \textbf{or} \quad 2x - 1 = 0$$
$$2x = -1 \qquad\qquad 2x = 1$$
$$x = -\tfrac{1}{2} \qquad\qquad x = \tfrac{1}{2}$$

35.
$$9y^2 - 4 = 0$$
$$(3y + 2)(3y - 2) = 0$$
$$3y + 2 = 0 \quad \textbf{or} \quad 3y - 2 = 0$$
$$3y = -2 \qquad\qquad 3y = 2$$
$$y = -\tfrac{2}{3} \qquad\qquad y = \tfrac{2}{3}$$

37.
$$x^2 = 49$$
$$x^2 - 49 = 0$$
$$(x + 7)(x - 7) = 0$$
$$x + 7 = 0 \quad \textbf{or} \quad x - 7 = 0$$
$$x = -7 \qquad\qquad x = 7$$

39.
$$4x^2 = 81$$
$$4x^2 - 81 = 0$$
$$(2x + 9)(2x - 9) = 0$$
$$2x + 9 = 0 \quad \textbf{or} \quad 2x - 9 = 0$$
$$2x = -9 \qquad\qquad 2x = 9$$
$$x = -\tfrac{9}{2} \qquad\qquad x = \tfrac{9}{2}$$

41.
$$x^2 - 13x + 12 = 0$$
$$(x - 1)(x - 12) = 0$$
$$x - 1 = 0 \quad \textbf{or} \quad x - 12 = 0$$
$$x = 1 \qquad\qquad x = 12$$

43.
$$x^2 - 2x - 15 = 0$$
$$(x + 3)(x - 5) = 0$$
$$x + 3 = 0 \quad \textbf{or} \quad x - 5 = 0$$
$$x = -3 \qquad\qquad x = 5$$

45.
$$x^2 - 4x - 21 = 0$$
$$(x + 3)(x - 7) = 0$$
$$x + 3 = 0 \quad \textbf{or} \quad x - 7 = 0$$
$$x = -3 \qquad\qquad x = 7$$

47.
$$x^2 + 8 - 9x = 0$$
$$x^2 - 9x + 8 = 0$$
$$(x - 8)(x - 1) = 0$$
$$x - 8 = 0 \quad \textbf{or} \quad x - 1 = 0$$
$$x = 8 \qquad\qquad x = 1$$

49.
$$a^2 + 8a = -15$$
$$a^2 + 8a + 15 = 0$$
$$(a + 3)(a + 5) = 0$$
$$a + 3 = 0 \quad \textbf{or} \quad a + 5 = 0$$
$$a = -3 \qquad\qquad a = -5$$

51.
$$2y - 8 = -y^2$$
$$y^2 + 2y - 8 = 0$$
$$(y + 4)(y - 2) = 0$$
$$y + 4 = 0 \quad \textbf{or} \quad y - 2 = 0$$
$$y = -4 \qquad\qquad y = 2$$

53.
$$2x^2 - 5x + 2 = 0$$
$$(2x - 1)(x - 2) = 0$$
$$2x - 1 = 0 \quad \textbf{or} \quad x - 2 = 0$$
$$2x = 1 \qquad\qquad x = 2$$
$$x = \tfrac{1}{2} \qquad\qquad x = 2$$

55.
$$6x^2 + x - 2 = 0$$
$$(2x - 1)(3x + 2) = 0$$
$$2x - 1 = 0 \quad \textbf{or} \quad 3x + 2 = 0$$
$$2x = 1 \qquad\qquad 3x = -2$$
$$x = \tfrac{1}{2} \qquad\qquad x = -\tfrac{2}{3}$$

57.
$$5p^2 - 6p + 1 = 0$$
$$(5p - 1)(p - 1) = 0$$
$$5p - 1 = 0 \quad \textbf{or} \quad p - 1 = 0$$
$$5p = 1 \qquad\qquad p = 1$$
$$p = \tfrac{1}{5} \qquad\qquad p = 1$$

59.
$$14m^2 + 23m + 3 = 0$$
$$(7m + 1)(2m + 3) = 0$$
$$7m + 1 = 0 \quad \textbf{or} \quad 2m + 3 = 0$$
$$7m = -1 \qquad\qquad 2m = -3$$
$$m = -\tfrac{1}{7} \qquad\qquad m = -\tfrac{3}{2}$$

61. $(x-1)(x^2+5x+6)=0$
$(x-1)(x+2)(x+3)=0$
$x-1=0$ **or** $x+2=0$ **or** $x+3=0$
$\quad x=1 \qquad\qquad x=-2 \qquad\qquad x=-3$

63. $(x+3)(x^2+2x-15)=0$
$(x+3)(x+5)(x-3)=0$
$x+3=0$ **or** $x+5=0$ **or** $x-3=0$
$\quad x=-3 \qquad\qquad x=-5 \qquad\qquad x=3$

65. $(x-5)(2x^2+x-3)=0$
$(x-5)(2x+3)(x-1)=0$
$x-5=0$ **or** $2x+3=0$ **or** $x-1=0$
$\quad x=5 \qquad\qquad 2x=-3 \qquad\qquad x=1$
$\qquad\qquad\qquad\quad x=-\frac{3}{2}$

67. $(p^2-81)(p+2)=0$
$(p+9)(p-9)(p+2)=0$
$p+9=0$ **or** $p-9=0$ **or** $p+2=0$
$\quad p=-9 \qquad\qquad p=9 \qquad\qquad p=-2$

69. $x^3+3x^2+2x=0$
$x(x^2+3x+2)=0$
$x(x+2)(x+1)=0$
$x=0$ **or** $x+2=0$ **or** $x+1=0$
$x=0 \qquad\qquad x=-2 \qquad\qquad x=-1$

71. $x^3-27x-6x^2=0$
$x(x^2-27-6x)=0$
$x(x^2-6x-27)=0$
$x(x-9)(x+3)=0$
$x=0$ **or** $x-9=0$ **or** $x+3=0$
$x=0 \qquad\qquad x=9 \qquad\qquad x=-3$

73.
$$3x^2-8x=3$$
$$3x^2-8x-3=0$$
$$(3x+1)(x-3)=0$$
$3x+1=0$ **or** $x-3=0$
$\quad 3x=-1 \qquad\qquad x=3$
$\quad x=-\frac{1}{3} \qquad\qquad x=3$

75.
$$15x^2-2=7x$$
$$15x^2-7x-2=0$$
$$(3x-2)(5x+1)=0$$
$3x-2=0$ **or** $5x+1=0$
$\quad 3x=2 \qquad\qquad 5x=-1$
$\quad x=\frac{2}{3} \qquad\qquad x=-\frac{1}{5}$

77.
$$x(6x + 5) = 6$$
$$6x^2 + 5x = 6$$
$$6x^2 + 5x - 6 = 0$$
$$(3x - 2)(2x + 3) = 0$$
$$3x - 2 = 0 \quad \textbf{or} \quad 2x + 3 = 0$$
$$3x = 2 \qquad\qquad 2x = -3$$
$$x = \tfrac{2}{3} \qquad\qquad x = -\tfrac{3}{2}$$

79.
$$(x + 1)(8x + 1) = 18x$$
$$8x^2 + 9x + 1 = 18x$$
$$8x^2 - 9x + 1 = 0$$
$$(x - 1)(8x - 1) = 0$$
$$x - 1 = 0 \quad \textbf{or} \quad 8x - 1 = 0$$
$$x = 1 \qquad\qquad 8x = 1$$
$$x = 1 \qquad\qquad x = \tfrac{1}{8}$$

81.
$$2x(3x^2 + 10x) = -6x$$
$$6x^3 + 20x^2 = -6x$$
$$6x^3 + 20x^2 + 6x = 0$$
$$2x(3x^2 + 10x + 3) = 0$$
$$2x(3x + 1)(x + 3) = 0$$
$$2x = 0 \quad \textbf{or} \quad 3x + 1 = 0 \quad \textbf{or} \quad x + 3 = 0$$
$$x = 0 \qquad\qquad 3x = -1 \qquad\qquad x = -3$$
$$x = 0 \qquad\qquad x = -\tfrac{1}{3} \qquad\qquad x = -3$$

83.
$$x^3 + 7x^2 = x^2 - 9x$$
$$x^3 + 6x^2 + 9x = 0$$
$$x(x^2 + 6x + 9) = 0$$
$$x(x + 3)(x + 3) = 0$$
$$x = 0 \quad \textbf{or} \quad x + 3 = 0 \quad \textbf{or} \quad x + 3 = 0$$
$$x = 0 \qquad\qquad x = -3 \qquad\qquad x = -3$$

85. Answers may vary.

87. Answers may vary.

89.
$$3a^2 + 9a - 2a - 6 = 0$$
$$3a(a + 3) - 2(a + 3) = 0$$
$$(a + 3)(3a - 2) = 0$$
$$a + 3 = 0 \quad \textbf{or} \quad 3a - 2 = 0$$
$$a = -3 \qquad\qquad 3a = 2$$
$$\qquad\qquad\qquad a = \tfrac{2}{3}$$

$$3a^2 + 9a - 2a - 6 = 0$$
$$3a^2 + 7a - 6 = 0$$
$$(a + 3)(3a - 2) = 0$$
$$a + 3 = 0 \quad \textbf{or} \quad 3a - 2 = 0$$
$$a = -3 \qquad\qquad 3a = 2$$
$$\qquad\qquad\qquad a = \tfrac{2}{3}$$

Exercise 5.8 (page 337)

1.
$$-2(5z + 2) = 3(2 - 3z)$$
$$-10z - 4 = 6 - 9z$$
$$-4 = 6 + z$$
$$-10 = z$$

3. Let $w =$ the width and $3w =$ the length.
$$\text{Perimeter} = 120$$
$$2(w) + 2(3w) = 120$$
$$2w + 6w = 120$$
$$8w = 120$$
$$w = 15$$
The area $= 15(45) = 675 \text{ cm}^2$.

5. analyze

7. Let x = the first positive integer. Then $x + 2$ = the other positive integer.

$$x(x + 2) = 35$$
$$x^2 + 2x = 35$$
$$x^2 + 2x - 35 = 0$$
$$(x + 7)(x - 5) = 0$$
$$x + 7 = 0 \quad \text{or} \quad x - 5 = 0$$
$$x = -7 \qquad\qquad x = 5$$

The answer of -7 is not a positive integer, so the integers are 5 and 7.

9. Let x = the composite integer.

$$x^2 + 4 = 10x - 5$$
$$x^2 - 10x + 9 = 0$$
$$(x - 9)(x - 1) = 0$$
$$x - 9 = 0 \quad \text{or} \quad x - 1 = 0$$
$$x = 9 \qquad\qquad x = 1$$

The answer of 1 is not a composite integer, so the integer is 9.

11. Let $v = 144$ and $h = 0$:

$$h = vt - 16t^2$$
$$0 = 144t - 16t^2$$
$$0 = 16t(9 - t)$$
$$16t = 0 \quad \text{or} \quad 9 - t = 0$$
$$t = 0 \qquad\qquad 9 = t$$

Since $t = 0$ is when the object was first thrown, it will hit the ground in 9 seconds.

13. Let $v = 220$ and $h = 600$:

$$h = vt - 16t^2$$
$$600 = 220t - 16t^2$$
$$16t^2 - 220t + 600 = 0$$
$$4\left(4t^2 - 55t + 150\right) = 0$$
$$4(4t - 15)(t - 10) = 0$$
$$4t - 15 = 0 \quad \text{or} \quad t - 10 = 0$$
$$4t = 15 \qquad\qquad t = 10$$
$$t = \tfrac{15}{4} \qquad\qquad t = 10$$

The cannonball will be at a height of 600 feet after $\frac{15}{4}$ seconds and after 10 seconds.

15.
$$h = -16t^2 + 64$$
$$0 = -16t^2 + 64$$
$$16t^2 - 64 = 0$$
$$16\left(t^2 - 4\right) = 0$$
$$16(t + 2)(t - 2) = 0$$
$$t + 2 = 0 \quad \text{or} \quad t - 2 = 0$$
$$t = -2 \qquad\qquad t = 2$$

The value of $t = -2$ does not make sense, so the dive lasts 2 seconds.

17.
$$h^2 + 72^2 = 75^2$$
$$h^2 + 5184 = 5625$$
$$h^2 - 441 = 0$$
$$(h + 21)(h - 21) = 0$$
$$h + 21 = 0 \quad \text{or} \quad h - 21 = 0$$
$$h = -21 \qquad\qquad h = 21$$

The value of $h = -21$ does not make sense, so the camper started at a height of 21 feet.

19. Area = Length · Width

$$36 = (2w + 1)w$$
$$36 = 2w^2 + w$$
$$0 = 2w^2 + w - 36$$
$$0 = (2w + 9)(w - 4)$$

$2w + 9 = 0 \quad \textbf{or} \quad w - 4 = 0$

$\quad 2w = -9 \qquad\qquad w = 4$

$\quad w = -\frac{9}{2} \qquad\qquad w = 4$

Since the answer $w = -\frac{9}{2}$ does not make sense, the dimensions are 4 m by 9 m.

21. Let w = the width and $w + 2$ = the length.

Area = Length · Width

$$143 = (w + 2)w$$
$$143 = w^2 + 2w$$
$$0 = w^2 + 2w - 143$$
$$0 = (w + 13)(w - 11)$$

$w + 13 = 0 \quad \textbf{or} \quad w - 11 = 0$

$\quad w = -13 \qquad\qquad w = 11$

Since the answer $w = -13$ does not make sense, the dimensions are 11 ft by 13 ft and the perimeter is 48 ft.

23. Let b = the base and $5b - 2$ = the height.

$$A = \frac{1}{2}bh$$
$$36 = \frac{1}{2}b(5b - 2)$$
$$72 = b(5b - 2)$$
$$72 = 5b^2 - 2b$$
$$0 = 5b^2 - 2b - 72$$
$$0 = (5b + 18)(b - 4)$$

$5b + 18 = 0 \quad \textbf{or} \quad b - 4 = 0$

$\quad 5b = -18 \qquad\qquad b = 4$

$\quad h = -\frac{18}{5} \qquad\qquad b = 4$

Since the answer $b = -\frac{18}{5}$ does not make sense, the base is 4 in. and the height is 18 in.

25. Let x = the base and the height.

| Base | + | Height | = | Area | − 6 |

$$x + x = \frac{1}{2}(x)(x) - 6$$
$$2x = \frac{1}{2}x^2 - 6$$
$$0 = \frac{1}{2}x^2 - 2x - 6$$
$$0 = x^2 - 4x - 12$$
$$0 = (x - 6)(x + 2)$$

$x - 6 = 0 \quad \textbf{or} \quad x + 2 = 0$

$\quad x = 6 \qquad\qquad x = -2$

The answer $x = -2$ does not make sense, so the area is $\frac{1}{2}(6)(6) = 18$ square units.

27.

| Large area | − | Small area | = | Border area |

$$(10 + 2w)(25 + 2w) - (10)(25) = 74$$
$$250 + 70w + 4w^2 - 250 = 74$$
$$4w^2 + 70w - 74 = 0$$
$$2(2w^2 + 35w - 37) = 0$$
$$2(2w + 37)(w - 1) = 0$$

$2w + 37 = 0 \quad \textbf{or} \quad w - 1 = 0$

$\quad 2w = -37 \qquad\qquad w = 1$

$\quad w = -\frac{37}{2} \qquad\qquad w = 1$

Since the answer $w = -\frac{37}{2}$ does not make sense, the width should be 1 meter.

29. Let w = the width and $2w + 1$ = the height.

$$V = lwh$$
$$210 = 10w(2w + 1)$$
$$210 = 20w^2 + 10w$$
$$0 = 20w^2 + 10w - 210$$
$$0 = 10(2w^2 + w - 21)$$
$$0 = 10(2w + 7)(w - 3)$$

$2w + 7 = 0 \quad \textbf{or} \quad w - 3 = 0$

$\quad 2w = -7 \qquad\qquad w = 3$

$\quad w = -\frac{7}{2} \qquad\qquad w = 3$

Since the answer $w = -\frac{7}{2}$ does not make sense, the width is 3 cm.

31. Let $x =$ one edge and $x - 3 =$ the other.

$$V = \frac{Bh}{3}$$
$$84 = \frac{x(x-3)(9)}{3}$$
$$252 = 9x^2 - 27x$$
$$0 = 9x^2 - 27x - 252$$
$$0 = 9(x+4)(x-7)$$
$$x + 4 = 0 \quad \textbf{or} \quad x - 7 = 0$$
$$x = -4 \qquad\qquad x = 7$$

Since the answer $x = -4$ does not make sense, the base is 7 cm by 4 cm.

33.
$$C = \tfrac{1}{2}(n^2 - n)$$
$$6 = \tfrac{1}{2}(n^2 - n)$$
$$2(66) = 2 \cdot \tfrac{1}{2}(n^2 - n)$$
$$132 = n^2 - n$$
$$0 = n^2 - n - 132$$
$$0 = (n + 11)(n - 12)$$
$$n + 11 = 0 \quad \textbf{or} \quad n - 12 = 0$$
$$n = -11 \qquad\qquad n = 12$$

The value of $n = -11$ does not make sense, so 12 telephones are needed.

35. Area #1 $= \pi r^2 = \pi(20 \text{ m})^2 = 400\pi \text{ m}^2$
Area #2 $= \pi r^2 = \pi(21 \text{ m})^2 = 441\pi \text{ m}^2$
Total Area $= 400\pi + 441\pi \text{ m}^2 = 841\pi \text{ m}^2$

37. Answers may vary.

39. Let $w =$ the width and $w + 2 =$ the length.
$$\text{Area} = 18$$
$$w(w + 2) = 18$$
$$w^2 + 2w = 18$$
$$w^2 + 2w - 18 = 0$$
This is a prime trinomial, so it cannot be factored in order to solve.

Chapter 5 Summary (page 341)

1. $35 = 5 \cdot 7$

2. $45 = 9 \cdot 5 = 3^2 \cdot 5$

3. $96 = 12 \cdot 8 = 4 \cdot 3 \cdot 2^3 = 2^2 \cdot 3 \cdot 2^3 = 2^5 \cdot 3$

4. $102 = 2 \cdot 51 = 2 \cdot 3 \cdot 17$

5. $87 = 3 \cdot 29$

6. $99 = 9 \cdot 11 = 3^2 \cdot 11$

7. $2{,}050 = 50 \cdot 41 = 25 \cdot 2 \cdot 41 = 2 \cdot 5^2 \cdot 41$

8. $4{,}096 = 64 \cdot 64 = 2^6 \cdot 2^6 = 2^{12}$

9. $3x + 9y = 3(x + 3y)$

10. $5ax^2 + 15a = 5a(x^2 + 3)$

11. $7x^2 + 14x = 7x(x + 2)$

12. $3x^2 - 3x = 3x(x - 1)$

13. $2x^3 + 4x^2 - 8x = 2x(x^2 + 2x - 4)$

14. $ax + ay - az = a(x + y - z)$

15. $ax + ay + a = a(x + y + 1)$

16. $x^2yz + xy^2z = xyz(x + y)$

17. $(x + y)a + (x + y)b = (x + y)(a + b)$

18. $(x + y)^2 + (x + y) = (x + y)(x + y) + 1(x + y) = (x + y)(x + y + 1)$

19. $2x^2(x+2) + 6x(x+2) = (x+2)(2x^2 + 6x) = (x+2)(2x)(x+3) = 2x(x+2)(x+3)$

20. $3x(y+z) - 9x(y+z)^2 = 3x[(y+z) - 3(y+z)(y+z)] = 3x(y+z)(1 - 3(y+z))$
$$= 3x(y+z)(1 - 3y - 3z)$$

21. $3p + 9q + ap + 3aq = 3(p+3q) + a(p+3q) = (p+3q)(3+a)$

22. $ar - 2as + 7r - 14s = a(r-2s) + 7(r-2s) = (r-2s)(a+7)$

23. $x^2 + ax + bx + ab = x(x+a) + b(x+a)$ **24.** $xy + 2x - 2y - 4 = x(y+2) - 2(y+2)$
$$\qquad\qquad\quad = (x+a)(x+b) \qquad\qquad\qquad\qquad\quad = (y+2)(x-2)$$

25. $xa + yb + ya + xb = xa + ya + xb + yb = a(x+y) + b(x+y) = (x+y)(a+b)$

26. $x^2 - 9 = x^2 - 3^2 = (x+3)(x-3)$ **27.** $x^2y^2 - 16 = (xy+4)(xy-4)$

28. $(x+2)^2 - y^2 = (x+2+y)(x+2-y)$

29. $z^2 - (x+y)^2 = [z+(x+y)][z-(x+y)] = (z+x+y)(z-x-y)$

30. $6x^2y - 24y^3 = 6y(x^2 - 4y^2) = 6y[x^2 - (2y)^2] = 6y(x+2y)(x-2y)$

31. $(x+y)^2 - z^2 = [(x+y)+z][(x+y)-z] = (x+y+z)(x+y-z)$

32. $x^2 + 10x + 21 = (x+3)(x+7)$ **33.** $x^2 + 4x - 21 = (x+7)(x-3)$

34. $x^2 + 2x - 24 = (x+6)(x-4)$ **35.** $x^2 - 4x - 12 = (x-6)(x+2)$

36. $2x^2 - 5x - 3 = (2x+1)(x-3)$ **37.** $3x^2 - 14x - 5 = (3x+1)(x-5)$

38. $6x^2 + 7x - 3 = (2x+3)(3x-1)$ **39.** $6x^2 + 3x - 3 = 3(2x^2 + x - 1)$
$$\qquad\qquad\qquad\qquad\qquad\qquad\qquad = 3(2x-1)(x+1)$$

40. $6x^3 + 17x^2 - 3x = x(6x^2 + 17x - 3)$ **41.** $4x^3 - 5x^2 - 6x = x(4x^2 - 5x - 6)$
$$\qquad\qquad\quad = x(x+3)(6x-1) \qquad\qquad\qquad\qquad = x(x-2)(4x+3)$$

42. $c^3 - 27 = c^3 - 3^3 = (c-3)(c^2 + 3c + 9)$ **43.** $d^3 + 8 = d^3 + 2^3 = (d+2)(d^2 - 2d + 4)$

44. $2x^3 + 54 = 2(x^3 + 27)$ **45.** $2ab^4 - 2ab = 2ab(b^3 - 1)$
$$\qquad = 2(x^3 + 3^3) \qquad\qquad\qquad\qquad = 2ab(b^3 - 1^3)$$
$$\qquad = 2(x+3)(x^2 - 3x + 9) \qquad\qquad = 2ab(b-1)(b^2 + b + 1)$$

46. $3x^2y - xy^2 - 6xy + 2y^2 = y(3x^2 - xy - 6x + 2y) = y[x(3x-y) - 2(3x-y)]$
$$= y(3x-y)(x-2)$$

47. $5x^2 + 10x - 15xy - 30y = 5(x^2 + 2x - 3xy - 6y) = 5[x(x+2) - 3y(x+2)]$
$$= 5(x+2)(x-3y)$$

48. $2a^2x + 2abx + a^3 + a^2b = a(2ax + 2bx + a^2 + ab) = a[2x(a+b) + a(a+b)]$
$$= a(a+b)(2x+a)$$

49. $x^2 + 2ax + a^2 - y^2 = (x^2 + 2ax + a^2) - y^2 = (x+a)(x+a) - y^2$
$$= (x+a)^2 - y^2 = (x+a+y)(x+a-y)$$

50. $x^2 - 4 + bx + 2b = x^2 - 2^2 + bx + 2b = (x+2)(x-2) + b(x+2) = (x+2)(x-2+b)$

51. $ax^6 - ay^6 = a(x^6 - y^6) = a\left[\left(x^3\right)^2 - \left(y^3\right)^2\right] = a\left(x^3 + y^3\right)\left(x^3 - y^3\right)$
$$= a(x+y)\left(x^2 - xy + y^2\right)(x-y)\left(x^2 + xy + y^2\right)$$

52. $x^2 + 2x = 0$
$x(x+2) = 0$
$x = 0$ **or** $x + 2 = 0$
$x = 0 \qquad\qquad x = -2$

53. $2x^2 - 6x = 0$
$2x(x-3) = 0$
$2x = 0$ **or** $x - 3 = 0$
$x = 0 \qquad\qquad x = 3$

54. $3x^2 = 2x$
$3x^2 - 2x = 0$
$x(3x-2) = 0$
$x = 0$ **or** $3x - 2 = 0$
$\qquad\qquad 3x = 2$
$\qquad\qquad x = \frac{2}{3}$

55. $5x^2 + 25x = 0$
$5x(x+5) = 0$
$5x = 0$ **or** $x + 5 = 0$
$x = \frac{0}{5} \qquad\qquad x = -5$
$x = 0$

56. $x^2 - 9 = 0$
$(x+3)(x-3) = 0$
$x + 3 = 0$ **or** $x - 3 = 0$
$x = -3 \qquad\qquad x = 3$

57. $x^2 - 25 = 0$
$(x+5)(x-5) = 0$
$x + 5 = 0$ **or** $x - 5 = 0$
$x = -5 \qquad\qquad x = 5$

58. $a^2 - 7a + 12 = 0$
$(a-3)(a-4) = 0$
$a - 3 = 0$ **or** $a - 4 = 0$
$a = 3 \qquad\qquad a = 4$

59. $x^2 - 2x - 15 = 0$
$(x+3)(x-5) = 0$
$x + 3 = 0$ **or** $x - 5 = 0$
$x = -3 \qquad\qquad x = 5$

60. $2x - x^2 + 24 = 0$
$x^2 - 2x - 24 = 0$
$(x+4)(x-6) = 0$
$x + 4 = 0$ **or** $x - 6 = 0$
$x = -4 \qquad\qquad x = 6$

61. $16 + x^2 - 10x = 0$
$x^2 - 10x + 16 = 0$
$(x-2)(x-8) = 0$
$x - 2 = 0$ **or** $x - 8 = 0$
$x = 2 \qquad\qquad x = 8$

62.
$$2x^2 - 5x - 3 = 0$$
$$(2x + 1)(x - 3) = 0$$
$$2x + 1 = 0 \quad \textbf{or} \quad x - 3 = 0$$
$$2x = -1 \qquad x = 3$$
$$x = -\tfrac{1}{2} \qquad x = 3$$

63.
$$2x^2 + x - 3 = 0$$
$$(2x + 3)(x - 1) = 0$$
$$2x + 3 = 0 \quad \textbf{or} \quad x - 1 = 0$$
$$2x = -3 \qquad x = 1$$
$$x = -\tfrac{3}{2} \qquad x = 1$$

64.
$$4x^2 = 1$$
$$4x^2 - 1 = 0$$
$$(2x + 1)(2x - 1) = 0$$
$$2x + 1 = 0 \quad \textbf{or} \quad 2x - 1 = 0$$
$$2x = -1 \qquad 2x = 1$$
$$x = -\tfrac{1}{2} \qquad x = \tfrac{1}{2}$$

65.
$$9x^2 = 4$$
$$9x^2 - 4 = 0$$
$$(3x + 2)(3x - 2) = 0$$
$$3x + 2 = 0 \quad \textbf{or} \quad 3x - 2 = 0$$
$$3x = -2 \qquad 3x = 2$$
$$x = -\tfrac{2}{3} \qquad x = \tfrac{2}{3}$$

66.
$$x^3 - 7x^2 + 12x = 0$$
$$x(x^2 - 7x + 12) = 0$$
$$x(x - 3)(x - 4) = 0$$
$$x = 0 \ \textbf{or} \ x - 3 = 0 \ \textbf{or} \ x - 4 = 0$$
$$x = 0 \qquad x = 3 \qquad x = 4$$

67.
$$x^3 + 5x^2 + 6x = 0$$
$$x(x^2 + 5x + 6) = 0$$
$$x(x + 2)(x + 3) = 0$$
$$x = 0 \ \textbf{or} \ x + 2 = 0 \quad \textbf{or} \ x + 3 = 0$$
$$x = 0 \qquad x = -2 \qquad x = -3$$

68.
$$2x^3 + 5x^2 = 3x$$
$$2x^3 + 5x^2 - 3x = 0$$
$$x(2x^2 + 5x - 3) = 0$$
$$x(2x - 1)(x + 3) = 0$$
$$x = 0 \ \textbf{or} \ 2x - 1 = 0 \ \textbf{or} \ x + 3 = 0$$
$$x = 0 \qquad 2x = 1 \qquad x = -3$$
$$x = 0 \qquad x = \tfrac{1}{2} \qquad x = -3$$

69.
$$3x^3 - 2x = x^2$$
$$3x^3 - x^2 - 2x = 0$$
$$x(3x^2 - x - 2) = 0$$
$$x(3x + 2)(x - 1) = 0$$
$$x = 0 \ \textbf{or} \ 3x + 2 = 0 \quad \textbf{or} \ x - 1 = 0$$
$$x = 0 \qquad 3x = -2 \qquad x = 1$$
$$x = 0 \qquad x = -\tfrac{2}{3} \qquad x = 1$$

70. Let $x =$ one number. Then
$12 - x =$ the other number.
$$x(12 - x) = 35$$
$$12x - x^2 = 35$$
$$0 = x^2 - 12x + 35$$
$$0 = (x - 5)(x - 7)$$
$$x - 5 = 0 \ \textbf{or} \ x - 7 = 0$$
$$x = 5 \qquad x = 7$$
The numbers are 5 and 7.

71. Let $x =$ the positive number.
$$3x^2 + 5x = 2$$
$$3x^2 + 5x - 2 = 0$$
$$(3x - 1)(x + 2) = 0$$
$$3x - 1 = 0 \quad \textbf{or} \quad x + 2 = 0$$
$$3x = 1 \qquad x = -2$$
$$x = \tfrac{1}{3} \qquad x = -2$$
The answer of -2 is not positive,
so the number is $\tfrac{1}{3}$.

72. Let w = the width and $w + 2$ = the length.

Area = Length · Width

$$48 = (w + 2)w$$
$$0 = w^2 + 2w - 48$$
$$0 = (w + 8)(w - 6)$$
$$w + 8 = 0 \quad \text{or} \quad w - 6 = 0$$
$$w = -8 \qquad\qquad w = 6$$

Since the answer $w = -8$ does not make sense, the dimensions are 6 ft by 8 ft.

73. Let w = the width and $2w + 3$ = the length.

Area = Length · Width

$$27 = (2w + 3)w$$
$$0 = 2w^2 + 3w - 27$$
$$0 = (2w + 9)(w - 3)$$
$$2w + 9 = 0 \quad \text{or} \quad w - 3 = 0$$
$$2w = -9 \qquad\qquad w = 3$$
$$w = -\tfrac{9}{2} \qquad\qquad w = 3$$

Since the answer $w = -\frac{9}{2}$ does not make sense, the dimensions are 3 ft by 9 ft.

74. Let w = the width and $w + 3$ = the length.

Area = Perimeter

$$(w + 3)w = 2w + 2(w + 3)$$
$$w^2 + 3w = 2w + 2w + 6$$
$$w^2 + 3w = 4w + 6$$
$$w^2 - w - 6 = 0$$
$$(w + 2)(w - 3) = 0$$
$$w + 2 = 0 \quad \text{or} \quad w - 3 = 0$$
$$w = -2 \qquad\qquad w = 3$$

Since the answer $w = -2$ does not make sense, the dimensions are 3 ft by 6 ft.

Chapter 5 Test (page 344)

1. $196 = 14 \cdot 14 = 2 \cdot 7 \cdot 2 \cdot 7 = 2^2 \cdot 7^2$

2. $111 = 3 \cdot 37$

3. $60ab^2c^3 + 30a^3b^2c - 25a = 5a(12b^2c^3 + 6a^2b^2c - 5)$

4. $3x^2(a + b) - 6xy(a + b) = 3x[x(a + b) - 2y(a + b)] = 3x(a + b)(x - 2y)$

5. $ax + ay + bx + by = a(x + y) + b(x + y)$
$$= (x + y)(a + b)$$

6. $x^2 - 25 = x^2 - 5^2 = (x + 5)(x - 5)$

7. $3a^2 - 27b^2 = 3(a^2 - 9b^2) = 3\left(a^2 - (3b)^2\right) = 3(a + 3b)(a - 3b)$

8. $16x^4 - 81y^4 = \left(4x^2\right)^2 - \left(9y^2\right)^2 = \left(4x^2 + 9y^2\right)\left(4x^2 - 9y^2\right)$
$$= \left(4x^2 + 9y^2\right)\left((2x)^2 - (3y)^2\right)$$
$$= \left(4x^2 + 9y^2\right)(2x + 3y)(2x - 3y)$$

9. $x^2 + 4x + 3 = (x + 3)(x + 1)$

10. $x^2 - 9x - 22 = (x - 11)(x + 2)$

11. $x^2 + 10xy + 9y^2 = (x + y)(x + 9y)$

12. $6x^2 - 30xy + 24y^2 = 6\left(x^2 - 5xy + 4y^2\right)$
$$= 6(x - 4y)(x - y)$$

13. $3x^2 + 13x + 4 = (3x + 1)(x + 4)$

14. $2a^2 + 5a - 12 = (2a - 3)(a + 4)$

15. $2x^2 + 3xy - 2y^2 = (2x - y)(x + 2y)$

16. $12 - 25x + 12x^2 = 12x^2 - 25x + 12$
$$= (4x - 3)(3x - 4)$$

17. $12a^2 + 6ab - 36b^2 = 6\left(2a^2 + ab - 6b^2\right)$
$$= 6(2a - 3b)(a + 2b)$$

18. $x^3 - 64 = x^3 - 4^3 = (x - 4)\left(x^2 + 4x + 16\right)$

19. $216 + 8a^3 = 8(27 + a^3) = 8(3^3 + a^3) = 8(3 + a)(9 - 3a + a^2)$

20. $x^9 z^3 - y^3 z^6 = z^3\left(x^9 - y^3 z^3\right) = z^3\left(\left(x^3\right)^3 - (yz)^3\right) = z^3\left(x^3 - yz\right)\left(\left(x^3\right)^2 + x^3 yz + (yz)^2\right)$
$$= z^3\left(x^3 - yz\right)\left(x^6 + x^3 yz + y^2 z^2\right)$$

21.
$$x^2 + 3x = 0$$
$$x(x + 3) = 0$$
$$x = 0 \quad \textbf{or} \quad x + 3 = 0$$
$$x = 0 \qquad\qquad x = -3$$

22.
$$2x^2 + 5x + 3 = 0$$
$$(2x + 3)(x + 1) = 0$$
$$2x + 3 = 0 \quad \textbf{or} \quad x + 1 = 0$$
$$2x = -3 \qquad\qquad x = -1$$
$$x = -\tfrac{3}{2} \qquad\qquad x = -1$$

23.
$$9y^2 - 81 = 0$$
$$9\left(y^2 - 9\right) = 0$$
$$9(y + 3)(y - 3) = 0$$
$$y + 3 = 0 \quad \textbf{or} \quad y - 3 = 0$$
$$y = -3 \qquad\qquad y = 3$$

24.
$$-3(y - 6) + 2 = y^2 + 2$$
$$-3y + 18 + 2 = y^2 + 2$$
$$-3y + 20 = y^2 + 2$$
$$0 = y^2 + 3y - 18$$
$$0 = (y + 6)(y - 3)$$
$$y + 6 = 0 \quad \textbf{or} \quad y - 3 = 0$$
$$y = -6 \qquad\qquad y = 3$$

25.
$$10x^2 - 13x = 9$$
$$10x^2 - 13x - 9 = 0$$
$$(2x + 1)(5x - 9) = 0$$
$$2x + 1 = 0 \quad \textbf{or} \quad 5x - 9 = 0$$
$$2x = -1 \qquad\qquad 5x = 9$$
$$x = -\tfrac{1}{2} \qquad\qquad x = \tfrac{9}{5}$$

26.
$$10x^2 - x = 9$$
$$10x^2 - x - 9 = 0$$
$$(10x + 9)(x - 1) = 0$$
$$10x + 9 = 0 \quad \textbf{or} \quad x - 1 = 0$$
$$10x = -9 \qquad\qquad x = 1$$
$$x = -\tfrac{9}{10} \qquad\qquad x = 1$$

27.
$$10x^2 + 43x = 9$$
$$10x^2 + 43x - 9 = 0$$
$$(2x + 9)(5x - 1) = 0$$

$2x + 9 = 0$ **or** $5x - 1 = 0$
$2x = -9$ $5x = 1$
$x = -\frac{9}{2}$ $x = \frac{1}{5}$

28.
$$10x^2 - 89x = 9$$
$$10x^2 - 89x - 9 = 0$$
$$(10x + 1)(x - 9) = 0$$

$10x + 1 = 0$ **or** $x - 9 = 0$
$10x = -1$ $x = 9$
$x = -\frac{1}{10}$ $x = 9$

29. Let $v = 192$ and $h = 0$:
$$h = vt - 16t^2$$
$$0 = 192t - 16t^2$$
$$0 = 16t(12 - t)$$

$16t = 0$ **or** $12 - t = 0$
$t = 0$ $12 = t$

Since $t = 0$ is when the cannonball was fired, it will hit the ground in 12 seconds.

30. Let $h =$ the height and $h + 2 =$ the base.
$$A = \tfrac{1}{2}bh$$
$$40 = \tfrac{1}{2}(h + 2)h$$
$$80 = (h + 2)h$$
$$0 = h^2 + 2h - 80$$
$$0 = (h + 10)(h - 8)$$

$h + 10 = 0$ **or** $h - 8 = 0$
$h = -10$ $h = 8$

Since the answer $h = -10$ does not make sense, the height is 8 m and the base is 10 m.

Exercise 6.1   (page 351)

1.
$$2x + 4 = 38$$
$$2x = 34$$
$$x = 17$$

3.
$$3(x + 2) = 24$$
$$3x + 6 = 24$$
$$3x = 18$$
$$x = 6$$

5. $\quad 2x + 6 = 2(x + 3)$

7. $\quad 2x^2 - x - 6 = (2x + 3)(x - 2)$

9. quotient

11. equal

13. Answers may vary.

15. $\dfrac{5}{7}$

17. $\dfrac{17}{34} = \dfrac{1}{2}$

19. $\dfrac{22}{33} = \dfrac{2}{3}$

21. $\dfrac{7}{24.5} = \dfrac{14}{49} = \dfrac{2}{7}$

23. $\dfrac{4 \text{ oz}}{12 \text{ oz}} = \dfrac{1}{3}$

25. $\dfrac{12 \text{ min}}{1 \text{ hr}} = \dfrac{12 \text{ min}}{60 \text{ min}} = \dfrac{1}{5}$

27. $\dfrac{3 \text{ days}}{1 \text{ week}} = \dfrac{3 \text{ days}}{7 \text{ days}} = \dfrac{3}{7}$

29. $\dfrac{18 \text{ months}}{2 \text{ years}} = \dfrac{18 \text{ months}}{24 \text{ months}} = \dfrac{3}{4}$

31. $750 + 652 + 188 + 125 + 110 = \$1{,}825$

33. $\dfrac{110}{1825} = \dfrac{22}{365}$

35. $995 + 1245 + 1680 + 4580 + 225 = \$8{,}725$

37. $\dfrac{1680}{8725} = \dfrac{336}{1745}$

39. $\dfrac{125}{2000} = \dfrac{1}{16}$

41. $\dfrac{\$30.09}{17 \text{ gal}} = \$1.77/\text{gal}$

43. $\dfrac{84¢}{12 \text{ oz}} = 7¢/\text{oz}$

45. $\dfrac{89¢}{6 \text{ oz}} \approx 14.873¢/\text{oz}; \dfrac{119¢}{8 \text{ oz}} = 14.875¢/\text{oz}$
The 6-oz can is the better buy.

47. $\dfrac{54 \text{ pg}}{40 \text{ min}} = 1.35 \dfrac{\text{pg}}{\text{min}}; \dfrac{80 \text{ pg}}{62 \text{ min}} \approx 1.29 \dfrac{\text{pg}}{\text{min}}$
The 1st student read faster.

49. $\dfrac{11880 \text{ gal}}{27 \text{ min}} = 440 \text{ gal/min}$

51. $\dfrac{36.75 - 35}{35} = \dfrac{1.75}{35} = 0.05 = 5\%$

53. $\dfrac{325 \text{ mi}}{5 \text{ hr}} = 65 \text{ mi/hr}$

55. $\dfrac{1235 \text{ mi}}{51.3 \text{ gal}} \approx 24 \dfrac{\text{mi}}{\text{gal}}; \dfrac{1456 \text{ mi}}{55.78 \text{ gal}} \approx 26 \dfrac{\text{mi}}{\text{gal}}$
The 2nd car had the better mpg rating.

57. Answers may vary.

59. $\dfrac{17}{19} = \dfrac{17 \cdot 21}{19 \cdot 21} = \dfrac{357}{399}; \dfrac{19}{21} = \dfrac{19 \cdot 19}{21 \cdot 19} = \dfrac{361}{399}$

Exercise 6.2 (page 361)

1. $\dfrac{9}{10} = 0.9 = 90\%$

3. $33\dfrac{1}{3}\% = \dfrac{100}{3}\% = \dfrac{100}{3} \cdot \dfrac{1}{100} = \dfrac{1}{3}$

5.
$$rb = a$$
$$0.30(1600) = a$$
$$480 = a$$

7.
$$rb = a$$
$$0.25(98) = a$$
$$24.50 = a$$
$$98 - 24.50 = \$73.50$$

9. proportion; ratios   **11.** means   **13.** shape   **15.** $ad; bc$

17. triangle

19.
$$\dfrac{9}{7} \overset{?}{=} \dfrac{81}{70}$$
$$9 \cdot 70 \overset{?}{=} 7 \cdot 81$$
$$630 \neq 567$$
not a proportion

21.
$$\dfrac{-7}{3} \overset{?}{=} \dfrac{14}{-6}$$
$$-7(-6) \overset{?}{=} 3 \cdot 14$$
$$42 = 42$$
proportion

23.
$$\dfrac{9}{19} \overset{?}{=} \dfrac{38}{80}$$
$$9 \cdot 80 \overset{?}{=} 19 \cdot 38$$
$$720 \neq 722$$
not a proportion

25.
$$\dfrac{10.4}{3.6} \overset{?}{=} \dfrac{41.6}{14.4}$$
$$10.4(14.4) \overset{?}{=} 3.6(41.6)$$
$$149.76 = 149.76$$
proportion

27.
$$\dfrac{2}{3} = \dfrac{x}{6}$$
$$2(6) = 3x$$
$$12 = 3x$$
$$4 = x$$

29.
$$\dfrac{5}{10} = \dfrac{3}{c}$$
$$5c = 10(3)$$
$$5c = 30$$
$$c = 6$$

31.
$$\frac{-6}{x} = \frac{8}{4}$$
$$-6(4) = 8x$$
$$-24 = 8x$$
$$-3 = x$$

33.
$$\frac{x}{3} = \frac{9}{3}$$
$$x(3) = 3(9)$$
$$3x = 27$$
$$x = 9$$

35.
$$\frac{x+1}{5} = \frac{3}{15}$$
$$15(x+1) = 5(3)$$
$$15x + 15 = 15$$
$$15x = 0$$
$$x = 0$$

37.
$$\frac{x+3}{12} = \frac{-7}{6}$$
$$6(x+3) = 12(-7)$$
$$6x + 18 = -84$$
$$6x = -102$$
$$x = -17$$

39.
$$\frac{4-x}{13} = \frac{11}{26}$$
$$26(4-x) = 13(11)$$
$$104 - 26x = 143$$
$$-26x = 39$$
$$x = -\frac{39}{26} = -\frac{3}{2}$$

41.
$$\frac{2x+1}{18} = \frac{14}{3}$$
$$3(2x+1) = 18(14)$$
$$6x + 3 = 252$$
$$6x = 249$$
$$x = \frac{249}{6} = \frac{83}{2}$$

43.
$$\frac{3p-2}{12} = \frac{p+1}{3}$$
$$3(3p-2) = 12(p+1)$$
$$9p - 6 = 12p + 12$$
$$-3p = 18$$
$$p = -\frac{18}{3} = -6$$

45. Let c = the cost of 51 pints of yogurt.
$$\frac{3 \text{ pints}}{51 \text{ pints}} = \frac{\text{cost of 3 pints}}{\text{cost of 51 pints}}$$
$$\frac{3}{51} = \frac{1}{c}$$
$$c \cdot 3 = 51 \cdot 1$$
$$3c = 51$$
$$c = 17$$
The 51 pints will cost \$17.

47. Let c = the cost of 39 packets.
$$\frac{3 \text{ packets}}{39 \text{ packets}} = \frac{\text{cost of 3 packets}}{\text{cost of 39 packets}}$$
$$\frac{3}{39} = \frac{50}{c}$$
$$c \cdot 3 = 39 \cdot 50$$
$$3c = 1{,}950$$
$$c = 650 \text{ cents} = \$6.50$$
The 39 packets will cost \$6.50.

49. Let d = the # of drops of pure essence needed with 56 drops of alcohol.
$$\frac{3 \text{ pure}}{7 \text{ alcohol}} = \frac{\text{drops of pure}}{56 \text{ drops of alcohol}}$$
$$\frac{3}{7} = \frac{d}{56}$$
$$3 \cdot 56 = 7d$$
$$168 = 7d$$
$$24 = d$$
24 drops of pure essence will be needed.

51. Let $f =$ the cups of flour needed for 12 dozen cookies.

$$\frac{1\frac{1}{4} \text{ C. flour}}{3\frac{1}{2} \text{ dozen}} = \frac{\text{C. flour}}{12 \text{ dozen}}$$

$$\frac{1\frac{1}{4}}{3\frac{1}{2}} = \frac{f}{12}$$

$$1\frac{1}{4} \cdot 12 = 3\frac{1}{2}f$$

$$\frac{5}{4} \cdot 12 = \frac{7}{2}f$$

$$\frac{5}{4} \cdot 12 = \frac{7}{2}f$$

$$15 = \frac{7}{2}f$$

$$30 = 7f$$

$$\frac{30}{7} = f$$

$$4\frac{2}{7} = f$$

$4\frac{2}{7} \left(\approx 4\frac{1}{4}\right)$ cups of flour will be needed.

53. Let $n =$ the number defective.

$$\frac{5 \text{ defective}}{100 \text{ parts}} = \frac{\text{number defective}}{940 \text{ parts}}$$

$$\frac{5}{100} = \frac{n}{940}$$

$$5 \cdot 940 = 100n$$

$$4700 = 100n$$

$$47 = n$$

There will be 47 defective parts.

55. Let $g =$ gallons of gas needed for 315 miles.

$$\frac{42 \text{ miles}}{315 \text{ miles}} = \frac{\text{gas for 42 miles}}{\text{gas for 315 miles}}$$

$$\frac{42}{315} = \frac{1}{g}$$

$$g \cdot 42 = 315 \cdot 1$$

$$42g = 315$$

$$g = \frac{315}{42} = \frac{15}{2} = 7\frac{1}{2}$$

To go 315 miles, $7\frac{1}{2}$ gallons of gas are needed.

57. Let $p =$ the amount he was paid last week.

$$\frac{30 \text{ hours}}{40 \text{ hours}} = \frac{\text{pay for 30 hours}}{\text{pay for 40 hours}}$$

$$\frac{30}{40} = \frac{p}{412}$$

$$412 \cdot 30 = 40 \cdot p$$

$$12,360 = 40p$$

$$309 = p$$

He was paid $309 last week.

59. Let $l =$ the length of a real caboose.

$$\frac{169 \text{ feet}}{1 \text{ foot}} = \frac{\text{length of a real caboose}}{\text{length of a model caboose}}$$

$$\frac{169}{1} = \frac{l}{3.5}$$

$$169(3.5) = 1 \cdot l$$

$$591.5 = l$$

A real engine is 591.5 inches $(49'\ 3.5'')$ long.

61. Let $t =$ the number of teachers required.

$$\frac{3 \text{ teachers}}{50 \text{ students}} = \frac{\text{number of teachers}}{\text{number of students}}$$

$$\frac{3}{50} = \frac{t}{2,700}$$

$$2,700 \cdot 3 = 50 \cdot t$$

$$8,100 = 50t$$

$$162 = t$$

A total of 162 teachers are needed.

63. Let $x =$ the amount of oil to be added.

$$\frac{50}{1} = \frac{\text{ounces of gasoline}}{\text{ounces of oil}}$$

$$\frac{50}{1} = \frac{6 \cdot 128}{x}$$

$$\frac{50}{1} = \frac{768}{x}$$

$$x \cdot 50 = 1 \cdot 768$$

$$50x = 768$$

$$x = \frac{768}{50} \approx 15.36 \text{ ounces}$$

The directions are close.

65. $\dfrac{\text{height of man}}{\text{man's shadow}} = \dfrac{\text{height of tree}}{\text{tree's shadow}}$

$$\frac{6}{4} = \frac{h}{26}$$
$$6 \cdot 26 = 4h$$
$$156 = 4h$$
$$39 = h$$

The tree is 39 feet tall.

67. $\dfrac{75}{32} = \dfrac{w}{20}$

$$75 \cdot 20 = 32w$$
$$1500 = 32w$$
$$\frac{1500}{32} = w, \text{ or } w = 46\frac{7}{8}$$

The river is $46\frac{7}{8}$ feet wide.

69. $\dfrac{x}{1350} = \dfrac{5}{1}$

$$1 \cdot x = 1350 \cdot 5$$
$$x = 6750$$

The plane will lose 6,750 feet in altitude.

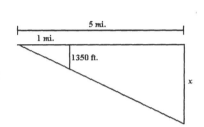

71. $\dfrac{x}{750} = \dfrac{52800}{2500}$

$$2500 \cdot x = 750 \cdot 52800$$
$$2500x = 39,600,000$$
$$x = 15,840$$

The road will rise 15,840 ft.

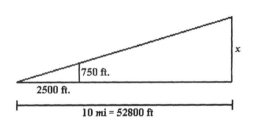

73. Answers may vary.

75. If $\dfrac{a}{b} = \dfrac{c}{d}$, then $ad = bc$. Thus, $\dfrac{a}{b} = \dfrac{a(b+d)}{b(b+d)} = \dfrac{ab+ad}{b(b+d)} = \dfrac{ab+bc}{b(b+d)} = \dfrac{b(a+c)}{b(b+d)} = \dfrac{a+c}{b+d}$.

Exercise 6.3 (page 370)

1. If a, b and c are real numbers, then $(a+b)+c = a+(b+c)$.

3. 0 is the additive identity.

5. $\frac{5}{3}$ is the additive inverse of $-\frac{5}{3}$.

7. numerator

9. 0

11. negatives (or opposites)

13. $\dfrac{a}{b}$

15. factor; common

17. The expression is undefined when the denominator equals 0:
$$y - 2 = 0$$
$$y = 2$$

19. The expression is undefined when the denominator equals 0:
$$x^2 - x - 2 = 0$$
$$(x + 1)(x - 2) = 0$$
$$x + 1 = 0 \quad \text{or} \quad x - 2 = 0$$
$$x = -1 \qquad\qquad x = 2$$

21. The expression is undefined when the denominator equals 0:
$$2m^2 - m - 3 = 0$$
$$(m + 1)(2m - 3) = 0$$
$$m + 1 = 0 \quad \text{or} \quad 2m - 3 = 0$$
$$m = -1 \qquad\qquad 2m = 3$$
$$m = \tfrac{3}{2}$$

23. $\dfrac{8}{10} = \dfrac{2 \cdot 4}{2 \cdot 5} = \dfrac{4}{5}$

25. $\dfrac{28}{35} = \dfrac{7 \cdot 4}{7 \cdot 5} = \dfrac{4}{5}$

27. $\dfrac{8}{52} = \dfrac{4 \cdot 2}{4 \cdot 13} = \dfrac{2}{13}$

29. $\dfrac{10}{45} = \dfrac{5 \cdot 2}{5 \cdot 9} = \dfrac{2}{9}$

31. $\dfrac{-18}{54} = -\dfrac{18 \cdot 1}{18 \cdot 3} = -\dfrac{1}{3}$

33. $\dfrac{4x}{2} = \dfrac{2 \cdot 2x}{2 \cdot 1} = 2x$

35. $\dfrac{-6x}{18} = -\dfrac{6 \cdot x}{6 \cdot 3} = -\dfrac{x}{3}$

37. $\dfrac{45}{9a} = \dfrac{9 \cdot 5}{9 \cdot a} = \dfrac{5}{a}$

39. $\dfrac{7 + 3}{5z} = \dfrac{10}{5z} = \dfrac{5 \cdot 2}{5 \cdot z} = \dfrac{2}{z}$

41. $\dfrac{(3 + 4)a}{24 - 3} = \dfrac{7a}{21} = \dfrac{7 \cdot a}{7 \cdot 3} = \dfrac{a}{3}$

43. $\dfrac{2x}{3x} = \dfrac{x \cdot 2}{x \cdot 3} = \dfrac{2}{3}$

45. $\dfrac{6x^2}{4x^2} = \dfrac{2x^2 \cdot 3}{2x^2 \cdot 2} = \dfrac{3}{2}$

47. $\dfrac{2x^2}{3y} \Rightarrow$ simplest form

49. $\dfrac{15x^2y}{5xy^2} = \dfrac{5xy \cdot 3x}{5xy \cdot y} = \dfrac{3x}{y}$

51. $\dfrac{28x}{32y} = \dfrac{4 \cdot 7x}{4 \cdot 8y} = \dfrac{7x}{8y}$

53. $\dfrac{x + 3}{3(x + 3)} = \dfrac{1(x + 3)}{3(x + 3)} = \dfrac{1}{3}$

55. $\dfrac{5x + 35}{x + 7} = \dfrac{5(x + 7)}{1(x + 7)} = 5$

57. $\dfrac{x^2 + 3x}{2x + 6} = \dfrac{x(x + 3)}{2(x + 3)} = \dfrac{x}{2}$

59. $\dfrac{15x - 3x^2}{25y - 5xy} = \dfrac{3x(5 - x)}{5y(5 - x)} = \dfrac{3x}{5y}$

61. $\dfrac{6a - 6b + 6c}{9a - 9b + 9c} = \dfrac{6(a - b + c)}{9(a - b + c)} = \dfrac{6}{9} = \dfrac{2}{3}$

63. $\dfrac{x - 7}{7 - x} = \dfrac{-(7 - x)}{7 - x} = -1$

65. $\dfrac{6x - 3y}{3y - 6x} = \dfrac{-(3y - 6x)}{3y - 6x} = -1$

67. $\dfrac{a + b - c}{c - a - b} = \dfrac{-(c - a - b)}{c - a - b} = -1$

69. $\dfrac{x^2 + 3x + 2}{x^2 + x - 2} = \dfrac{(x + 2)(x + 1)}{(x + 2)(x - 1)} = \dfrac{x + 1}{x - 1}$

71. $\dfrac{x^2 - 8x + 15}{x^2 - x - 6} = \dfrac{(x - 5)(x - 3)}{(x + 2)(x - 3)} = \dfrac{x - 5}{x + 2}$

73. $\dfrac{2x^2 - 8x}{x^2 - 6x + 8} = \dfrac{2x(x - 4)}{(x - 2)(x - 4)} = \dfrac{2x}{x - 2}$

75. $\dfrac{xy + 2x^2}{2xy + y^2} = \dfrac{x(y + 2x)}{y(2x + y)} = \dfrac{x}{y}$

77. $\dfrac{x^2 + 3x + 2}{x^3 + x^2} = \dfrac{(x + 2)(x + 1)}{x^2(x + 1)} = \dfrac{x + 2}{x^2}$

79. $\dfrac{x^2 - 8x + 16}{x^2 - 16} = \dfrac{(x - 4)(x - 4)}{(x + 4)(x - 4)} = \dfrac{x - 4}{x + 4}$

81. $\dfrac{2x^2 - 8}{x^2 - 3x + 2} = \dfrac{2(x^2 - 4)}{(x - 1)(x - 2)} = \dfrac{2(x + 2)(x - 2)}{(x - 1)(x - 2)} = \dfrac{2(x + 2)}{x - 1}$

83. $\dfrac{x^2 - 2x - 15}{x^2 + 2x - 15} = \dfrac{(x - 5)(x + 3)}{(x + 5)(x - 3)}$

85. $\dfrac{x^2 - 3(2x - 3)}{9 - x^2} = \dfrac{x^2 - 6x + 9}{3^2 - x^2} = \dfrac{(x - 3)(x - 3)}{(3 + x)(3 - x)} = -\dfrac{x - 3}{x + 3}$

87. $\dfrac{4(x + 3) + 4}{3(x + 2) + 6} = \dfrac{4x + 12 + 4}{3x + 6 + 6} = \dfrac{4x + 16}{3x + 12} = \dfrac{4(x + 4)}{3(x + 4)} = \dfrac{4}{3}$

89. $\dfrac{x^2 - 9}{(2x + 3) - (x + 6)} = \dfrac{x^2 - 9}{2x + 3 - x - 6} = \dfrac{x^2 - 9}{x - 3} = \dfrac{(x + 3)(x - 3)}{x - 3} = x + 3$

91. $\dfrac{x^3 + 1}{x^2 - x + 1} = \dfrac{x^3 + 1^3}{x^2 - x + 1} = \dfrac{(x + 1)(x^2 - 1x + 1^2)}{x^2 - x + 1} = \dfrac{(x + 1)(x^2 - x + 1)}{x^2 - x + 1} = x + 1$

93. $\dfrac{2a^3 - 16}{2a^2 + 4a + 8} = \dfrac{2(a^3 - 8)}{2(a^2 + 2a + 4)} = \dfrac{a^3 - 2^3}{a^2 + 2a + 4} = \dfrac{(a - 2)(a^2 + 2a + 2^2)}{a^2 + 2a + 4}$

$= \dfrac{(a - 2)(a^2 + 2a + 4)}{a^2 + 2a + 4} = a - 2$

95. $\dfrac{ab + b + 2a + 2}{ab + a + b + 1} = \dfrac{b(a + 1) + 2(a + 1)}{a(b + 1) + 1(b + 1)} = \dfrac{(a + 1)(b + 2)}{(b + 1)(a + 1)} = \dfrac{b + 2}{b + 1}$

97. $\dfrac{xy + 3y + 3x + 9}{x^2 - 9} = \dfrac{y(x + 3) + 3(x + 3)}{(x + 3)(x - 3)} = \dfrac{(x + 3)(y + 3)}{(x + 3)(x - 3)} = \dfrac{y + 3}{x - 3}$

99. **Answers may vary.**

101. $\dfrac{3 - x}{3 + x} = \dfrac{-(x - 3)}{x + 3} = -\dfrac{x - 3}{x + 3}$

Exercise 6.4   (page 378)

1. $2x^3y^2(-3x^2y^4z) = -6x^5y^6z$

3. $(3y)^{-4} = \dfrac{1}{(3y)^4} = \dfrac{1}{81y^4}$

5. $\dfrac{x^{3m}}{x^{4m}} = x^{-m} = \dfrac{1}{x^m}$

7. $-4\left(y^3 - 4y^2 + 3y - 2\right) + 6\left(-2y^2 + 4\right) - 4\left(-2y^3 - y\right)$
$$= -4y^3 + 16y^2 - 12y + 8 - 12y^2 + 24 + 8y^3 + 4y$$
$$= 4y^3 + 4y^2 - 8y + 32$$

9. numerator   **11.** numerators; denominators   **13.** 1   **15.** divisor; multiply

17. $\dfrac{5}{7} \cdot \dfrac{9}{13} = \dfrac{5 \cdot 9}{7 \cdot 13} = \dfrac{45}{91}$

19. $\dfrac{25}{35} \cdot \dfrac{-21}{55} = \dfrac{-5 \cdot 5 \cdot 3 \cdot 7}{7 \cdot 5 \cdot 5 \cdot 11} = -\dfrac{3}{11}$

21. $\dfrac{2}{3} \cdot \dfrac{15}{2} \cdot \dfrac{1}{7} = \dfrac{2 \cdot 3 \cdot 5}{3 \cdot 2 \cdot 7} = \dfrac{5}{7}$

23. $\dfrac{3x}{y} \cdot \dfrac{y}{2} = \dfrac{3x \cdot y}{y \cdot 2} = \dfrac{3x}{2}$

25. $\dfrac{5y}{7} \cdot \dfrac{7x}{5z} = \dfrac{5 \cdot y \cdot 7 \cdot x}{7 \cdot 5 \cdot z} = \dfrac{xy}{z}$

27. $\dfrac{7z}{9z} \cdot \dfrac{4z}{2z} = \dfrac{7 \cdot z \cdot 2 \cdot 2 \cdot z}{9 \cdot z \cdot 2 \cdot z} = \dfrac{14}{9}$

29. $\dfrac{2x^2 y}{3xy} \cdot \dfrac{3xy^2}{2} = \dfrac{6x^3 y^3}{6xy} = x^2 y^2$

31. $\dfrac{8x^2 y^2}{4x^2} \cdot \dfrac{2xy}{2y} = \dfrac{16x^3 y^3}{8x^2 y} = 2xy^2$

33. $\dfrac{-2xy}{x^2} \cdot \dfrac{3xy}{2} = -\dfrac{6x^2 y^2}{2x^2} = -3y^2$

35. $\dfrac{ab^2}{a^2 b} \cdot \dfrac{b^2 c^2}{abc} \cdot \dfrac{abc^2}{a^3 c^2} = \dfrac{a^2 b^5 c^4}{a^6 b^2 c^3} = \dfrac{b^3 c}{a^4}$

37. $\dfrac{10r^2 st^3}{6rs^2} \cdot \dfrac{3r^3 t}{2rst} \cdot \dfrac{2s^3 t^4}{5s^2 t^3} = \dfrac{60r^5 s^4 t^8}{60r^2 s^5 t^4} = \dfrac{r^3 t^4}{s}$

39. $\dfrac{z+7}{7} \cdot \dfrac{z+2}{z} = \dfrac{(z+7)(z+2)}{7z}$

41. $\dfrac{x-2}{2} \cdot \dfrac{2x}{x-2} = \dfrac{2x(x-2)}{2(x-2)} = x$

43. $\dfrac{x+5}{5} \cdot \dfrac{x}{x+5} = \dfrac{x(x+5)}{5(x+5)} = \dfrac{x}{5}$

45. $\dfrac{(x+1)^2}{x+1} \cdot \dfrac{x+2}{x+1} = \dfrac{(x+2)(x+1)^2}{(x+1)^2}$
$$= x+2$$

47. $\dfrac{2x+6}{x+3} \cdot \dfrac{3}{4x} = \dfrac{2(x+3)}{x+3} \cdot \dfrac{3}{4x}$
$$= \dfrac{6(x+3)}{4x(x+3)} = \dfrac{3}{2x}$$

49. $\dfrac{x^2 - x}{x} \cdot \dfrac{3x-6}{3x-3} = \dfrac{x(x-1)}{x} \cdot \dfrac{3(x-2)}{3(x-1)} = \dfrac{3x(x-1)(x-2)}{3x(x-1)} = x-2$

51. $\dfrac{7y-14}{y-2} \cdot \dfrac{x^2}{7x} = \dfrac{7(y-2)}{y-2} \cdot \dfrac{x^2}{7x} = \dfrac{7x^2(y-2)}{7x(y-2)} = x$

53. $\dfrac{x^2 + x - 6}{5x} \cdot \dfrac{5x-10}{x+3} = \dfrac{(x+3)(x-2)}{5x} \cdot \dfrac{5(x-2)}{x+3} = \dfrac{5(x+3)(x-2)^2}{5x(x+3)} = \dfrac{(x-2)^2}{x}$

55. $\dfrac{m^2 - 2m - 3}{2m+4} \cdot \dfrac{m^2 - 4}{m^2 + 3m + 2} = \dfrac{(m-3)(m+1)}{2(m+2)} \cdot \dfrac{(m+2)(m-2)}{(m+2)(m+1)}$
$$= \dfrac{(m-3)(m+1)(m+2)(m-2)}{2(m+2)(m+2)(m+1)} = \dfrac{(m-3)(m-2)}{2(m+2)}$$

57. $\dfrac{x^2 + 7xy + 12y^2}{x^2 + 2xy - 8y^2} \cdot \dfrac{x^2 - xy - 2y^2}{x^2 + 4xy + 3y^2} = \dfrac{(x+4y)(x+3y)}{(x+4y)(x-2y)} \cdot \dfrac{(x-2y)(x+y)}{(x+3y)(x+y)}$

$$= \dfrac{(x+4y)(x+3y)(x-2y)(x+y)}{(x+4y)(x-2y)(x+3y)(x+y)} = 1$$

59. $\dfrac{ax + bx + ay + by}{x^3 - y^3} \cdot \dfrac{x^2 + xy + y^2}{ax + bx} = \dfrac{x(a+b) + y(a+b)}{(x-y)(x^2+xy+y^2)} \cdot \dfrac{x^2+xy+y^2}{x(a+b)}$

$$= \dfrac{(a+b)(x+y)}{(x-y)(x^2+xy+y^2)} \cdot \dfrac{x^2+xy+y^2}{x(a+b)}$$

$$= \dfrac{(a+b)(x+y)(x^2+xy+y^2)}{(x-y)(x^2+xy+y^2) \cdot x(a+b)} = \dfrac{x+y}{x(x-y)}$$

61. $\dfrac{x^2 - y^2}{y^2 - xy} \cdot \dfrac{yx^3 - y^4}{ax + ay + bx + by} = \dfrac{(x+y)(x-y)}{y(y-x)} \cdot \dfrac{y(x^3 - y^3)}{a(x+y) + b(x+y)}$

$$= -\dfrac{(x+y)}{y} \cdot \dfrac{y(x-y)(x^2+xy+y^2)}{(x+y)(a+b)}$$

$$= -\dfrac{(x+y) \cdot y(x-y)(x^2+xy+y^2)}{y(x+y)(a+b)}$$

$$= -\dfrac{(x-y)(x^2+xy+y^2)}{a+b}$$

63. $\dfrac{abc^2}{a+1} \cdot \dfrac{c}{a^2b^2} \cdot \dfrac{a^2 + a}{ac} = \dfrac{abc^2}{a+1} \cdot \dfrac{c}{a^2b^2} \cdot \dfrac{a(a+1)}{ac} = \dfrac{a^2bc^3(a+1)}{a^3b^2c(a+1)} = \dfrac{c^2}{ab}$

65. $\dfrac{3x^2 + 5x + 2}{x^2 - 9} \cdot \dfrac{x - 3}{x^2 - 4} \cdot \dfrac{x^2 + 5x + 6}{6x + 4} = \dfrac{(3x+2)(x+1)}{(x+3)(x-3)} \cdot \dfrac{x-3}{(x+2)(x-2)} \cdot \dfrac{(x+2)(x+3)}{2(3x+2)}$

$$= \dfrac{(3x+2)(x+1)(x-3)(x+2)(x+3)}{2(x+3)(x-3)(x+2)(x-2)(3x+2)} = \dfrac{x+1}{2(x-2)}$$

67. $\dfrac{1}{3} \div \dfrac{1}{2} = \dfrac{1}{3} \cdot \dfrac{2}{1} = \dfrac{2}{3}$

69. $\dfrac{21}{14} \div \dfrac{5}{2} = \dfrac{21}{14} \cdot \dfrac{2}{5} = \dfrac{3 \cdot 7 \cdot 2}{2 \cdot 7 \cdot 5} = \dfrac{3}{5}$

71. $\dfrac{2}{y} \div \dfrac{4}{3} = \dfrac{2}{y} \cdot \dfrac{3}{4} = \dfrac{2 \cdot 3}{y \cdot 2 \cdot 2} = \dfrac{3}{2y}$

73. $\dfrac{3x}{2} \div \dfrac{x}{2} = \dfrac{3x}{2} \cdot \dfrac{2}{x} = \dfrac{3 \cdot x \cdot 2}{2 \cdot x} = 3$

75. $\dfrac{3x}{y} \div \dfrac{2x}{4} = \dfrac{3x}{y} \cdot \dfrac{4}{2x} = \dfrac{3 \cdot x \cdot 2 \cdot 2}{y \cdot 2 \cdot x} = \dfrac{6}{y}$

77. $\dfrac{4x}{3x} \div \dfrac{2y}{9y} = \dfrac{4x}{3x} \cdot \dfrac{9y}{2y} = \dfrac{2 \cdot 2 \cdot x \cdot 3 \cdot 3 \cdot y}{3 \cdot x \cdot 2 \cdot y}$

$$= 6$$

79. $\dfrac{x^2}{3} \div \dfrac{2x}{4} = \dfrac{x^2}{3} \cdot \dfrac{4}{2x} = \dfrac{x \cdot x \cdot 2 \cdot 2}{3 \cdot 2 \cdot x} = \dfrac{2x}{3}$

81. $\dfrac{x^2y}{3xy} \div \dfrac{xy^2}{6y} = \dfrac{x^2y}{3xy} \cdot \dfrac{6y}{xy^2} = \dfrac{6x^2y^2}{3x^2y^3} = \dfrac{2}{y}$

83. $\dfrac{x+2}{3x} \div \dfrac{x+2}{2} = \dfrac{x+2}{3x} \cdot \dfrac{2}{x+2}$

$\qquad\qquad = \dfrac{2(x+2)}{3x(x+2)} = \dfrac{2}{3x}$

85. $\dfrac{(z-2)^2}{3z^2} \div \dfrac{z-2}{6z} = \dfrac{(z-2)^2}{3z^2} \cdot \dfrac{6z}{z-2}$

$\qquad\qquad = \dfrac{6z(z-2)^2}{3z^2(z-2)} = \dfrac{2(z-2)}{z}$

87. $\dfrac{(z-7)^2}{z+2} \div \dfrac{z(z-7)}{5z^2} = \dfrac{(z-7)^2}{z+2} \cdot \dfrac{5z^2}{z(z-7)} = \dfrac{5z^2(z-7)^2}{z(z+2)(z-7)} = \dfrac{5z(z-7)}{z+2}$

89. $\dfrac{x^2-4}{3x+6} \div \dfrac{x-2}{x+2} = \dfrac{x^2-4}{3x+6} \cdot \dfrac{x+2}{x-2} = \dfrac{(x+2)(x-2)(x+2)}{3(x+2)(x-2)} = \dfrac{x+2}{3}$

91. $\dfrac{x^2-1}{3x-3} \div \dfrac{x+1}{3} = \dfrac{x^2-1}{3x-3} \cdot \dfrac{3}{x+1} = \dfrac{(x+1)(x-1)3}{3(x-1)(x+1)} = 1$

93. $\dfrac{5x^2+13x-6}{x+3} \div \dfrac{5x^2-17x+6}{x-2} = \dfrac{5x^2+13x-6}{x+3} \cdot \dfrac{x-2}{5x^2-17x+6}$

$\qquad\qquad = \dfrac{(5x-2)(x+3)(x-2)}{(x+3)(5x-2)(x-3)} = \dfrac{x-2}{x-3}$

95. $\dfrac{2x^2+8x-42}{x-3} \div \dfrac{2x^2+14x}{x^2+5x} = \dfrac{2(x^2+4x-21)}{x-3} \cdot \dfrac{x^2+5x}{2x^2+14x}$

$\qquad\qquad = \dfrac{2(x+7)(x-3)(x)(x+5)}{(x-3)(2x)(x+7)} = x+5$

97. $\dfrac{ab+4a+2b+8}{b^2+4b+16} \div \dfrac{b^2-16}{b^3-64} = \dfrac{ab+4a+2b+8}{b^2+4b+16} \cdot \dfrac{b^3-64}{b^2-16}$

$\qquad\qquad = \dfrac{a(b+4)+2(b+4)}{b^2+4b+16} \cdot \dfrac{b^3-4^3}{(b+4)(b-4)}$

$\qquad\qquad = \dfrac{(b+4)(a+2)}{b^2+4b+16} \cdot \dfrac{(b-4)(b^2+4b+16)}{(b+4)(b-4)}$

$\qquad\qquad = \dfrac{(b+4)(a+2)(b-4)(b^2+4b+16)}{(b^2+4b+16)(b+4)(b-4)} = a+2$

99. $\dfrac{p^3-p^2q+pq^2}{mp-mq+np-nq} \div \dfrac{q^3+p^3}{q^2-p^2} = \dfrac{p^3-p^2q+pq^2}{mp-mq+np-nq} \cdot \dfrac{q^2-p^2}{q^3+p^3}$

$\qquad\qquad = \dfrac{p(p^2-pq+q^2)}{m(p-q)+n(p-q)} \cdot \dfrac{(q+p)(q-p)}{(q+p)(q^2-pq+p^2)}$

$\qquad\qquad = \dfrac{p(p^2-pq+q^2)}{(p-q)(m+n)} \cdot \dfrac{(q+p)(q-p)}{(q+p)(q^2-pq+p^2)}$

$\qquad\qquad = \dfrac{p(p^2-pq+q^2)(q+p)(q-p)}{(p-q)(m+n)(q+p)(q^2-pq+p^2)} = -\dfrac{p}{m+n}$

101. $\dfrac{x}{3} \cdot \dfrac{9}{4} \div \dfrac{x^2}{6} = \dfrac{x}{3} \cdot \dfrac{9}{4} \cdot \dfrac{6}{x^2}$

$\qquad = \dfrac{x \cdot 3 \cdot 3 \cdot 2 \cdot 3}{3 \cdot 2 \cdot 2 \cdot x \cdot x} = \dfrac{9}{2x}$

103. $\dfrac{x^2}{18} \div \dfrac{x^3}{6} \div \dfrac{12}{x^2} = \dfrac{x^2}{18} \cdot \dfrac{6}{x^3} \cdot \dfrac{x^2}{12}$

$\qquad = \dfrac{6x^4}{216x^3} = \dfrac{x}{36}$

105. $\dfrac{x^2 - 1}{x^2 - 9} \cdot \dfrac{x + 3}{x + 2} \div \dfrac{5}{x + 2} = \dfrac{x^2 - 1}{x^2 - 9} \cdot \dfrac{x + 3}{x + 2} \cdot \dfrac{x + 2}{5} = \dfrac{(x + 1)(x - 1)(x + 3)(x + 2)}{5(x + 3)(x - 3)(x + 2)}$

$\qquad\qquad = \dfrac{(x + 1)(x - 1)}{5(x - 3)}$

107. $\dfrac{x^2 - 4}{2x + 6} \div \dfrac{x + 2}{4} \cdot \dfrac{x + 3}{x - 2} = \dfrac{x^2 - 4}{2x + 6} \cdot \dfrac{4}{x + 2} \cdot \dfrac{x + 3}{x - 2} = \dfrac{4(x + 2)(x - 2)(x + 3)}{2(x + 3)(x + 2)(x - 2)} = 2$

109. $\dfrac{x - x^2}{x^2 - 4}\left(\dfrac{2x + 4}{x + 2} \div \dfrac{5}{x + 2} \right) = \dfrac{x - x^2}{x^2 - 4}\left(\dfrac{2x + 4}{x + 2} \cdot \dfrac{x + 2}{5} \right) = \dfrac{x - x^2}{x^2 - 4}\left(\dfrac{2(x + 2)(x + 2)}{5(x + 2)} \right)$

$\qquad\qquad = \dfrac{x(1 - x)}{(x + 2)(x - 2)} \cdot \dfrac{2(x + 2)}{5}$

$\qquad\qquad = \dfrac{2x(1 - x)}{5(x - 2)}$

111. $\dfrac{y^2}{x + 1} \cdot \dfrac{x^2 + 2x + 1}{x^2 - 1} \div \dfrac{3y}{xy - y} = \dfrac{y^2}{x + 1} \cdot \dfrac{(x + 1)(x + 1)}{(x + 1)(x - 1)} \cdot \dfrac{xy - y}{3y} = \dfrac{y^2(x + 1)(x + 1)(y)(x - 1)}{(x + 1)(x + 1)(x - 1)(3y)}$

$\qquad\qquad = \dfrac{y^2}{3}$

113. $\dfrac{x^2 + x - 6}{x^2 - 4} \cdot \dfrac{x^2 + 2x}{x - 2} \div \dfrac{x^2 + 3x}{x + 2} = \dfrac{x^2 + x - 6}{x^2 - 4} \cdot \dfrac{x^2 + 2x}{x - 2} \cdot \dfrac{x + 2}{x^2 + 3x}$

$\qquad\qquad = \dfrac{(x + 3)(x - 2)(x)(x + 2)(x + 2)}{(x + 2)(x - 2)(x - 2)(x)(x + 3)} = \dfrac{x + 2}{x - 2}$

115. Answers may vary.

117. Answers may vary.

119. You always get the original value of x after simplifying.

Exercise 6.5 (page 390)

1. $49 = 7 \cdot 7 = 7^2$

3. $136 = 4 \cdot 34 = 2 \cdot 2 \cdot 2 \cdot 17 = 2^3 \cdot 17$

5. $102 = 2 \cdot 51 = 2 \cdot 3 \cdot 17$

7. $144 = 16 \cdot 9 = 2 \cdot 2 \cdot 2 \cdot 2 \cdot 3 \cdot 3 = 2^4 \cdot 3^2$

9. LCD

11. numerators; common denominator

13. $\dfrac{1}{3} + \dfrac{1}{3} = \dfrac{1 + 1}{3} = \dfrac{2}{3}$

15. $\dfrac{2}{9} + \dfrac{1}{9} = \dfrac{2 + 1}{9} = \dfrac{3}{9} = \dfrac{1}{3}$

17. $\dfrac{2x}{y} + \dfrac{2x}{y} = \dfrac{2x + 2x}{y} = \dfrac{4x}{y}$

19. $\dfrac{4}{7y} + \dfrac{10}{7y} = \dfrac{4 + 10}{7y} = \dfrac{14}{7y} = \dfrac{2}{y}$

21. $\dfrac{y+2}{5z} + \dfrac{y+4}{5z} = \dfrac{y+2+y+4}{5z} = \dfrac{2y+6}{5z}$

23. $\dfrac{3x-5}{x-2} + \dfrac{6x-13}{x-2} = \dfrac{3x-5+6x-13}{x-2} = \dfrac{9x-18}{x-2} = \dfrac{9(x-2)}{x-2} = 9$

25. $\dfrac{5}{7} - \dfrac{4}{7} = \dfrac{5-4}{7} = \dfrac{1}{7}$

27. $\dfrac{35}{72} - \dfrac{44}{72} = \dfrac{35-44}{72} = \dfrac{-9}{72} = -\dfrac{1}{8}$

29. $\dfrac{2x}{y} - \dfrac{x}{y} = \dfrac{2x-x}{y} = \dfrac{x}{y}$

31. $\dfrac{9y}{3x} - \dfrac{6y}{3x} = \dfrac{9y-6y}{3x} = \dfrac{3y}{3x} = \dfrac{y}{x}$

33. $\dfrac{6x-5}{3xy} - \dfrac{3x-5}{3xy} = \dfrac{6x-5-(3x-5)}{3xy} = \dfrac{6x-5-3x+5}{3xy} = \dfrac{3x}{3xy} = \dfrac{1}{y}$

35. $\dfrac{3y-2}{y+3} - \dfrac{2y-5}{y+3} = \dfrac{3y-2-(2y-5)}{y+3} = \dfrac{3y-2-2y+5}{y+3} = \dfrac{y+3}{y+3} = 1$

37. $\dfrac{13x}{15} + \dfrac{12x}{15} - \dfrac{5x}{15} = \dfrac{13x+12x-5x}{15} = \dfrac{20x}{15} = \dfrac{4x}{3}$

39. $\dfrac{x}{3y} + \dfrac{2x}{3y} - \dfrac{x}{3y} = \dfrac{x+2x-x}{3y} = \dfrac{2x}{3y}$

41. $\dfrac{3x}{y+2} - \dfrac{3y}{y+2} + \dfrac{x+y}{y+2} = \dfrac{3x-3y+x+y}{y+2} = \dfrac{4x-2y}{y+2} = \dfrac{2(2x-y)}{y+2}$

43. $\dfrac{x+1}{x-2} - \dfrac{2(x-3)}{x-2} + \dfrac{3(x+1)}{x-2} = \dfrac{x+1-2(x-3)+3(x+1)}{x-2} = \dfrac{x+1-2x+6+3x+3}{x-2}$
$= \dfrac{2x+10}{x-2} = \dfrac{2(x+5)}{x-2}$

45. $\dfrac{25}{4} = \dfrac{25 \cdot 5}{4 \cdot 5} = \dfrac{125}{20}$

47. $\dfrac{8}{x} = \dfrac{8 \cdot xy}{x \cdot xy} = \dfrac{8xy}{x^2 y}$

49. $\dfrac{3x}{x+1} = \dfrac{3x(x+1)}{(x+1)(x+1)} = \dfrac{3x(x+1)}{(x+1)^2}$

51. $\dfrac{2y}{x} = \dfrac{2y(x+1)}{x(x+1)} = \dfrac{2y(x+1)}{x^2+x}$

53. $\dfrac{z}{z-1} = \dfrac{z(z+1)}{(z-1)(z+1)} = \dfrac{z(z+1)}{z^2-1}$

55. $\dfrac{2}{x+1} = \dfrac{2(x+2)}{(x+1)(x+2)} = \dfrac{2(x+2)}{x^2+3x+2}$

57. $2x = 2 \cdot x$
$6x = 2 \cdot 3 \cdot x$
$\text{LCD} = 2 \cdot 3 \cdot x = 6x$

59. $3x = 3 \cdot x$
$6y = 2 \cdot 3 \cdot y$
$9xy = 3^2 \cdot x \cdot y$
$\text{LCD} = 2 \cdot 3^2 \cdot x \cdot y = 18xy$

61. $x^2 - 1 = (x+1)(x-1)$
   $x + 1 = x + 1$
   $\text{LCD} = (x+1)(x-1) = x^2 - 1$

63. $x^2 + 6x = x(x+6)$
   $x + 6 = x + 6$
   $x = x$
   $\text{LCD} = x(x+6) = x^2 + 6x$

65. $x^2 - 4x - 5 = (x+1)(x-5)$
   $x^2 - 25 = (x+5)(x-5)$
   $\text{LCD} = (x+1)(x-5)(x+5)$

67. $\dfrac{1}{2} + \dfrac{2}{3} = \dfrac{1 \cdot 3}{2 \cdot 3} + \dfrac{2 \cdot 2}{3 \cdot 2} = \dfrac{3}{6} + \dfrac{4}{6} = \dfrac{7}{6}$

69. $\dfrac{2y}{9} + \dfrac{y}{3} = \dfrac{2y}{9} + \dfrac{y \cdot 3}{3 \cdot 3} = \dfrac{2y}{9} + \dfrac{3y}{9} = \dfrac{5y}{9}$

71. $\dfrac{21x}{14} - \dfrac{5x}{21} = \dfrac{21x \cdot 3}{14 \cdot 3} - \dfrac{5x \cdot 2}{21 \cdot 2}$
   $= \dfrac{63x}{42} - \dfrac{10x}{42} = \dfrac{53x}{42}$

73. $\dfrac{4x}{3} + \dfrac{2x}{y} = \dfrac{4x \cdot y}{3 \cdot y} + \dfrac{2x \cdot 3}{y \cdot 3} = \dfrac{4xy}{3y} + \dfrac{6x}{3y} = \dfrac{4xy + 6x}{3y}$

75. $\dfrac{2}{x} - 3x = \dfrac{2}{x} - \dfrac{3x}{1} = \dfrac{2}{x} - \dfrac{3x \cdot x}{1 \cdot x} = \dfrac{2}{x} - \dfrac{3x^2}{x} = \dfrac{2 - 3x^2}{x}$

77. $\dfrac{y+2}{5y} + \dfrac{y+4}{15y} = \dfrac{(y+2)3}{5y \cdot 3} + \dfrac{y+4}{15y} = \dfrac{3y+6}{15y} + \dfrac{y+4}{15y} = \dfrac{4y+10}{15y}$

79. $\dfrac{x+5}{xy} - \dfrac{x-1}{x^2 y} = \dfrac{(x+5)x}{xy \cdot x} - \dfrac{x-1}{x^2 y} = \dfrac{x^2 + 5x}{x^2 y} - \dfrac{x-1}{x^2 y} = \dfrac{x^2 + 4x + 1}{x^2 y}$

81. $\dfrac{x}{x+1} + \dfrac{x-1}{x} = \dfrac{x \cdot x}{(x+1)x} + \dfrac{(x-1)(x+1)}{x(x+1)} = \dfrac{x^2}{x(x+1)} + \dfrac{x^2 - 1}{x(x+1)} = \dfrac{2x^2 - 1}{x(x+1)}$

83. $\dfrac{x-1}{x} + \dfrac{y+1}{y} = \dfrac{(x-1)y}{x \cdot y} + \dfrac{(y+1)x}{y \cdot x} = \dfrac{xy - y}{xy} + \dfrac{xy + x}{xy} = \dfrac{2xy + x - y}{xy}$

85. $\dfrac{x}{x-2} + \dfrac{4+2x}{x^2 - 4} = \dfrac{x}{x-2} + \dfrac{2(x+2)}{(x+2)(x-2)} = \dfrac{x}{x-2} + \dfrac{2}{x-2} = \dfrac{x+2}{x-2}$

87. $\dfrac{x+1}{x-1} + \dfrac{x-1}{x+1} = \dfrac{(x+1)(x+1)}{(x-1)(x+1)} + \dfrac{(x-1)(x-1)}{(x+1)(x-1)} = \dfrac{x^2 + 2x + 1}{(x+1)(x-1)} + \dfrac{x^2 - 2x + 1}{(x+1)(x-1)}$
   $= \dfrac{2x^2 + 2}{(x+1)(x-1)}$

89. $\dfrac{2x+2}{x-2} - \dfrac{2x}{2-x} = \dfrac{2x+2}{x-2} + \dfrac{2x}{x-2} = \dfrac{4x+2}{x-2} = \dfrac{2(2x+1)}{x-2}$

91. $\dfrac{2x}{x^2 - 3x + 2} + \dfrac{2x}{x - 1} - \dfrac{x}{x - 2} = \dfrac{2x}{(x - 2)(x - 1)} + \dfrac{2x}{x - 1} - \dfrac{x}{x - 2}$

$$= \dfrac{2x}{(x - 2)(x - 1)} + \dfrac{2x(x - 2)}{(x - 1)(x - 2)} - \dfrac{x(x - 1)}{(x - 2)(x - 1)}$$

$$= \dfrac{2x + 2x(x - 2) - x(x - 1)}{(x - 2)(x - 1)}$$

$$= \dfrac{2x + 2x^2 - 4x - x^2 + x}{(x - 2)(x - 1)}$$

$$= \dfrac{x^2 - x}{(x - 2)(x - 1)} = \dfrac{x(x - 1)}{(x - 2)(x - 1)} = \dfrac{x}{x - 2}$$

93. $\dfrac{2x}{x - 1} + \dfrac{3x}{x + 1} - \dfrac{x + 3}{x^2 - 1} = \dfrac{2x}{x - 1} + \dfrac{3x}{x + 1} - \dfrac{x + 3}{(x + 1)(x - 1)}$

$$= \dfrac{2x(x + 1)}{(x - 1)(x + 1)} + \dfrac{3x(x - 1)}{(x + 1)(x - 1)} - \dfrac{x + 3}{(x + 1)(x - 1)}$$

$$= \dfrac{2x(x + 1) + 3x(x - 1) - (x + 3)}{(x + 1)(x - 1)}$$

$$= \dfrac{2x^2 + 2x + 3x^2 - 3x - x - 3}{(x + 1)(x - 1)}$$

$$= \dfrac{5x^2 - 2x - 3}{(x + 1)(x - 1)} = \dfrac{(5x + 3)(x - 1)}{(x + 1)(x - 1)} = \dfrac{5x + 3}{x + 1}$$

95. $\dfrac{x + 1}{2x + 4} - \dfrac{x^2}{2x^2 - 8} = \dfrac{x + 1}{2(x + 2)} - \dfrac{x^2}{2(x + 2)(x - 2)} = \dfrac{(x + 1)(x - 2)}{2(x + 2)(x - 2)} - \dfrac{x^2}{2(x + 2)(x - 2)}$

$$= \dfrac{x^2 - 2x + x - 2 - x^2}{2(x + 2)(x - 2)}$$

$$= \dfrac{-x - 2}{2(x + 2)(x - 2)}$$

$$= \dfrac{-(x + 2)}{2(x + 2)(x - 2)} = -\dfrac{1}{2(x - 2)}$$

97. Answers may vary.

99. Answers may vary.

101. The subtraction needs to be distributed in the numerator of the 2nd fraction.

103. $\dfrac{a}{b} + \dfrac{c}{d} = \dfrac{a(d)}{b(d)} + \dfrac{c(b)}{d(b)}$

$$= \dfrac{ad}{bd} + \dfrac{bc}{bd} = \dfrac{ad + bc}{bd}$$

Exercise 6.6 (page 397)

1. $t^3 t^4 t^2 = t^{3+4+2} = t^9$

3. $-2r(r^3)^2 = -2rr^6 = -2r^7$

5. $\left(\dfrac{3r}{4r^3}\right)^4 = \left(\dfrac{3}{4r^2}\right)^4 = \dfrac{3^4}{(4r^2)^4} = \dfrac{81}{256r^8}$

7. $\left(\dfrac{6r^{-2}}{2r^3}\right)^{-2} = \left(3r^{-5}\right)^{-2} = \dfrac{r^{10}}{9}$

9. complex fraction

11. single; divide

13. $\dfrac{\frac{2}{3}}{\frac{3}{4}} = \dfrac{2}{3} \div \dfrac{3}{4} = \dfrac{2}{3} \cdot \dfrac{4}{3} = \dfrac{8}{9}$

15. $\dfrac{\frac{4}{5}}{\frac{32}{15}} = \dfrac{4}{5} \div \dfrac{32}{15} = \dfrac{4}{5} \cdot \dfrac{15}{32} = \dfrac{3}{8}$

17. $\dfrac{\frac{2}{3}+1}{\frac{1}{3}+1} = \dfrac{\left(\frac{2}{3}+1\right)3}{\left(\frac{1}{3}+1\right)3} = \dfrac{\frac{2}{3}\cdot 3 + 1\cdot 3}{\frac{1}{3}\cdot 3 + 1 \cdot 3}$
$= \dfrac{2+3}{1+3} = \dfrac{5}{4}$

19. $\dfrac{\frac{1}{2}+\frac{3}{4}}{\frac{3}{2}+\frac{1}{4}} = \dfrac{\left(\frac{1}{2}+\frac{3}{4}\right)4}{\left(\frac{3}{2}+\frac{1}{4}\right)4} = \dfrac{\frac{1}{2}\cdot 4 + \frac{3}{4}\cdot 4}{\frac{3}{2}\cdot 4 + \frac{1}{4}\cdot 4}$
$= \dfrac{2+3}{6+1} = \dfrac{5}{7}$

21. $\dfrac{\frac{x}{y}}{\frac{1}{x}} = \dfrac{x}{y} \div \dfrac{1}{x} = \dfrac{x}{y} \cdot \dfrac{x}{1} = \dfrac{x^2}{y}$

23. $\dfrac{\frac{5t^2}{9x^2}}{\frac{3t}{x^2 t}} = \dfrac{5t^2}{9x^2} \div \dfrac{3t}{x^2 t} = \dfrac{5t^2}{9x^2} \cdot \dfrac{x^2 t}{3t}$
$= \dfrac{5x^2 t^3}{27 x^2 t} = \dfrac{5t^2}{27}$

25. $\dfrac{\frac{1}{x}-3}{\frac{5}{x}+2} = \dfrac{\left(\frac{1}{x}-3\right)x}{\left(\frac{5}{x}+2\right)x} = \dfrac{\frac{1}{x}\cdot x - 3\cdot x}{\frac{5}{x}\cdot x + 2\cdot x} = \dfrac{1-3x}{5+2x}$

27. $\dfrac{\frac{2}{x}+2}{\frac{4}{x}+2} = \dfrac{\left(\frac{2}{x}+2\right)x}{\left(\frac{4}{x}+2\right)x} = \dfrac{\frac{2}{x}\cdot x + 2\cdot x}{\frac{4}{x}\cdot x + 2\cdot x} = \dfrac{2+2x}{4+2x} = \dfrac{2(1+x)}{2(2+x)} = \dfrac{1+x}{2+x}$

29. $\dfrac{\frac{3y}{x}-y}{y-\frac{y}{x}} = \dfrac{\left(\frac{3y}{x}-y\right)x}{\left(y-\frac{y}{x}\right)x} = \dfrac{\frac{3y}{x}\cdot x - y\cdot x}{y\cdot x - \frac{y}{x}\cdot x} = \dfrac{3y-xy}{xy-y} = \dfrac{y(3-x)}{y(x-1)} = \dfrac{3-x}{x-1}$

31. $\dfrac{\frac{1}{x+1}}{1+\frac{1}{x+1}} = \dfrac{\left(\frac{1}{x+1}\right)(x+1)}{\left(1+\frac{1}{x+1}\right)(x+1)} = \dfrac{\frac{1}{x+1}(x+1)}{1(x+1)+\frac{1}{x+1}(x+1)} = \dfrac{1}{x+1+1} = \dfrac{1}{x+2}$

33. $\dfrac{\frac{x}{x+2}}{\frac{x}{x+2}+x} = \dfrac{\left(\frac{x}{x+2}\right)(x+2)}{\left(\frac{x}{x+2}+x\right)(x+2)} = \dfrac{\frac{x}{x+2}(x+2)}{\frac{x}{x+2}(x+2)+x(x+2)} = \dfrac{x}{x+x^2+2x} = \dfrac{x}{x^2+3x}$
$= \dfrac{x}{x(x+3)} = \dfrac{1}{x+3}$

35. $\dfrac{1}{\frac{1}{x}+\frac{1}{y}} = \dfrac{1(xy)}{\left(\frac{1}{x}+\frac{1}{y}\right)xy} = \dfrac{xy}{\frac{1}{x}\cdot xy + \frac{1}{y}\cdot xy} = \dfrac{xy}{y+x}$

37. $\dfrac{\frac{2}{x}}{\frac{2}{y}-\frac{4}{x}} = \dfrac{\frac{2}{x}(xy)}{\left(\frac{2}{y}-\frac{4}{x}\right)(xy)} = \dfrac{\frac{2}{x}(xy)}{\frac{2}{y}(xy)-\frac{4}{x}(xy)} = \dfrac{2y}{2x-4y} = \dfrac{2y}{2(x-2y)} = \dfrac{y}{x-2y}$

39. $\dfrac{3 + \frac{3}{x-1}}{3 - \frac{3}{x}} = \dfrac{\left(3 + \frac{3}{x-1}\right)(x)(x-1)}{\left(3 - \frac{3}{x}\right)(x)(x-1)} = \dfrac{3x(x-1) + \frac{3}{x-1}(x)(x-1)}{3x(x-1) - \frac{3}{x}(x)(x-1)} = \dfrac{3x^2 - 3x + 3x}{3x^2 - 3x - 3(x-1)}$

$$= \dfrac{3x^2}{3x^2 - 6x + 3}$$

$$= \dfrac{3x^2}{3(x^2 - 2x + 1)}$$

$$= \dfrac{x^2}{x^2 - 2x + 1} = \dfrac{x^2}{(x-1)^2}$$

41. $\dfrac{\frac{3}{x} + \frac{4}{x+1}}{\frac{2}{x+1} - \frac{3}{x}} = \dfrac{\left(\frac{3}{x} + \frac{4}{x+1}\right)(x)(x+1)}{\left(\frac{2}{x+1} - \frac{3}{x}\right)(x)(x+1)} = \dfrac{\frac{3}{x}(x)(x+1) + \frac{4}{x+1}(x)(x+1)}{\frac{2}{x+1}(x)(x+1) - \frac{3}{x}(x)(x+1)}$

$$= \dfrac{3(x+1) + 4x}{2x - 3(x+1)} = \dfrac{3x + 3 + 4x}{2x - 3x - 3} = \dfrac{7x + 3}{-x - 3}$$

43. $\dfrac{\frac{2}{x} - \frac{3}{x+1}}{\frac{2}{x+1} - \frac{3}{x}} = \dfrac{\left(\frac{2}{x} - \frac{3}{x+1}\right)(x)(x+1)}{\left(\frac{2}{x+1} - \frac{3}{x}\right)(x)(x+1)} = \dfrac{2(x+1) - 3x}{2x - 3(x+1)}$

$$= \dfrac{2x + 2 - 3x}{2x - 3x - 3} = \dfrac{-x + 2}{-x - 3} = \dfrac{-(x-2)}{-(x+3)} = \dfrac{x-2}{x+3}$$

45. $\dfrac{\frac{1}{y^2+y} - \frac{1}{xy+x}}{\frac{1}{xy+x} - \frac{1}{y^2+y}} = \dfrac{\frac{1}{y(y+1)} - \frac{1}{x(y+1)}}{\frac{1}{x(y+1)} - \frac{1}{y(y+1)}} = \dfrac{\left(\frac{1}{y(y+1)} - \frac{1}{x(y+1)}\right)(x)(y)(y+1)}{\left(\frac{1}{x(y+1)} - \frac{1}{y(y+1)}\right)(x)(y)(y+1)}$

$$= \dfrac{1x - 1y}{1y - 1x} = \dfrac{x - y}{-(x - y)} = -1$$

47. $\dfrac{x^{-2}}{y^{-1}} = \dfrac{\frac{1}{x^2}}{\frac{1}{y}} = \dfrac{1}{x^2} \div \dfrac{1}{y} = \dfrac{1}{x^2} \cdot \dfrac{y}{1} = \dfrac{y}{x^2}$

49. $\dfrac{1 + x^{-1}}{x^{-1} - 1} = \dfrac{1 + \frac{1}{x}}{\frac{1}{x} - 1} = \dfrac{\left(1 + \frac{1}{x}\right)x}{\left(\frac{1}{x} - 1\right)x}$

$$= \dfrac{1x + \frac{1}{x} \cdot x}{\frac{1}{x} \cdot x - 1x} = \dfrac{x + 1}{1 - x}$$

51. $\dfrac{a^{-2} + a}{a + 1} = \dfrac{\frac{1}{a^2} + a}{a + 1} = \dfrac{\left(\frac{1}{a^2} + a\right)a^2}{(a+1)a^2} = \dfrac{1 + a^3}{a^3 + a^2} = \dfrac{(1+a)(1-a+a^2)}{a^2(a+1)} = \dfrac{a^2 - a + 1}{a^2}$

53. $\dfrac{2x^{-1} + 4x^{-2}}{2x^{-2} + x^{-1}} = \dfrac{\frac{2}{x} + \frac{4}{x^2}}{\frac{2}{x^2} + \frac{1}{x}} = \dfrac{\left(\frac{2}{x} + \frac{4}{x^2}\right)x^2}{\left(\frac{2}{x^2} + \frac{1}{x}\right)x^2} = \dfrac{2x + 4}{2 + x} = \dfrac{2(x+2)}{x + 2} = 2$

55. $\dfrac{1 - 25y^{-2}}{1 + 10y^{-1} + 25y^{-2}} = \dfrac{1 - \frac{25}{y^2}}{1 + \frac{10}{y} + \frac{25}{y^2}} = \dfrac{\left(1 - \frac{25}{y^2}\right)y^2}{\left(1 + \frac{10}{y} + \frac{25}{y^2}\right)y^2} = \dfrac{y^2 - 25}{y^2 + 10y + 25}$

$$= \dfrac{(y+5)(y-5)}{(y+5)(y+5)} = \dfrac{y-5}{y+5}$$

57. **Answers may vary.**

59. $\dfrac{1}{1+1} = \dfrac{1}{2}; \dfrac{1}{1+\frac{1}{2}} = \dfrac{1}{\frac{3}{2}} = \dfrac{2}{3}; \dfrac{1}{1+\frac{1}{1+\frac{1}{2}}} = \dfrac{1}{1+\frac{1}{\frac{3}{2}}} = \dfrac{1}{1+\frac{2}{3}} = \dfrac{1}{\frac{5}{3}} = \dfrac{3}{5}$

$\dfrac{1}{1+\frac{1}{1+\frac{1}{1+\frac{1}{2}}}} = \dfrac{1}{1+\frac{1}{1+\frac{1}{\frac{3}{2}}}} = \dfrac{1}{1+\frac{1}{1+\frac{2}{3}}} = \dfrac{1}{1+\frac{1}{\frac{5}{3}}} = \dfrac{1}{1+\frac{3}{5}} = \dfrac{1}{\frac{8}{5}} = \dfrac{5}{8}$

Exercise 6.7 (page 404)

1. $x^2 + 4x = x(x+4)$ **3.** $2x^2 + x - 3 = (2x+3)(x-1)$

5. $x^4 - 16 = \left(x^2\right)^2 - 4^2 = (x^2+4)(x^2-4) = (x^2+4)(x+2)(x-2)$

7. extraneous   **9.** LCD   **11.** xy

13.
$$\dfrac{x}{2} + 4 = \dfrac{3x}{2}$$
$$2\left(\dfrac{x}{2}+4\right) = 2\left(\dfrac{3x}{2}\right)$$
$$x + 8 = 3x$$
$$8 = 2x$$
$$4 = x$$

15.
$$\dfrac{2y}{5} - 8 = \dfrac{4y}{5}$$
$$5\left(\dfrac{2y}{5}-8\right) = 5\left(\dfrac{4y}{5}\right)$$
$$2y - 40 = 4y$$
$$-40 = 2y$$
$$-20 = y$$

17.
$$\dfrac{x}{3} + 1 = \dfrac{x}{2}$$
$$6\left(\dfrac{x}{3}+1\right) = 6\left(\dfrac{x}{2}\right)$$
$$2x + 6 = 3x$$
$$6 = x$$

19.
$$\dfrac{x}{5} - \dfrac{x}{3} = -8$$
$$15\left(\dfrac{x}{5}-\dfrac{x}{3}\right) = 15(-8)$$
$$3x - 5x = -120$$
$$-2x = -120$$
$$x = 60$$

21.
$$\dfrac{3a}{2} + \dfrac{a}{3} = -22$$
$$6\left(\dfrac{3a}{2}+\dfrac{a}{3}\right) = 6(-22)$$
$$9a + 2a = -132$$
$$11a = -132$$
$$a = -12$$

23.
$$\dfrac{x-3}{3} + 2x = -1$$
$$3\left(\dfrac{x-3}{3}+2x\right) = 3(-1)$$
$$x - 3 + 6x = -3$$
$$7x = 0$$
$$x = 0$$

25.
$$\dfrac{z-3}{2} = z + 2$$
$$2\left(\dfrac{z-3}{2}\right) = 2(z+2)$$
$$z - 3 = 2z + 4$$
$$-7 = z$$

27.
$$\dfrac{5(x+1)}{8} = x + 1$$
$$8\left(\dfrac{5(x+1)}{8}\right) = 8(x+1)$$
$$5(x+1) = 8(x+1)$$
$$5x + 5 = 8x + 8$$
$$-3x = 3$$
$$x = -1$$

29.
$$\dfrac{c-4}{4} = \dfrac{c+4}{8}$$
$$8\left(\dfrac{c-4}{4}\right) = 8\left(\dfrac{c+4}{8}\right)$$
$$2(c-4) = c + 4$$
$$2c - 8 = c + 4$$
$$c = 12$$

31.
$$\frac{x+1}{3} + \frac{x-1}{5} = \frac{2}{15}$$
$$15\left(\frac{x+1}{3} + \frac{x-1}{5}\right) = 15\left(\frac{2}{15}\right)$$
$$5(x+1) + 3(x-1) = 2$$
$$5x + 5 + 3x - 3 = 2$$
$$8x = 0$$
$$x = 0$$

33.
$$\frac{3x-1}{6} - \frac{x+3}{2} = \frac{3x+4}{3}$$
$$12\left(\frac{3x-1}{6} - \frac{x+3}{2}\right) = 12\left(\frac{3x+4}{3}\right)$$
$$2(3x-1) - 6(x+3) = 4(3x+4)$$
$$6x - 2 - 6x - 18 = 12x + 16$$
$$-36 = 12x$$
$$-3 = x$$

35.
$$\frac{3}{x} + 2 = 3$$
$$x\left(\frac{3}{x} + 2\right) = x(3)$$
$$3 + 2x = 3x$$
$$3 = x \Rightarrow \text{The answer checks.}$$

37.
$$\frac{5}{a} - \frac{4}{a} = 8 + \frac{1}{a}$$
$$a\left(\frac{5}{a} - \frac{4}{a}\right) = a\left(8 + \frac{1}{a}\right)$$
$$5 - 4 = 8a + 1$$
$$0 = 8a$$
$$0 = a$$
0 does not check. $\Rightarrow$ no solutions

39.
$$\frac{2}{y+1} + 5 = \frac{12}{y+1}$$
$$(y+1)\left(\frac{2}{y+1} + 5\right) = (y+1)\left(\frac{12}{y+1}\right)$$
$$2 + 5(y+1) = 12$$
$$2 + 5y + 5 = 12$$
$$5y = 5$$
$$y = 1$$
The answer checks.

41.
$$\frac{1}{x-1} + \frac{3}{x-1} = 1$$
$$(x-1)\left(\frac{1}{x-1} + \frac{3}{x-1}\right) = (x-1)(1)$$
$$1 + 3 = x - 1$$
$$5 = x$$
The answer checks.

43.
$$\frac{a^2}{a+2} - \frac{4}{a+2} = a$$
$$(a+2)\left(\frac{a^2}{a+2} - \frac{4}{a+2}\right) = (a+2)a$$
$$a^2 - 4 = a^2 + 2a$$
$$-4 = 2a$$
$$-2 = a$$
-2 does not check $\Rightarrow$ no solutions

45.
$$\frac{x}{x-5} - \frac{5}{x-5} = 3$$
$$(x-5)\left(\frac{x}{x-5} - \frac{5}{x-5}\right) = (x-5)3$$
$$x - 5 = 3x - 15$$
$$-2x = -10$$
$$x = 5$$
5 does not check $\Rightarrow$ no solutions

47.
$$\frac{3r}{2} - \frac{3}{r} = \frac{3r}{2} + 3$$
$$2r\left(\frac{3r}{2} - \frac{3}{r}\right) = 2r\left(\frac{3r}{2} + 3\right)$$
$$3r^2 - 6 = 3r^2 + 6r$$
$$-6 = 6r$$
$$-1 = r \Rightarrow \text{The answer checks.}$$

49.
$$\frac{1}{3} + \frac{2}{x-3} = 1$$
$$3(x-3)\left(\frac{1}{3} + \frac{2}{x-3}\right) = 3(x-3)(1)$$
$$x - 3 + 3(2) = 3x - 9$$
$$x - 3 + 6 = 3x - 9$$
$$12 = 2x$$
$$6 = x$$

The answer checks.

51.
$$\frac{u}{u-1} + \frac{1}{u} = \frac{u^2+1}{u^2-u}$$
$$\frac{u}{u-1} + \frac{1}{u} = \frac{u^2+1}{u(u-1)}$$
$$u(u-1)\left(\frac{u}{u-1} + \frac{1}{u}\right) = u(u-1)\left(\frac{u^2+1}{u(u-1)}\right)$$
$$u^2 + u - 1 = u^2 + 1$$
$$u - 1 = 1$$
$$u = 2 \Rightarrow \text{The answer checks.}$$

53.
$$\frac{3}{x-2} + \frac{1}{x} = \frac{2(3x+2)}{x^2-2x}$$
$$\frac{3}{x-2} + \frac{1}{x} = \frac{6x+4}{x(x-2)}$$
$$x(x-2)\left(\frac{3}{x-2} + \frac{1}{x}\right) = x(x-2)\left(\frac{6x+4}{x(x-2)}\right)$$
$$3x + x - 2 = 6x + 4$$
$$4x - 2 = 6x + 4$$
$$-2x = 6$$
$$x = -3 \Rightarrow \text{The answer checks.}$$

55.
$$\frac{7}{q^2-q-2} + \frac{1}{q+1} = \frac{3}{q-2}$$
$$\frac{7}{(q-2)(q+1)} + \frac{1}{q} \qquad \frac{3}{1-2}$$
$$(q-2)(q+1)\left(\frac{7}{(q-2)(q+1)} + \frac{1}{q+1}\right) = (q-2)(q+1)\left(\frac{3}{q-2}\right)$$
$$7 + q - 2 = 3(q+1)$$
$$q + 5 = 3q + 3$$
$$-2q - -2$$
$$q = 1 \Rightarrow \text{The answer checks.}$$

57.

$$\frac{3y}{3y-6} + \frac{8}{y^2-4} = \frac{2y}{2y+4}$$

$$\frac{3y}{3(y-2)} + \frac{8}{(y+2)(y-2)} = \frac{2y}{2(y+2)}$$

$$\frac{y}{y-2} + \frac{8}{(y+2)(y-2)} = \frac{y}{y+2}$$

$$(y+2)(y-2)\left(\frac{y}{y-2} + \frac{8}{(y+2)(y-2)}\right) = (y+2)(y-2)\left(\frac{y}{y+2}\right)$$

$$y(y+2) + 8 = y(y-2)$$

$$y^2 + 2y + 8 = y^2 - 2y$$

$$8 = -4y$$

$$-2 = y: \quad -2 \text{ does not check} \Rightarrow \text{no solutions}$$

59.

$$y + \frac{2}{3} = \frac{2y-12}{3y-9}$$

$$y + \frac{2}{3} = \frac{2(y-6)}{3(y-3)}$$

$$3(y-3)\left(y + \frac{2}{3}\right) = 3(y-3)\left(\frac{2y-12}{3(y-3)}\right)$$

$$3y(y-3) + 2(y-3) = 2y - 12$$

$$3y^2 - 9y + 2y - 6 = 2y - 12$$

$$3y^2 - 9y + 6 = 0$$

$$3(y^2 - 3y + 2) = 0$$

$$3(y-2)(y-1) = 0$$

$$y - 2 = 0 \quad \textbf{or} \quad y - 1 = 0$$

$$y = 2 \qquad\qquad y = 1$$

Both answers check.

61.

$$\frac{5}{4y+12} - \frac{3}{4} = \frac{5}{4y+12} - \frac{y}{4}$$

$$\frac{5}{4(y+3)} - \frac{3}{4} = \frac{5}{4(y+3)} - \frac{y}{4}$$

$$4(y+3)\left(\frac{5}{4(y+3)} - \frac{3}{4}\right) = 4(y+3)\left(\frac{5}{4(y+3)} - \frac{y}{4}\right)$$

$$5 - 3(y+3) = 5 - y(y+3)$$

$$5 - 3y - 9 = 5 - y^2 - 3y$$

$$y^2 - 9 = 0$$

$$(y+3)(y-3) = 0$$

$$y + 3 = 0 \quad \textbf{or} \quad y - 3 = 0 \quad \Rightarrow y = -3 \text{ does not check, so the only solution is } y = 3.$$

$$y = -3 \qquad\qquad y = 3$$

63.

$$\frac{x}{x-1} - \frac{12}{x^2-x} = \frac{-1}{x-1}$$

$$\frac{x}{x-1} - \frac{12}{x(x-1)} = \frac{-1}{x-1}$$

$$x(x-1)\left(\frac{x}{x-1} - \frac{12}{x(x-1)}\right) = x(x-1)\left(\frac{-1}{x-1}\right)$$

$$x^2 - 12 = -x$$

$$x^2 + x - 12 = 0$$

$$(x+4)(x-3) = 0$$

$$x + 4 = 0 \quad \textbf{or} \quad x - 3 = 0 \quad \Rightarrow \text{ Both answers check.}$$
$$x = -4 \qquad\qquad x = 3$$

65.

$$\frac{z-4}{z-3} = \frac{z+2}{z+1}$$

$$(z-3)(z+1)\left(\frac{z-4}{z-3}\right) = (z-3)(z+1)\left(\frac{z+2}{z+1}\right)$$

$$(z+1)(z-4) = (z-3)(z+2)$$

$$z^2 - 3z - 4 = z^2 - z - 6$$

$$-2z = -2$$

$$z = 1 \Rightarrow \text{ The answer checks.}$$

67.

$$\frac{n}{n^2-9} + \frac{n+8}{n+3} = \frac{n-8}{n-3}$$

$$\frac{n}{(n+3)(n-3)} + \frac{n+8}{n+3} = \frac{n-8}{n-3}$$

$$(n+3)(n-3)\left(\frac{n}{(n+3)(n-3)} + \frac{n+8}{n+3}\right) = (n+3)(n-3)\left(\frac{n-8}{n-3}\right)$$

$$n + (n-3)(n+8) = (n+3)(n-8)$$

$$n + n^2 + 5n - 24 = n^2 - 5n - 24$$

$$11n = 0$$

$$n = 0 \Rightarrow \text{ The answer checks.}$$

69.

$$\frac{b+2}{b+3} + 1 = \frac{-7}{b-5}$$

$$(b+3)(b-5)\left(\frac{b+2}{b+3} + 1\right) = (b+3)(b-5)\left(\frac{-7}{b-5}\right)$$

$$(b-5)(b+2) + (b+3)(b-5) = -7(b+3)$$

$$b^2 - 3b - 10 + b^2 - 2b - 15 = -7b - 21$$

$$2b^2 + 2b - 4 = 0$$

$$2\left(b^2 + b - 2\right) = 0$$

$$2(b+2)(b-1) = 0$$

$$b + 2 = 0 \quad \textbf{or} \quad b - 1 = 0 \quad \Rightarrow \text{ Both answers check.}$$
$$b = -2 \qquad\qquad b = 1$$

71.
$$\frac{1}{a} + \frac{1}{b} = 1$$
$$ab\left(\frac{1}{a} + \frac{1}{b}\right) = ab(1)$$
$$b + a = ab$$
$$b = ab - a$$
$$b = a(b - 1)$$
$$\frac{b}{b-1} = a$$

73.
$$\frac{1}{f} = \frac{1}{d_1} + \frac{1}{d_2}$$
$$fd_1d_2 \cdot \frac{1}{f} = fd_1d_2\left(\frac{1}{d_1} + \frac{1}{d_2}\right)$$
$$d_1d_2 = fd_2 + fd_1$$
$$d_1d_2 = f(d_2 + d_1)$$
$$\frac{d_1d_2}{d_2 + d_1} = f$$

75. **Answers may vary.**

77.
$$x = \frac{1}{x}$$
$$x(x) = x\left(\frac{1}{x}\right)$$
$$x^2 = 1$$
$$x^2 - 1 = 0$$
$$(x + 1)(x - 1) = 0$$
$$x + 1 = 0 \quad \textbf{or} \quad x - 1 = 0$$
$$x = -1 \qquad\qquad x = 1$$
The numbers 1 and -1 are equal to their own reciprocals.

Exercise 6.8 (page 409)

1.
$$x^2 - 5x - 6 = 0$$
$$(x - 6)(x + 1) = 0$$
$$x - 6 = 0 \quad \textbf{or} \quad x + 1 = 0$$
$$x = 6 \qquad\qquad x = -1$$

3.
$$(t + 2)\left(t^2 + 7t + 12\right) = 0$$
$$(t + 2)(t + 4)(t + 3) = 0$$
$$t + 2 = 0 \quad \textbf{or} \quad t + 4 = 0 \quad \textbf{or} \quad t + 3 = 0$$
$$t = -2 \qquad\qquad t = -4 \qquad\qquad t = -3$$

5.
$$y^3 - y^2 = 0$$
$$y^2(y - 1) = 0$$
$$y = 0 \quad \textbf{or} \quad y = 0 \quad \textbf{or} \quad y - 1 = 0$$
$$y = 0 \qquad\qquad y = 0 \qquad\qquad y = 1$$

7.
$$\left(x^2 - 1\right)\left(x^2 - 4\right) = 0$$
$$(x + 1)(x - 1)(x + 2)(x - 2) = 0$$
$$x + 1 = 0 \quad \textbf{or} \quad x - 1 = 0 \quad \textbf{or} \quad x + 2 = 0 \quad \textbf{or} \quad x - 2 = 0$$
$$x = -1 \qquad\qquad x = 1 \qquad\qquad x = -2 \qquad\qquad x = 2$$

9. Answers may vary.

11. Let x = the number.

$$\frac{2(3)}{4+x} = 1$$

$$(4+x)\left(\frac{6}{4+x}\right) = (4+x)(1)$$

$$6 = 4+x$$

$$2 = x \Rightarrow \text{The number is 2.}$$

13. Let x = the number.

$$\frac{3+x}{4+2x} = \frac{4}{7}$$

$$7(4+2x)\left(\frac{3+x}{4+2x}\right) = 7(4+2x)\left(\frac{4}{7}\right)$$

$$7(x+3) = 4(4+2x)$$

$$7x+21 = 16+8x$$

$$5 = x$$

The number is 5.

15. Let x = the number. $\Rightarrow x + \dfrac{1}{x} = \dfrac{13}{6}$

$$6x\left(x + \frac{1}{x}\right) = 6x\left(\frac{13}{6}\right)$$

$$6x^2 + 6 = 13x$$

$$6x^2 - 13x + 6 = 0$$

$$(2x - 3)(3x - 2) = 0$$

$$2x - 3 = 0 \quad \textbf{or} \quad 3x - 2 = 0$$

$$2x = 3 \qquad\qquad 3x = 2$$

$$x = \tfrac{3}{2} \qquad\qquad x = \tfrac{2}{3} \quad \text{The numbers are } \tfrac{3}{2} \text{ and } \tfrac{2}{3}.$$

17. Let x = hours for both pipes to fill the pool.

1st pipe in 1 hour	+	2nd pipe in 1 hour	=	Total in 1 hour

$$\frac{1}{5} + \frac{1}{4} = \frac{1}{x}$$

$$20x\left(\frac{1}{5} + \frac{1}{4}\right) = 20x\left(\frac{1}{x}\right)$$

$$4x + 5x = 20$$

$$9x = 20$$

$$x = \frac{20}{9}$$

The pool can be filled in $2\frac{2}{9}$ hours.

19. Let x = days for both working together.

Roofer in 1 day	+	Owner in 1 day	=	Total in 1 day

$$\frac{1}{4} + \frac{1}{7} = \frac{1}{x}$$

$$28x\left(\frac{1}{4} + \frac{1}{7}\right) = 28x\left(\frac{1}{x}\right)$$

$$7x + 4x = 28$$

$$11x = 28$$

$$x = \frac{28}{11}$$

The pool can be filled in $2\frac{6}{11}$ days.

21. Let r = the rate at which he walks.

$$\boxed{\begin{array}{c}\text{Time he}\\\text{walks}\end{array}} = \boxed{\begin{array}{c}\text{Time he}\\\text{rides}\end{array}}$$

$$\frac{8}{r} = \frac{28}{r+10}$$

$$r(r+10)\left(\frac{8}{r}\right) = r(r+10)\left(\frac{28}{r+10}\right)$$

$$8(r+10) = 28r$$
$$8r+80 = 28r$$
$$80 = 20r$$
$$4 = r$$

	d	r	t
Walks	8	r	$\dfrac{8}{r}$
Rides	28	$r+10$	$\dfrac{28}{r+10}$

He walks 4 miles per hour, so it will take him $\frac{30}{4} = 7\frac{1}{2}$ hours to walk 30 miles.

23. Let r = the speed of the current.

$$\boxed{\begin{array}{c}\text{Time}\\\text{downstream}\end{array}} = \boxed{\begin{array}{c}\text{Time}\\\text{upstream}\end{array}}$$

$$\frac{22}{18+r} = \frac{14}{18-r}$$

$$(18+r)(18-r)\left(\frac{22}{18+r}\right) = (18+r)(18-r)\left(\frac{14}{18-r}\right)$$

$$22(18-r) = 14(18+r)$$
$$396 - 22r = 252 + 14r$$
$$144 = 36r$$
$$4 = r$$

	d	r	t
Downstream	22	$18+r$	$\dfrac{22}{18+r}$
Upstream	14	$18-r$	$\dfrac{14}{18-r}$

The current has a speed of 4 miles per hour.

25. Let r = the lower interest rate.

	I	P	r
Lower rate CD	175	$\dfrac{175}{r}$	r
Higher rate CD	200	$\dfrac{200}{r+.01}$	$r+.01$

$$\boxed{\begin{array}{c}\text{Lower rate}\\\text{principal}\end{array}} = \boxed{\begin{array}{c}\text{Higher rate}\\\text{principal}\end{array}}$$

$$\frac{175}{r} = \frac{200}{r+.01}$$

$$r(r+.01)\left(\frac{175}{r}\right) = r(r+.01)\left(\frac{200}{r+.01}\right)$$

$$175(r+.01) = 200r$$
$$175r + 1.75 = 200r$$
$$1.75 = 25r$$
$$0.07 = r$$

The rates are 7% and 8%.

27. Let x = the number who contributed.

$$\boxed{\begin{array}{c}\text{Original}\\\text{share}\end{array}} = \boxed{\begin{array}{c}\text{Share with}\\\text{more workers}\end{array}} + 2$$

$$\frac{35}{x} = \frac{35}{x+2} + 2$$

$$x(x+2)\left(\frac{35}{x}\right) = x(x+2)\left(\frac{35}{x+2} + 2\right)$$

$$35(x+2) = 35x + 2x(x+2)$$
$$35x + 70 = 35x + 2x^2 + 4x$$
$$0 = 2x^2 + 4x - 70$$
$$0 = 2\left(x^2 + 2x - 35\right)$$
$$0 = 2(x+7)(x-5)$$

$$x + 7 = 0 \quad \textbf{or} \quad x - 5 = 0$$
$$x = -7 \qquad\qquad x = 5$$

Since the answer cannot be negative, 5 workers must have contributed.

29. Let $x =$ the number bought. Then each cost $\frac{120}{x}$.

$$\boxed{\begin{array}{c}\text{Number}\\\text{bought}\end{array}} \cdot \boxed{\begin{array}{c}\text{Amount charged}\\\text{for each}\end{array}} = 120$$

$$(x + 10) \cdot \left(\frac{120}{x} - 1\right) = 120$$

$$120 - x + \frac{1200}{x} - 10 = 120$$

$$\frac{1200}{x} - x - 10 = 0$$

$$x\left(\frac{1200}{x} - x - 10\right) = 0$$

$$-x^2 - 10x + 1200 = 0$$

$$-\left(x^2 + 10x - 1200\right) = 0$$

$$-(x + 40)(x - 30) = 0$$

$$x + 40 = 0 \quad \textbf{or} \quad x - 30 = 0 \qquad \text{The store can buy 30 at the regular price.}$$
$$x = -40 \qquad\qquad x = 30$$

31. Let $r =$ the still-water speed.

$$\boxed{\begin{array}{c}\text{Time}\\\text{upstream}\end{array}} + \boxed{\begin{array}{c}\text{Time}\\\text{downstream}\end{array}} = 5$$

$$\frac{60}{r - 5} + \frac{60}{r + 5} = 5$$

$$(r + 5)(r - 5)\left(\frac{60}{r - 5} + \frac{60}{r + 5}\right) = 5(r + 5)(r - 5)$$

$$60(r + 5) + 60(r - 5) = 5\left(r^2 - 25\right)$$

$$60r + 300 + 60r - 300 = 5r^2 - 125$$

$$0 = 5r^2 - 120r - 125$$

$$0 = 5\left(r^2 - 24r - 25\right)$$

$$0 = 5(r + 1)(r - 25)$$

$$r + 1 = 0 \quad \textbf{or} \quad r - 25 = 0$$
$$r = -1 \qquad\qquad r = 25$$

	d	r	t
Upstream	60	$r - 5$	$\dfrac{60}{r - 5}$
Downstream	60	$r + 5$	$\dfrac{60}{r + 5}$

The still-water speed should be 25 miles per hour.

33. **Answers may vary.**

35. **Answers may vary.**

Chapter 6 Summary   (page 412)

1. $\dfrac{3}{6} = \dfrac{1}{2}$

2. $\dfrac{12x}{15x} = \dfrac{4}{5}$

3. $\dfrac{2 \text{ ft}}{1 \text{ yd}} = \dfrac{2 \text{ ft}}{3 \text{ ft}} = \dfrac{2}{3}$

4. $\dfrac{5 \text{ pt}}{3 \text{ qt}} = \dfrac{5 \text{ pt}}{6 \text{ pt}} = \dfrac{5}{6}$

5. $\dfrac{\$8.79}{3 \text{ lb}} = \$2.93/\text{lb}$

6. $\dfrac{2275 \text{ kwh}}{4 \text{ weeks}} = \$568.75 \text{ kwh/week}$

7.
$$\frac{4}{7} \overset{?}{=} \frac{20}{34}$$
$$4 \cdot 34 \overset{?}{=} 7 \cdot 20$$
$$136 \neq 140$$
not a proportion

8.
$$\frac{5}{7} \overset{?}{=} \frac{30}{42}$$
$$5 \cdot 42 \overset{?}{=} 7 \cdot 30$$
$$210 = 210$$
proportion

9.
$$\frac{3}{x} = \frac{6}{9}$$
$$3(9) = 6x$$
$$27 = 6x$$
$$\frac{27}{6} = x$$
$$\frac{9}{2} = x$$

10.
$$\frac{x}{3} = \frac{x}{5}$$
$$5x = 3x$$
$$2x = 0$$
$$x = 0$$

11.
$$\frac{x-2}{5} = \frac{x}{7}$$
$$7(x-2) = 5x$$
$$7x - 14 = 5x$$
$$2x = 14$$
$$x = 7$$

12.
$$\frac{4x-1}{18} = \frac{x}{6}$$
$$6(4x-1) = 18x$$
$$24x - 6 = 18x$$
$$6x = 6$$
$$x = 1$$

13. Let h = the height of the pole.
$$\frac{\text{height of pole}}{\text{shadow of pole}} = \frac{\text{height of man}}{\text{shadow of man}}$$
$$\frac{h}{12} = \frac{6}{3.6}$$
$$3.6h = 12(6)$$
$$3.6h = 72$$
$$h = \frac{72}{3.6} = 20 \text{ ft}$$

14. The expression is undefined when the denominator equals 0:
$$(x+3)(x-3) = 0$$
$$x + 3 = 0 \quad \textbf{or} \quad x - 3 = 0$$
$$x = -3 \qquad\qquad x = 3$$

15. $\dfrac{10}{25} = \dfrac{5 \cdot 2}{5 \cdot 5} = \dfrac{2}{5}$

16. $\dfrac{-12}{18} = -\dfrac{6 \cdot 2}{6 \cdot 3} = -\dfrac{2}{3}$

17. $\dfrac{-51}{153} = -\dfrac{51 \cdot 1}{51 \cdot 3} = -\dfrac{1}{3}$

18. $\dfrac{105}{45} = \dfrac{15 \cdot 7}{15 \cdot 3} = \dfrac{7}{3}$

19. $\dfrac{3x^2}{6x^3} = \dfrac{1}{2x}$

20. $\dfrac{5xy^2}{2x^2y^2} = \dfrac{5}{2x}$

21. $\dfrac{x^2}{x^2+x} = \dfrac{x^2}{x(x+1)} = \dfrac{x}{x+1}$

22. $\dfrac{x+2}{x^2+2x} = \dfrac{x+2}{x(x+2)} = \dfrac{1}{x}$

23. $\dfrac{6xy}{3xy} = 2$

24. $\dfrac{8x^2y}{2x(4xy)} = \dfrac{8x^2y}{8x^2y} = 1$

25. $\dfrac{3p-2}{2-3p} = -1$

26. $\dfrac{x^2-x-56}{x^2-5x-24} = \dfrac{(x-8)(x+7)}{(x-8)(x+3)} = \dfrac{x+7}{x+3}$

27. $\dfrac{2x^2-16x}{2x^2-18x+16} = \dfrac{2x(x-8)}{2(x^2-9x+8)} = \dfrac{2x(x-8)}{2(x-8)(x-1)} = \dfrac{x}{x-1}$

28. $\dfrac{a^2+2a+ab+2b}{a^2+2ab+b^2} = \dfrac{a(a+2)+b(a+2)}{(a+b)(a+b)} = \dfrac{(a+2)(a+b)}{(a+b)(a+b)} = \dfrac{a+2}{a+b}$

29. $\dfrac{3xy}{2x} \cdot \dfrac{4x}{2y^2} = \dfrac{12x^2y}{4xy^2} = \dfrac{3x}{y}$

30. $\dfrac{3x}{x^2 - x} \cdot \dfrac{2x - 2}{x^2} = \dfrac{3x(2x - 2)}{(x^2 - x)x^2} = \dfrac{6x(x - 1)}{x^3(x - 1)}$
$$= \dfrac{6}{x^2}$$

31. $\dfrac{x^2 + 3x + 2}{x^2 + 2x} \cdot \dfrac{x}{x + 1} = \dfrac{(x^2 + 3x + 2)x}{(x^2 + 2x)(x + 1)} = \dfrac{x(x + 2)(x + 1)}{x(x + 2)(x + 1)} = 1$

32. $\dfrac{x^2 + x}{3x - 15} \cdot \dfrac{6x - 30}{x^2 + 2x + 1} = \dfrac{(x^2 + x)(6x - 30)}{(3x - 15)(x^2 + 2x + 1)} = \dfrac{x(x + 1) \cdot 6(x - 5)}{3(x - 5) \cdot (x + 1)(x + 1)} = \dfrac{2x}{x + 1}$

33. $\dfrac{3x^2}{5x^2y} \div \dfrac{6x}{15xy^2} = \dfrac{3x^2}{5x^2y} \cdot \dfrac{15xy^2}{6x} = \dfrac{45x^3y^2}{30x^3y} = \dfrac{3y}{2}$

34. $\dfrac{x^2 + 5x}{x^2 + 4x - 5} \div \dfrac{x^2}{x - 1} = \dfrac{x^2 + 5x}{x^2 + 4x - 5} \cdot \dfrac{x - 1}{x^2} = \dfrac{x(x + 5)(x - 1)}{x^2(x + 5)(x - 1)} = \dfrac{1}{x}$

35. $\dfrac{x^2 - x - 6}{2x - 1} \div \dfrac{x^2 - 2x - 3}{2x^2 + x - 1} = \dfrac{x^2 - x - 6}{2x - 1} \cdot \dfrac{2x^2 + x - 1}{x^2 - 2x - 3} = \dfrac{(x - 3)(x + 2)(2x - 1)(x + 1)}{(2x - 1)(x - 3)(x + 1)}$
$$= x + 2$$

36. $\dfrac{x^2 - 3x}{x^2 - x - 6} \div \dfrac{x^2 - x}{x^2 + x - 2} = \dfrac{x^2 - 3x}{x^2 - x - 6} \cdot \dfrac{x^2 + x - 2}{x^2 - x} = \dfrac{x(x - 3) \cdot (x + 2)(x - 1)}{(x + 2)(x - 3) \cdot x(x - 1)} = 1$

37. $\dfrac{x^2 + 4x + 4}{x^2 + x - 6} \left(\dfrac{x - 2}{x - 1} \div \dfrac{x + 2}{x^2 + 2x - 3} \right) = \dfrac{x^2 + 4x + 4}{x^2 + x - 6} \left(\dfrac{x - 2}{x - 1} \cdot \dfrac{x^2 + 2x - 3}{x + 2} \right)$
$$= \dfrac{(x + 2)(x + 2)}{(x + 3)(x - 2)} \left(\dfrac{(x - 2)(x + 3)(x - 1)}{(x - 1)(x + 2)} \right)$$
$$= \dfrac{(x + 2)(x + 2)(x - 2)(x + 3)(x - 1)}{(x + 3)(x - 2)(x - 1)(x + 2)} = x + 2$$

38. $\dfrac{x}{x + y} + \dfrac{y}{x + y} = \dfrac{x + y}{x + y} = 1$

39. $\dfrac{3x}{x - 7} - \dfrac{x - 2}{x - 7} = \dfrac{3x - x + 2}{x - 7} = \dfrac{2x + 2}{x - 7}$

40. $\dfrac{x}{x - 1} + \dfrac{1}{x} = \dfrac{(x)(x)}{(x - 1)(x)} + \dfrac{1(x - 1)}{x(x - 1)} = \dfrac{x^2}{x(x - 1)} + \dfrac{x - 1}{x(x - 1)} = \dfrac{x^2 + x - 1}{x(x - 1)}$

41. $\dfrac{1}{7} - \dfrac{1}{x} = \dfrac{1(x)}{7(x)} - \dfrac{1(7)}{x(7)} = \dfrac{x}{7x} - \dfrac{7}{7x} = \dfrac{x - 7}{7x}$

42. $\dfrac{3}{x + 1} - \dfrac{2}{x} = \dfrac{3(x)}{(x + 1)(x)} - \dfrac{2(x + 1)}{x(x + 1)} = \dfrac{3x}{x(x + 1)} - \dfrac{2x + 2}{x(x + 1)} = \dfrac{3x - 2x - 2}{x(x + 1)} = \dfrac{x - 2}{x(x + 1)}$

43. $\dfrac{x+2}{2x} - \dfrac{2-x}{x^2} = \dfrac{(x+2)x}{2x(x)} - \dfrac{(2-x)2}{x^2(2)} = \dfrac{x^2+2x}{2x^2} - \dfrac{4-2x}{2x^2} = \dfrac{x^2+2x-4+2x}{2x^2}$

$$= \dfrac{x^2+4x-4}{2x^2}$$

44. $\dfrac{x}{x+2} + \dfrac{3}{x} - \dfrac{4}{x^2+2x} = \dfrac{x}{x+2} + \dfrac{3}{x} - \dfrac{4}{x(x+2)} = \dfrac{x(x)}{x(x+2)} + \dfrac{3(x+2)}{x(x+2)} - \dfrac{4}{x(x+2)}$

$$= \dfrac{x^2}{x(x+2)} + \dfrac{3x+6}{x(x+2)} - \dfrac{4}{x(x+2)}$$

$$= \dfrac{x^2+3x+2}{x(x+2)} = \dfrac{(x+2)(x+1)}{x(x+2)} = \dfrac{x+1}{x}$$

45. $\dfrac{2}{x-1} - \dfrac{3}{x+1} + \dfrac{x-5}{x^2-1} = \dfrac{2}{x-1} - \dfrac{3}{x+1} + \dfrac{x-5}{(x+1)(x-1)}$

$$= \dfrac{2(x+1)}{(x-1)(x+1)} - \dfrac{3(x-1)}{(x+1)(x-1)} + \dfrac{x-5}{(x+1)(x-1)}$$

$$= \dfrac{2(x+1)-3(x-1)+x-5}{(x+1)(x-1)}$$

$$= \dfrac{2x+2-3x+3+x-5}{(x+1)(x-1)} = \dfrac{0}{(x+1)(x-1)} = 0$$

46. $\dfrac{\frac{3}{2}}{\frac{2}{3}} = \dfrac{3}{2} \div \dfrac{2}{3} = \dfrac{3}{2} \cdot \dfrac{3}{2} = \dfrac{9}{4}$

47. $\dfrac{\frac{3}{2}+1}{\frac{2}{3}+1} = \dfrac{\left(\frac{3}{2}+1\right)6}{\left(\frac{2}{3}+1\right)6} = \dfrac{\frac{3}{2}\cdot 6+1\cdot 6}{\frac{2}{3}\cdot 6+1\cdot 6} = \dfrac{9+6}{4+6} = \dfrac{15}{10} = \dfrac{3}{2}$

48. $\dfrac{\frac{1}{x}+1}{\frac{1}{x}-1} = \dfrac{\left(\frac{1}{x}+1\right)x}{\left(\frac{1}{x}-1\right)x} = \dfrac{\frac{1}{x}\cdot x+1\cdot x}{\frac{1}{x}\cdot x-1\cdot x}$

$$= \dfrac{1+x}{1-x}$$

49. $\dfrac{1+\frac{3}{x}}{2-\frac{1}{x^2}} = \dfrac{\left(1+\frac{3}{x}\right)x^2}{\left(2-\frac{1}{x^2}\right)x^2} = \dfrac{x^2+3x}{2x^2-1}$

50. $\dfrac{\frac{2}{x-1}+\frac{x-1}{x+1}}{\frac{1}{x^2-1}} = \dfrac{\frac{2}{x-1}+\frac{x-1}{x+1}}{\frac{1}{(x+1)(x-1)}} = \dfrac{\left(\frac{2}{x-1}+\frac{x-1}{x+1}\right)(x+1)(x-1)}{\left(\frac{1}{(x+1)(x-1)}\right)(x+1)(x-1)} = \dfrac{2(x+1)+(x-1)(x-1)}{1}$

$$= 2x+2+x^2-2x+1 = x^2+3$$

51. $\dfrac{\frac{a}{b}+c}{\frac{b}{a}+c} = \dfrac{\left(\frac{a}{b}+c\right)ab}{\left(\frac{b}{a}+c\right)ab} = \dfrac{a^2+abc}{b^2+abc} = \dfrac{a(a+bc)}{b(b+ac)}$

52.
$$\frac{3}{x} = \frac{2}{x-1}$$
$$x(x-1)\left(\frac{3}{x}\right) = x(x-1)\left(\frac{2}{x-1}\right)$$
$$3(x-1) = 2x$$
$$3x - 3 = 2x$$
$$x = 3 \Rightarrow \text{The answer checks.}$$

53.
$$\frac{5}{x+4} = \frac{3}{x+2}$$
$$(x+4)(x+2)\left(\frac{5}{x+4}\right) = (x+4)(x+2)\left(\frac{3}{x+2}\right)$$
$$5(x+2) = 3(x+4)$$
$$5x + 10 = 3x + 12$$
$$2x = 2$$
$$x = 1 \Rightarrow \text{The answer checks.}$$

54.
$$\frac{2}{3x} + \frac{1}{x} = \frac{5}{9}$$
$$9x\left(\frac{2}{3x} + \frac{1}{x}\right) = 9x\left(\frac{5}{9}\right)$$
$$6 + 9 = 5x$$
$$15 = 5x$$
$$3 = x \Rightarrow \text{The answer checks.}$$

55.
$$\frac{2x}{x+4} = \frac{3}{x-1}$$
$$(x+4)(x-1)\left(\frac{2x}{x+4}\right) = (x+4)(x-1)\left(\frac{3}{x-1}\right)$$
$$2x(x-1) = 3(x+4)$$
$$2x^2 - 2x = 3x + 12$$
$$2x^2 - 5x - 12 = 0$$
$$(2x+3)(x-4) = 0$$

$2x + 3 = 0 \quad \text{or} \quad x - 4 = 0 \Rightarrow \text{Both answers check.}$
$$2x = -3 \qquad\qquad x = 4$$
$$x = -\frac{3}{2} \qquad\qquad x = 4$$

56.
$$\frac{2}{x-1} + \frac{3}{x+4} = \frac{-5}{x^2+3x-4}$$
$$\frac{2}{x-1} + \frac{3}{x+4} = \frac{-5}{(x+4)(x-1)}$$
$$(x-1)(x+4)\left(\frac{2}{x-1} + \frac{3}{x+4}\right) = (x-1)(x+4)\left(\frac{-5}{(x+4)(x-1)}\right)$$
$$2(x+4) + 3(x-1) = -5$$
$$2x + 8 + 3x - 3 = -5$$
$$5x = -10$$
$$x = -2 \Rightarrow \text{The answer checks.}$$

57.
$$\frac{4}{x+2} - \frac{3}{x+3} = \frac{6}{x^2+5x+6}$$
$$\frac{4}{x+2} - \frac{3}{x+3} = \frac{6}{(x+2)(x+3)}$$
$$(x+2)(x+3)\left(\frac{4}{x+2} - \frac{3}{x+3}\right) = (x+2)(x+3)\left(\frac{6}{(x+2)(x+3)}\right)$$
$$4(x+3) - 3(x+2) = 6$$
$$4x + 12 - 3x - 6 = 6$$
$$x = 0 \Rightarrow \text{The answer checks.}$$

58.
$$\frac{1}{r} = \frac{1}{r_1} + \frac{1}{r_2}$$
$$rr_1r_2 \cdot \frac{1}{r} = rr_1r_2\left(\frac{1}{r_1} + \frac{1}{r_2}\right)$$
$$r_1r_2 = rr_2 + rr_1$$
$$r_1r_2 - rr_1 = rr_2$$
$$r_1(r_2 - r) = rr_2$$
$$r_1 = \frac{rr_2}{r_2 - r}$$

59.
$$E = 1 - \frac{T_2}{T_1}$$
$$T_1E = T_1\left(1 - \frac{T_2}{T_1}\right)$$
$$T_1E = T_1 - T_2$$
$$T_1E - T_1 = -T_2$$
$$T_1(E - 1) = -T_2$$
$$T_1 = -\frac{T_2}{E-1} = \frac{T_2}{1-E}$$

60.
$$H = \frac{RB}{R+B}$$
$$(R+B)H = (R+B)\left(\frac{RB}{R+B}\right)$$
$$RH + BH = RB$$
$$BH = RB - RH$$
$$BH = R(B - H)$$
$$\frac{BH}{B-H} = R$$

61. Let x = hours for both pipes to empty.

1st pipe in 1 hour	+	2nd pipe in 1 hour	=	Total in 1 hour

$$\frac{1}{18} + \frac{1}{20} = \frac{1}{x}$$
$$180x\left(\frac{1}{18} + \frac{1}{20}\right) = 180x\left(\frac{1}{x}\right)$$
$$10x + 9x = 180$$
$$19x = 180$$
$$x = \frac{180}{19}$$

It can be emptied in $9\frac{9}{19}$ hours.

62. Let $x =$ days for both working together.

$$\boxed{\text{Painter in 1 day}} + \boxed{\text{Owner in 1 day}} = \boxed{\text{Total in 1 day}}$$

$$\frac{1}{10} + \frac{1}{14} = \frac{1}{x}$$

$$70x\left(\frac{1}{10} + \frac{1}{14}\right) = 70x\left(\frac{1}{x}\right)$$

$$7x + 5x = 70$$

$$12x = 70$$

$$x = \frac{70}{12} = \frac{35}{6}$$

It can be painted in $5\frac{5}{6}$ days.

63. Let $r =$ the rate at which he jogs.

	d	r	t
Jogs	10	r	$\dfrac{10}{r}$
Rides	30	$r+10$	$\dfrac{30}{r+10}$

$$\boxed{\text{Time he jogs}} = \boxed{\text{Time he rides}}$$

$$\frac{10}{r} = \frac{30}{r+10}$$

$$r(r+10)\left(\frac{10}{r}\right) = r(r+10)\left(\frac{30}{r+10}\right)$$

$$10(r+10) = 30r$$

$$10r + 100 = 30r$$

$$100 = 20r$$

$$5 = r$$

He jogs 5 miles per hour.

64. Let $r =$ the speed of the wind.

	d	r	t
Downwind	400	$360+r$	$\dfrac{400}{360+r}$
Upwind	320	$360-r$	$\dfrac{320}{360-r}$

$$\boxed{\text{Time downwind}} = \boxed{\text{Time upwind}}$$

$$\frac{400}{360+r} = \frac{320}{360-r}$$

$$(360+r)(360-r)\left(\frac{400}{360+r}\right) = (360+r)(360-r)\left(\frac{320}{360-r}\right)$$

$$400(360-r) = 320(360+r)$$

$$144000 - 400r = 115200 + 320r$$

$$28800 = 720r$$

$$40 = r \Rightarrow \text{The wind has a speed of 40 miles per hour.}$$

Chapter 6 Test (page 416)

1. $\dfrac{6 \text{ ft}}{3 \text{ yd}} = \dfrac{6 \text{ ft}}{9 \text{ ft}} = \dfrac{2}{3}$

2. $\dfrac{3xy}{5xy} \overset{?}{=} \dfrac{3xt}{5xt}$

$3xy \cdot 5xt \overset{?}{=} 5xy \cdot 3xt$

$15x^2yt = 15x^2yt \Rightarrow \text{proportion}$

3. $\dfrac{y}{y-1} = \dfrac{y-2}{y}$

$y(y) = (y-1)(y-2)$

$y^2 = y^2 - 3y + 2$

$3y = 2$

$y = \dfrac{2}{3}$

4. Let $h = $ the height of the tree.

$\dfrac{\text{height of tree}}{\text{shadow of tree}} = \dfrac{\text{height of man}}{\text{shadow of man}}$

$\dfrac{h}{30} = \dfrac{6}{4}$

$4h = 30(6)$

$4h = 180$

$h = 45 \text{ ft}$

5. $\dfrac{48x^2y}{54xy^2} = \dfrac{6 \cdot 8x^2y}{6 \cdot 9xy^2} = \dfrac{8x}{9y}$

6. $\dfrac{2x^2 - x - 3}{4x^2 - 9} = \dfrac{(2x-3)(x+1)}{(2x+3)(2x-3)} = \dfrac{x+1}{2x+3}$

7. $\dfrac{3(x+2) - 3}{2x - 4 - (x-5)} = \dfrac{3x + 6 - 3}{2x - 4 - x + 5}$

$= \dfrac{3x+3}{x+1} = \dfrac{3(x+1)}{x+1} = 3$

8. $\dfrac{12x^2y}{15xyz} \cdot \dfrac{25y^2z}{16xt} = \dfrac{3 \cdot 4 \cdot 5 \cdot 5x^2y^3z}{3 \cdot 5 \cdot 4 \cdot 4x^2ytz} = \dfrac{5y^2}{4t}$

9. $\dfrac{x^2 + 3x + 2}{3x + 9} \cdot \dfrac{x+3}{x^2 - 4} = \dfrac{(x+2)(x+1)}{3(x+3)} \cdot \dfrac{x+3}{(x+2)(x-2)} = \dfrac{(x+2)(x+1)(x+3)}{3(x+3)(x+2)(x-2)} = \dfrac{x+1}{3(x-2)}$

10. $\dfrac{8x^2y}{25xt} \div \dfrac{16x^2y^3}{30xyt^3} = \dfrac{8x^2y}{25xt} \cdot \dfrac{30xyt^3}{16x^2y^3} = \dfrac{4 \cdot 2 \cdot 5 \cdot 2 \cdot 3x^3y^2t^3}{5 \cdot 5 \cdot 4 \cdot 4x^3y^3t} = \dfrac{3t^2}{5y}$

11. $\dfrac{x^2 - x}{3x^2 + 6x} \div \dfrac{3x - 3}{3x^3 + 6x^2} = \dfrac{x^2 - x}{3x^2 + 6x} \cdot \dfrac{3x^3 + 6x^2}{3x - 3} = \dfrac{x(x-1)}{3x(x+2)} \cdot \dfrac{3x^2(x+2)}{3(x-1)}$

$= \dfrac{3x^3(x-1)(x+2)}{9x(x-1)(x+2)} = \dfrac{x^2}{3}$

12. $\dfrac{x^2 + xy}{x - y} \cdot \dfrac{x^2 - y^2}{x^2 - 2x} \div \dfrac{x^2 + 2xy + y^2}{x^2 - 4} = \dfrac{x^2 + xy}{x - y} \cdot \dfrac{x^2 - y^2}{x^2 - 2x} \cdot \dfrac{x^2 - 4}{x^2 + 2xy + y^2}$

$= \dfrac{x(x+y)}{x - y} \cdot \dfrac{(x+y)(x-y)}{x(x-2)} \cdot \dfrac{(x+2)(x-2)}{(x+y)(x+y)}$

$= \dfrac{x(x+y)(x+y)(x-y)(x+2)(x-2)}{x(x-y)(x-2)(x+y)(x+y)} = x + 2$

13. $\dfrac{5x - 4}{x - 1} + \dfrac{5x + 3}{x - 1} = \dfrac{5x - 4 + 5x + 3}{x - 1} = \dfrac{10x - 1}{x - 1}$

14. $\dfrac{3y+7}{2y+3} - \dfrac{3(y-2)}{2y+3} = \dfrac{3y+7-3(y-2)}{2y+3} = \dfrac{3y+7-3y+6}{2y+3} = \dfrac{13}{2y+3}$

15. $\dfrac{x+1}{x} + \dfrac{x-1}{x+1} = \dfrac{(x+1)(x+1)}{x(x+1)} + \dfrac{x(x-1)}{x(x+1)} = \dfrac{(x+1)(x+1)+x(x-1)}{x(x+1)}$

$$= \dfrac{x^2+2x+1+x^2-x}{x(x+1)} = \dfrac{2x^2+x+1}{x(x+1)}$$

16. $\dfrac{5x}{x-2} - 3 = \dfrac{5x}{x-2} - \dfrac{3}{1} = \dfrac{5x}{x-2} - \dfrac{3(x-2)}{x-2} = \dfrac{5x-3(x-2)}{x-2} = \dfrac{5x-3x+6}{x-2} = \dfrac{2x+6}{x-2}$

17. $\dfrac{\frac{8x^2}{xy^3}}{\frac{4y^3}{x^2y^3}} = \dfrac{8x^2}{xy^3} \div \dfrac{4y^3}{x^2y^3} = \dfrac{8x^2}{xy^3} \cdot \dfrac{x^2y^3}{4y^3} = \dfrac{8x^4y^3}{4xy^6} = \dfrac{2x^3}{y^3}$

18. $\dfrac{1+\frac{y}{x}}{\frac{y}{x}-1} = \dfrac{\left(1+\frac{y}{x}\right)x}{\left(\frac{y}{x}-1\right)x} = \dfrac{1 \cdot x + \frac{y}{x} \cdot x}{\frac{y}{x} \cdot x - 1 \cdot x}$

$$= \dfrac{x+y}{y-x}$$

19. $\dfrac{x}{10} - \dfrac{1}{2} = \dfrac{x}{5}$

$10\left(\dfrac{x}{10} - \dfrac{1}{2}\right) = 10\left(\dfrac{x}{5}\right)$

$x - 5 = 2x$

$-5 = x$

20. $3x - \dfrac{2(x+3)}{3} = 16 - \dfrac{x+2}{2}$

$6\left(3x - \dfrac{2x+6}{3}\right) = 6\left(16 - \dfrac{x+2}{2}\right)$

$18x - 2(2x+6) = 96 - 3(x+2)$

$18x - 4x - 12 = 96 - 3x - 6$

$17x = 102$

$x = 6$

21. $\dfrac{7}{x+4} - \dfrac{1}{2} = \dfrac{3}{x+4}$

$2(x+4)\left(\dfrac{7}{x+4} - \dfrac{1}{2}\right) = 2(x+4)\left(\dfrac{3}{x+4}\right)$

$2(7) - (x+4) = 2(3)$

$14 - x - 4 = 6$

$-x = -4$

$x = 4$

The answer checks.

22. $H = \dfrac{RB}{R+B}$

$(R+B)H = (R+B)\left(\dfrac{RB}{R+B}\right)$

$RH + BH = RB$

$RH = RB - BH$

$RH = B(R-H)$

$\dfrac{RH}{R-H} = B$

23. Let $x =$ hours working together.

1st worker in 1 hour		2nd worker in 1 hour		Total in 1 hour

$$\dfrac{1}{7} + \dfrac{1}{9} = \dfrac{1}{x}$$

$$63x\left(\dfrac{1}{7} + \dfrac{1}{9}\right) = 63x\left(\dfrac{1}{x}\right)$$

$$9x + 7x = 63$$

$$16x = 63$$

$$x = \dfrac{63}{16}$$

They can finish in $3\frac{15}{16}$ hours.

24. Let $r =$ the speed of the current.

$$\boxed{\begin{array}{c}\text{Time} \\ \text{downstream}\end{array}} = \boxed{\begin{array}{c}\text{Time} \\ \text{upstream}\end{array}}$$

$$\frac{28}{23 + r} = \frac{18}{23 - r}$$

$$(23 + r)(23 - r)\left(\frac{28}{23 + r}\right) = (23 + r)(23 - r)\left(\frac{18}{23 - r}\right)$$

$$28(23 - r) = 18(23 + r)$$

$$644 - 28r = 414 + 18r$$

$$230 = 46r$$

$$5 = r$$

	d	r	t
Downstream	28	$23 + r$	$\dfrac{28}{23 + r}$
Upstream	18	$23 - r$	$\dfrac{18}{23 - r}$

The current has a speed of 5 miles per hour.

25.
$$\frac{575}{\frac{1}{2}} = \frac{x}{7}$$

$$7(575) = \frac{1}{2}x$$

$$4025 = \frac{1}{2}x$$

$$2(4025) = 2\left(\frac{1}{2}x\right)$$

$$8050 = x \Rightarrow \text{The plane will lose 8,050 feet of altitude.}$$

Cumulative Review Exercises (page 417)

1. $x^2 x^5 = x^{2+5} = x^7$

2. $\left(x^2\right)^5 = x^{2 \cdot 5} = x^{10}$

3. $\dfrac{x^5}{x^2} = x^{5-2} = x^3$

4. $\left(3x^5\right)^0 = 1$

5. $(3x^2 - 2x) + (6x^3 - 3x^2 - 1) = 3x^2 - 2x + 6x^3 - 3x^2 - 1 = 6x^3 - 2x - 1$

6. $(4x^3 - 2x) - (2x^3 - 2x^2 - 3x + 1) = 4x^3 - 2x - 2x^3 + 2x^2 + 3x - 1 = 2x^3 + 2x^2 + x - 1$

7. $3(5x^2 - 4x + 3) + 2(-x^2 + 2x - 4) = 15x^2 - 12x + 9 - 2x^2 + 4x - 8 = 13x^2 - 8x + 1$

8. $4(3x^2 - 4x - 1) - 2(-2x^2 + 4x - 3) = 12x^2 - 16x - 4 + 4x^2 - 8x + 6 = 16x^2 - 24x + 2$

9. $(3x^3 y^2)(-4x^2 y^3) = 3(-4)x^3 x^2 y^2 y^3 = -12x^5 y^5$

10. $-5x^2(7x^3 - 2x^2 - 2) = -35x^5 + 10x^4 + 10x^2$

11. $(3x + 1)(2x + 4) = 6x^2 + 12x + 2x + 4 = 6x^2 + 14x + 4$

12. $(5x - 4y)(3x + 2y) = 15x^2 + 10xy - 12xy - 8y^2 = 15x^2 - 2xy - 8y^2$

13.
$$\begin{array}{r} x+4 \\ x+3\overline{\smash{\big)}\,x^2+7x+12} \\ \underline{x^2+3x} \\ 4x+12 \\ \underline{4x+12} \\ 0 \end{array}$$

14.
$$\begin{array}{r} x^2+x+1 \\ 2x-3\overline{\smash{\big)}\,2x^3-x^2-x-3} \\ \underline{2x^3-3x^2} \\ 2x^2-x \\ \underline{2x^2-3x} \\ 2x-3 \\ \underline{2x-3} \\ 0 \end{array}$$

15. $3x^2y - 6xy^2 = 3xy(x - 2y)$

16. $3(a+b) + x(a+b) = (a+b)(3+x)$

17. $2a + 2b + ab + b^2 = 2(a+b) + b(a+b)$
$$= (a+b)(2+b)$$

18. $25p^4 - 16q^2 = \left(5p^2\right)^2 - (4q)^2$
$$= \left(5p^2 + 4q\right)\left(5p^2 - 4q\right)$$

19. $x^2 - 11x - 12 = (x - 12)(x + 1)$

20. $x^2 - xy - 6y^2 = (x - 3y)(x + 2y)$

21. $6a^2 - 7a - 20 = (2a - 5)(3a + 4)$

22. $8m^2 - 10mn - 3n^2 = (4m + n)(2m - 3n)$

23. $p^3 - 27q^3 = p^3 - (3q)^3 = (p - 3q)\left(p^2 + p(3q) + (3q)^2\right) = (p - 3q)(p^2 + 3pq + 9q^2)$

24. $8r^3 + 64s^3 = 8\left(r^3 + 8s^3\right) = 8\left[r^3 + (2s)^3\right] = 8(r + 2s)\left(r^2 - r \cdot 2s + (2s)^2\right)$
$$= 8(r + 2s)\left(r^2 - 2rs + 4s^2\right)$$

25.
$$\frac{4}{5}x + 6 = 18$$
$$\frac{4}{5}x = 12$$
$$5 \cdot \frac{4}{5}x = 5(12)$$
$$4x = 60$$
$$x = 15$$

26.
$$5 - \frac{x+2}{3} = 7 - x$$
$$3\left(5 - \frac{x+2}{3}\right) = 3(7 - x)$$
$$15 - (x + 2) = 21 - 3x$$
$$15 - x - 2 = 21 - 3x$$
$$2x = 8$$
$$x = 4$$

27.
$$6x^2 - x - 2 = 0$$
$$(2x + 1)(3x - 2) = 0$$
$$2x + 1 = 0 \quad \textbf{or} \quad 3x - 2 = 0$$
$$2x = -1 \qquad\qquad 3x = 2$$
$$x = -\tfrac{1}{2} \qquad\qquad x = \tfrac{2}{3}$$

28.
$$5x^2 = 10x$$
$$5x^2 - 10x = 0$$
$$5x(x - 2) = 0$$
$$5x = 0 \quad \textbf{or} \quad x - 2 = 0$$
$$x = 0 \qquad\qquad x = 2$$

29.
$$x^2 + 3x + 2 = 0$$
$$(x + 1)(x + 2) = 0$$
$$x + 1 = 0 \quad \textbf{or} \quad x + 2 = 0$$
$$x = -1 \qquad\qquad x = -2$$

30.
$$2y^2 + 5y - 12 = 0$$
$$(2y - 3)(y + 4) = 0$$
$$2y - 3 = 0 \quad \textbf{or} \quad y + 4 = 0$$
$$2y = 3 \qquad\qquad y = -4$$
$$y = \tfrac{3}{2} \qquad\qquad y = -4$$

31. $5x - 3 > 7$
$$5x > 10$$
$$x > 2$$

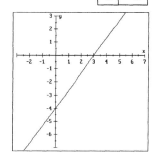

32. $7x - 9 < 5$
$$7x < 14$$
$$x < 2$$

33. $-2 < -x + 3 < 5$
$$-5 < \quad -x \quad < 2$$
$$\frac{-5}{-1} > \quad \frac{-x}{-1} \quad > \frac{2}{-1}$$
$$5 > \quad x \quad > -2$$
$$-2 < \quad x \quad < 5$$

34. $0 \le \dfrac{4 - x}{3} \le 2$
$$3(0) \le 3 \cdot \frac{4 - x}{3} \le 3(2)$$
$$0 \le \quad 4 - x \quad \le 6$$
$$-4 \le \quad -x \quad \le 2$$
$$\frac{-4}{-1} \ge \quad \frac{-x}{-1} \quad \ge \frac{2}{-1}$$
$$4 \ge \quad x \quad \ge -2$$
$$-2 \le \quad x \quad \le 4$$

35. $4x - 3y = 12$

x	y
0	-4
3	0

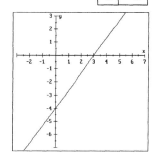

36. $3x + 4y = 4y + 12$
$$3x = 12$$
$$x = 4$$

x	y
4	0
4	2

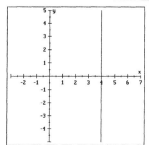

37. $f(0) = 2(0)^2 - 3 = 0 - 3 = -3$

38. $f(3) = 2(3)^2 - 3 = 2(9) - 3 = 18 - 3 = 15$

39. $f(-2) = 2(-2)^2 - 3 = 2(4) - 3$
$$= 8 - 3 = 5$$

40. $f(2x) = 2(2x)^2 - 3 = 2(4x^2) - 3$
$$= 8x^2 - 3$$

41. $\dfrac{x^2 + 2x + 1}{x^2 - 1} = \dfrac{(x+1)(x+1)}{(x+1)(x-1)} = \dfrac{x+1}{x-1}$

42. $\dfrac{x^2 + 2x - 15}{x^2 + 3x - 10} = \dfrac{(x+5)(x-3)}{(x+5)(x-2)} = \dfrac{x-3}{x-2}$

43. $\dfrac{x^2 + x - 6}{5x - 5} \cdot \dfrac{5x - 10}{x + 3} = \dfrac{(x+3)(x-2) \cdot 5(x-2)}{5(x-1) \cdot (x+3)} = \dfrac{(x-2)^2}{x-1}$

44. $\dfrac{p^2 - p - 6}{3p - 9} \div \dfrac{p^2 + 6p + 9}{p^2 - 9} = \dfrac{p^2 - p - 6}{3p - 9} \cdot \dfrac{p^2 - 9}{p^2 + 6p + 9} = \dfrac{(p-3)(p+2) \cdot (p+3)(p-3)}{3(p-3) \cdot (p+3)(p+3)}$

$$= \dfrac{(p+2)(p-3)}{3(p+3)}$$

45. $\dfrac{3x}{x+2} + \dfrac{5x}{x+2} - \dfrac{7x-2}{x+2} = \dfrac{3x + 5x - (7x - 2)}{x + 2} = \dfrac{3x + 5x - 7x + 2}{x + 2} = \dfrac{x+2}{x+2} = 1$

46. $\dfrac{x-1}{x+1} + \dfrac{x+1}{x-1} = \dfrac{(x-1)(x-1)}{(x+1)(x-1)} + \dfrac{(x+1)(x+1)}{(x-1)(x+1)} = \dfrac{x^2 - 2x + 1}{(x+1)(x-1)} + \dfrac{x^2 + 2x + 1}{(x+1)(x-1)}$

$$= \dfrac{2x^2 + 2}{(x+1)(x-1)}$$

47. $\dfrac{a+1}{2a+4} - \dfrac{a^2}{2a^2 - 8} = \dfrac{a+1}{2(a+2)} - \dfrac{a^2}{2(a+2)(a-2)} = \dfrac{(a+1)(a-2)}{2(a+2)(a-2)} - \dfrac{a^2}{2(a+2)(a-2)}$

$$= \dfrac{a^2 - 2a + a - 2 - a^2}{2(a+2)(a-2)}$$

$$= \dfrac{-a - 2}{2(a+2)(a-2)}$$

$$= \dfrac{-(a+2)}{2(a+2)(a-2)} = -\dfrac{1}{2(a-2)}$$

48. $\dfrac{\frac{1}{x} + \frac{1}{y}}{\frac{1}{x} - \frac{1}{y}} = \dfrac{\left(\frac{1}{x} + \frac{1}{y}\right)xy}{\left(\frac{1}{x} - \frac{1}{y}\right)xy} = \dfrac{\frac{1}{x} \cdot xy + \frac{1}{y} \cdot xy}{\frac{1}{x} \cdot xy - \frac{1}{y} \cdot xy} = \dfrac{y + x}{y - x}$

Exercise 7.1 (page 428)

1. $(-2)^4 = (-2)(-2)(-2)(-2) = 16$

3. $3x - x^2 = 3(-3) - (-3)^2 = -9 - (+9)$
$$= -9 - 9 = -18$$

5. system

7. independent

9. inconsistent

11. $\begin{array}{ll} x + y = 2 & 2x - y = 1 \\ 1 + 1 \stackrel{?}{=} 2 & 2(1) - 1 \stackrel{?}{=} 1 \\ 2 = 2 & 2 - 1 \stackrel{?}{=} 1 \\ & 1 = 1 \end{array}$

$(1, 1)$ is a solution to the system.

13. $\begin{array}{ll} 2x + y = 4 & x + y = 1 \\ 2(3) + (-2) \stackrel{?}{=} 4 & 3 + (-2) \stackrel{?}{=} 1 \\ 6 + (-2) \stackrel{?}{=} 4 & 1 = 1 \\ 4 = 4 & \end{array}$

$(3, -2)$ is a solution to the system.

15.
$$2x - 3y = -7 \qquad 4x - 5y = 25$$
$$2(4) - 3(5) \stackrel{?}{=} -7 \qquad 4(4) - 5(5) \stackrel{?}{=} 25$$
$$8 - 15 \stackrel{?}{=} -7 \qquad 16 - 25 \stackrel{?}{=} 25$$
$$-7 = -7 \qquad -9 \neq 25$$

$(4, 5)$ is not a solution to the system.

17.
$$4x + 5y = -23 \qquad -3x + 2y = 0$$
$$4(-2) + 5(-3) \stackrel{?}{=} -23 \qquad -3(-2) + 2(-3) \stackrel{?}{=} 0$$
$$-8 + (-15) \stackrel{?}{=} -23 \qquad 6 + (-6) \stackrel{?}{=} 0$$
$$-23 = -23 \qquad 0 = 0$$

$(-2, -3)$ is a solution to the system.

19.
$$2x + y = 4 \qquad 4x - 3y = 11$$
$$2\left(\tfrac{1}{2}\right) + 3 \stackrel{?}{=} 4 \qquad 4\left(\tfrac{1}{2}\right) - 3(3) \stackrel{?}{=} 11$$
$$1 + 3 \stackrel{?}{=} 4 \qquad 2 - 9 \stackrel{?}{=} 11$$
$$4 = 4 \qquad -7 \neq 11$$

$\left(\tfrac{1}{2}, 3\right)$ is not a solution to the system.

21.
$$5x - 4y = -6 \qquad 8y = 10x + 12$$
$$5\left(-\tfrac{2}{5}\right) - 4\left(\tfrac{1}{4}\right) \stackrel{?}{=} -6 \qquad 8\left(\tfrac{1}{4}\right) \stackrel{?}{=} 10\left(-\tfrac{2}{5}\right) + 12$$
$$-2 - 1 \stackrel{?}{=} -6 \qquad 2 \stackrel{?}{=} -4 + 12$$
$$-3 \neq -6 \qquad 2 \neq 8$$

$\left(-\tfrac{2}{5}, \tfrac{1}{4}\right)$ is not a solution to the system.

23. $x + y = 2 \qquad x - y = 0$

x	y
0	2
1	1

x	y
0	0
1	1

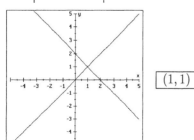

$(1, 1)$

25. $x + y = 2 \qquad x - y = 4$

x	y
0	2
1	1

x	y
0	-4
1	-3

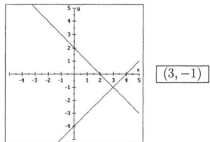

$(3, -1)$

27. $3x + 2y = -8$ $2x - 3y = -1$

x	y
0	-4
2	-7

x	y
1	1
-2	-1

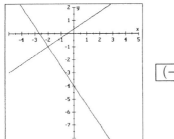

$\boxed{(-2, -1)}$

29. $4x - 2y = 8$ $y = 2x - 4$

x	y
0	-4
1	-2

x	y
0	-4
1	-2

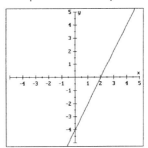

dependent equations

31. $2x - 3y = -18$ $3x + 2y = -1$

x	y
-3	4
-6	2

x	y
1	-2
-1	1

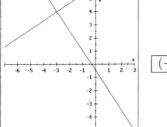

$\boxed{(-3, 4)}$

33. $4x = 3(4 - y)$ $2y = 4(3 - x)$
$4x = 12 - 3y$ $2y = 12 - 4x$

x	y
0	4
3	0

x	y
3	0
0	6

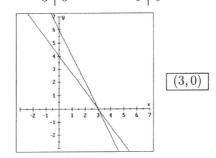

$\boxed{(3, 0)}$

35. $x + 2y = -4$ $x - \dfrac{1}{2}y = 6$
                 $2x - y = 12$

x	y
0	-2
2	-3

x	y
4	-4
6	0

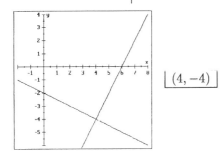

$\boxed{(4, -4)}$

37. $-\dfrac{3}{4}x + y = 3$ $\dfrac{1}{4}x + y = -1$
$-3x + 4y = 12$ $x + 4y = -4$

x	y
0	3
-4	0

x	y
0	-1
-4	0

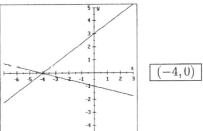

$\boxed{(-4, 0)}$

39. $\dfrac{1}{2}x + \dfrac{1}{4}y = 0 \qquad \dfrac{1}{4}x - \dfrac{3}{8}y = -2$

$\qquad\quad 2x + y = 0 \qquad\quad 2x - 3y = -16$

x	y
0	0
1	-2

x	y
-2	4
-5	2

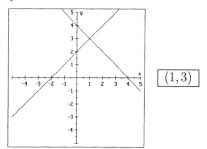

$(-2, 4)$

41. $\dfrac{1}{3}x - \dfrac{1}{2}y = \dfrac{1}{6} \qquad \dfrac{2}{5}x + \dfrac{1}{2}y = \dfrac{13}{10}$

$\qquad\quad 2x - 3y = 1 \qquad\quad 4x + 5y = 13$

x	y
5	3
2	1

x	y
-3	5
2	1

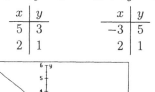

$(2, 1)$

43. $\begin{cases} y = 4 - x \\ y = 2 + x \end{cases}$

$(1, 3)$

45. $\begin{cases} 3x - 6y = 4 & \Rightarrow y = \dfrac{1}{2}x - \dfrac{2}{3} \\ 2x + y = 1 & \Rightarrow y = -2x + 1 \end{cases}$

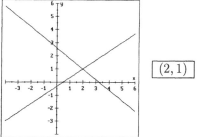

$(0.67, -0.33)$

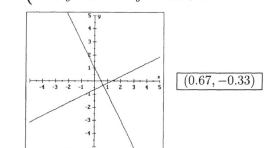

47. **a.** Donors outnumbered patients.
 b. 1999; about 4100
 c. Patients outnumber donors.

49. **a.** Houston, New Orleans, St. Augustine
 b. St. Louis, Memphis, New Orleans
 c. New Orleans

51. Answers may vary.

53. Answers may vary.

Exercise 7.2 (page 436)

1. $y^2 - x^2 = (3)^2 - (-2)^2 = 9 - 4 = 5$

3. $\dfrac{3x - 2y}{2x + y} = \dfrac{3(-2) - 2(3)}{2(-2) + 3} = \dfrac{-6 - 6}{-4 + 3}$

$\qquad\qquad\qquad\qquad\qquad = \dfrac{-12}{-1} = 12$

5. $-x(3y - 4) = -(-2)[3(3) - 4]$
$\qquad\qquad\quad = -(-2)(5) = 10$

7. y, terms

9. remove

11. infinitely many

13. $\begin{cases} (1) \quad y = 2x \\ (2) \quad x + y = 6 \end{cases}$

Substitute $y = 2x$ from (1) into (2):

$x + y = 6$

$x + 2x = 6$

$3x = 6$

$x = 2$

Substitute and solve for y:

$y = 2x$

$y = 2(2) = 4$

$\boxed{x = 2, y = 4}$

15. $\begin{cases} (1) \quad y = 2x - 6 \\ (2) \quad 2x + y = 6 \end{cases}$

Substitute $y = 2x - 6$ from (1) into (2):

$2x + y = 6$

$2x + 2x - 6 = 6$

$4x = 12$

$x = 3$

Substitute and solve for y:

$y = 2x - 6$

$y = 2(3) - 6 = 6 - 6 = 0$

$\boxed{x = 3, y = 0}$

17. $\begin{cases} (1) \quad y = 2x + 5 \\ (2) \quad x + 2y = -5 \end{cases}$

Substitute $y = 2x + 5$ from (1) into (2):

$x + 2y = -5$

$x + 2(2x + 5) = -5$

$x + 4x + 10 = -5$

$5x = -15$

$x = -3$

Substitute and solve for y:

$y = 2x + 5$

$y = 2(-3) + 5 = -6 + 5 = -1$

$\boxed{x = -3, y = -1}$

19. $\begin{cases} (1) \quad 2a + 4b = -24 \\ (2) \quad a = 20 - 2b \end{cases}$

Substitute $a = 20 - 2b$ from (2) into (1):

$2a + 4b = -24$

$2(20 - 2b) + 4b = -24$

$40 - 4b + 4b = -24$

$40 \neq -24$

$\boxed{\text{inconsistent system}}$

21. $\begin{cases} (1) \quad 2a = 3b - 13 \\ (2) \quad b = 2a + 7 \end{cases}$

Substitute $b = 2a + 7$ from (2) into (1):

$2a = 3b - 13$

$2a = 3(2a + 7) - 13$

$2a = 6a + 21 - 13$

$-4a = 8$

$a = -2$

Substitute and solve for b:

$b = 2a + 7$

$b = 2(-2) + 7 = -4 + 7 = 3$

$\boxed{a = -2, b = 3}$

23. $\begin{cases} 2x + y = 4 \quad \Rightarrow (1) \quad y = -2x + 4 \\ 4x + y = 5 \quad \Rightarrow (2) \quad 4x + y = 5 \end{cases}$

Substitute $y = -2x + 4$ from (1) into (2):

$4x + y = 5$

$4x + (-2x + 4) = 5$

$4x - 2x + 4 = 5$

$2x = 1$

$x = \dfrac{1}{2}$

Substitute and solve for y:

$y = -2x + 4$

$y = -2\left(\dfrac{1}{2}\right) + 4 = -1 + 4 = 3$

$\boxed{x = \frac{1}{2}, y = 3}$

25. $\begin{cases} r + 3s = 9 & \Rightarrow (1) \quad r = -3s + 9 \\ 3r + 2s = 13 & \Rightarrow (2) \quad 3r + 2s = 13 \end{cases}$

Substitute $r = -3s + 9$ from (1) into (2):

$$3r + 2s = 13$$
$$3(-3s + 9) + 2s = 13$$
$$-9s + 27 + 2s = 13$$
$$-7s = -14$$
$$s = 2$$

Substitute and solve for r:

$$r = -3s + 9$$
$$r = -3(2) + 9 = -6 + 9 = 3$$
$$\boxed{r = 3, s = 2}$$

27. $\begin{cases} 4x + 5y = 2 & \Rightarrow (1) \quad 4x + 5y = 2 \\ 3x - y = 11 & \Rightarrow (2) \quad y = 3x - 11 \end{cases}$

Substitute $y = 3x - 11$ from (2) into (1):

$$4x + 5y = 2$$
$$4x + 5(3x - 11) = 2$$
$$4x + 15x - 55 = 2$$
$$19x = 57$$
$$x = 3$$

Substitute and solve for y:

$$y = 3x - 11$$
$$y = 3(3) - 11 = 9 - 11 = -2$$
$$\boxed{x = 3, y = -2}$$

29. $\begin{cases} 2x + y = 0 & \Rightarrow (1) \quad y = -2x \\ 3x + 2y = 1 & \Rightarrow (2) \quad 3x + 2y = 1 \end{cases}$

Substitute $y = -2x$ from (1) into (2):

$$3x + 2y = 1$$
$$3x + 2(-2x) = 1$$
$$3x - 4x = 1$$
$$-x = 1$$
$$x = -1$$

Substitute and solve for y:

$$y = -2x$$
$$y = -2(-1) = 2$$
$$\boxed{x = -1, y = 2}$$

31. $\begin{cases} 3x + 4y = -7 & \Rightarrow (1) \quad 3x + 4y = -7 \\ 2y - x = -1 & \Rightarrow (2) \quad x = 2y + 1 \end{cases}$

Substitute $x = 2y + 1$ from (2) into (1):

$$3x + 4y = -7$$
$$3(2y + 1) + 4y = -7$$
$$6y + 3 + 4y = -7$$
$$10y = -10$$
$$y = -1$$

Substitute and solve for x:

$$x = 2y + 1$$
$$x = 2(-1) + 1 = -2 + 1 = -1$$
$$\boxed{x = -1, y = -1}$$

33. $\begin{cases} 9x = 3y + 12 & \Rightarrow (1) \quad 9x = 3y + 12 \\ 4 = 3x - y & \Rightarrow (2) \quad y = 3x - 4 \end{cases}$

Substitute $y = 3x - 4$ from (2) into (1):

$$9x = 3y + 12$$
$$9x = 3(3x - 4) + 12$$
$$9x = 9x - 12 + 12$$
$$0 = 0$$
$$\boxed{\text{dependent equations}}$$

35.
$$\begin{cases} (1) & 2x + 3y = 5 \quad \Rightarrow 2x = 5 - 3y \quad \Rightarrow x = \dfrac{5 - 3y}{2} \\ (2) & 3x + 2y = 5 \end{cases}$$

Substitute $x = \dfrac{5 - 3y}{2}$ from (1) into (2):

$$3x + 2y = 5$$
$$3 \cdot \dfrac{5 - 3y}{2} + 2y = 5$$
$$3(5 - 3y) + 4y = 10$$
$$15 - 9y + 4y = 10$$
$$-5y = -5$$
$$y = 1$$

Substitute and solve for x:

$$x = \dfrac{5 - 3y}{2}$$
$$x = \dfrac{5 - 3(1)}{2}$$
$$x = \dfrac{5 - 3}{2} = \dfrac{2}{2} = 1$$

Solution:

$$\boxed{x = 1, y = 1}$$

37.
$$\begin{cases} (1) & a = \dfrac{3}{2}b + 5 \\ (2) & 2a - 3b = 8 \end{cases}$$

Substitute $a = \dfrac{3}{2}b + 5$ from (1) into (2):

$$2a - 3b = 8$$
$$2\left(\dfrac{3}{2}b + 5\right) - 3b = 8$$
$$3b + 10 - 3b = 8$$
$$10 \neq 8$$

inconsistent system

39.
$$\begin{cases} (1) & 2x + 5y = -2 \quad \Rightarrow 2x = -2 - 5y \quad \Rightarrow x = \dfrac{-2 - 5y}{2} \\ (2) & 4x + 3y = 10 \end{cases}$$

Substitute $x = \dfrac{-2 - 5y}{2}$ from (1) into (2):

$$4x + 3y = 10$$
$$4 \cdot \dfrac{-2 - 5y}{2} + 3y = 10$$
$$4(-2 - 5y) + 6y = 20$$
$$-8 - 20y + 6y = 20$$
$$-14y = 28$$
$$y = -2$$

Substitute and solve for x:

$$x = \dfrac{-2 - 5y}{2}$$
$$x = \dfrac{-2 - 5(-2)}{2}$$
$$x = \dfrac{-2 + 10}{2} = \dfrac{8}{2} = 4$$

Solution:

$$\boxed{x = 4, y = -2}$$

41. $\begin{cases} (1) & 2x - 3y = -3 \quad \Rightarrow 2x = 3y - 3 \quad \Rightarrow x = \dfrac{3y - 3}{2} \\ (2) & 3x + 5y = -14 \end{cases}$

Substitute $x = \dfrac{3y - 3}{2}$ from (1) into (2):   Substitute and solve for x: Solution:

$$3x + 5y = -14$$
$$3 \cdot \frac{3y - 3}{2} + 5y = -14$$
$$3(3y - 3) + 10y = -28$$
$$9y - 9 + 10y = -28$$
$$19y = -19$$
$$y = -1$$

$$x = \frac{3y - 3}{2}$$
$$x = \frac{3(-1) - 3}{2}$$
$$x = \frac{-3 - 3}{2} = \frac{-6}{2} = -3$$

$\boxed{x = -3, y = -1}$

43. $\begin{cases} (1) & 7x - 2y = -1 \quad \Rightarrow 7x = 2y - 1 \quad \Rightarrow x = \dfrac{2y - 1}{7} \\ (2) & -5x + 2y = -1 \end{cases}$

Substitute $x = \dfrac{2y - 1}{7}$ from (1) into (2):   Substitute and solve for x: Solution:

$$-5x + 2y = -1$$
$$-5 \cdot \frac{2y - 1}{7} + 2y = -1$$
$$-5(2y - 1) + 14y = -7$$
$$-10y + 5 + 14y = -7$$
$$4y = -12$$
$$y = -3$$

$$x = \frac{2y - 1}{7}$$
$$x = \frac{2(-3) - 1}{7}$$
$$x = \frac{-6 - 1}{7} = \frac{-7}{7} = -1$$

$\boxed{x = -1, y = -3}$

45. $\begin{cases} (1) & 2a + 3b = 2 \quad \Rightarrow 2a = 2 - 3b \quad \Rightarrow a = \dfrac{2 - 3b}{2} \\ (2) & 8a - 3b = 3 \end{cases}$

Substitute $a = \dfrac{2 - 3b}{2}$ from (1) into (2):   Substitute and solve for a: Solution:

$$8a - 3b = 3$$
$$8 \cdot \frac{2 - 3b}{2} - 3b = 3$$
$$4(2 - 3b) - 3b = 3$$
$$8 - 12b - 3b = 3$$
$$-15b = -5$$
$$b = \frac{1}{3}$$

$$a = \frac{2 - 3b}{2}$$
$$a = \frac{2 - 3\left(\frac{1}{3}\right)}{2}$$
$$a = \frac{2 - 1}{2} = \frac{1}{2}$$

$\boxed{a = \tfrac{1}{2}, b = \tfrac{1}{3}}$

47. $\begin{cases} (1) & y - x = 3x \quad \Rightarrow y = 4x \\ (2) & 2(x+y) = 14 - y \quad \Rightarrow 2x + 2y = 14 - y \quad \Rightarrow 2x + 3y = 14 \end{cases}$

Substitute $y = 4x$ from (1) into (2):   Substitute and solve for y: Solution:

$2x + 3y = 14$ $y = 4x$ $\boxed{x = 1, y = 4}$

$2x + 3(4x) = 14$ $y = 4(1) = 4$

$2x + 12x = 14$

$14x = 14$

$x = 1$

49. $\begin{cases} (1) & 3(x-1) + 3 = 8 + 2y \quad \Rightarrow 3x - 3 + 3 = 8 + 2y \quad \Rightarrow 3x = 2y + 8 \quad \Rightarrow x = \dfrac{2y+8}{3} \\ (2) & 2(x+1) = 4 + 3y \quad \Rightarrow 2x + 2 = 4 + 3y \quad \Rightarrow 2x - 3y = 2 \end{cases}$

Substitute $x = \dfrac{2y+8}{3}$ from (1) into (2):   Substitute and solve for x: Solution:

$2x - 3y = 2$ $x = \dfrac{2y+8}{3}$ $\boxed{x = 4, y = 2}$

$2 \cdot \dfrac{2y+8}{3} - 3y = 2$ $x = \dfrac{2(2)+8}{3}$

$2(2y + 8) - 9y = 6$ $x = \dfrac{4+8}{3} = \dfrac{12}{3} = 4$

$4y + 16 - 9y = 6$

$-5y = -10$

$y = 2$

51. $\begin{cases} (1) & 6a = 5(3 + b + a) - a \quad \Rightarrow 6a = 15 + 5b + 5a - a \quad \Rightarrow 2a = 5b + 15 \quad \Rightarrow a = \dfrac{5b+15}{2} \\ (2) & 3(a - b) + 4b = 5(1 + b) \quad \Rightarrow 3a - 3b + 4b = 5 + 5b \quad \Rightarrow 3a - 4b = 5 \end{cases}$

Substitute $a = \dfrac{5b+15}{2}$ from (1) into (2):   Substitute and solve for a: Solution:

$3a - 4b = 5$ $a = \dfrac{5b+15}{2}$ $\boxed{a = -5, b = -5}$.

$3 \cdot \dfrac{5b+15}{2} - 4b = 5$ $a = \dfrac{5(-5)+15}{2}$

$3(5b + 15) - 8b = 10$ $a = \dfrac{-25+15}{2} = \dfrac{-10}{2} = -5$

$15b + 45 - 8b = 10$

$7b = -35$

$b = -5$

53. $\begin{cases} (1) & \dfrac{1}{2}x + \dfrac{1}{2}y = -1 \quad \Rightarrow x + y = -2 \quad \Rightarrow x = -y - 2 \\ (2) & \dfrac{1}{3}x - \dfrac{1}{2}y = -4 \quad \Rightarrow 2x - 3y = -24 \end{cases}$

Substitute $x = -y - 2$ from (1) into (2):   Substitute and solve for x: Solution:

$2x - 3y = -24$ $x = -y - 2$ $\boxed{x = -6, y = 4}$

$2(-y - 2) - 3y = -24$ $x = -4 - 2 = -6$

$-2y - 4 - 3y = -24$

$-5y = -20$

$y = 4$

55. $\begin{cases} (1) & 5x = \dfrac{1}{2}y - 1 \quad \Rightarrow 10x = y - 2 \\[2mm] (2) & \dfrac{1}{4}y = 10x - 1 \quad \Rightarrow y = 40x - 4 \end{cases}$

Substitute $y = 40x - 4$ from (2) into (1):

$10x = y - 2$

$10x = 40x - 4 - 2$

$-30x = -6$

$x = \dfrac{1}{5}$

Substitute and solve for y:

$y = 40x - 4$

$y = 40\left(\dfrac{1}{5}\right) - 4$

$y = 8 - 4 = 4$

Solution:

$\boxed{x = \tfrac{1}{5}, y = 4}$

57. $\begin{cases} (1) & \dfrac{6x - 1}{3} - \dfrac{5}{3} = \dfrac{3y + 1}{2} \quad \Rightarrow [\times 6] \Rightarrow \quad 2(6x - 1) - 2(5) = 3(3y + 1) \\ & \hspace{6cm} 12x - 2 - 10 = 9y + 3 \quad \Rightarrow 12x = 9y + 15 \\[2mm] (2) & \dfrac{1 + 5y}{4} + \dfrac{x + 3}{4} = \dfrac{17}{2} \quad \Rightarrow [\times 4] \Rightarrow \quad 1 + 5y + x + 3 = 2(17) \quad \Rightarrow x = 30 - 5y \end{cases}$

Substitute $x = 30 - 5y$ from (2) into (1):

$12x = 9y + 15$

$12(30 - 5y) = 9y + 15$

$360 - 60y = 9y + 15$

$-69y = -345$

$y = 5$

Substitute and solve for x:

$x = 30 - 5y$

$x = 30 - 5(5)$

$x = 30 - 25 = 5$

Solution:

$\boxed{x - 5, y = 5}$

59. Answers may vary.

61. Answers may vary.

Exercise 7.3 (page 443)

1. $8(3x - 5) - 12 = 4(2x + 3)$

$24x - 40 - 12 = 8x + 12$

$24x - 52 = 8x + 12$

$24x - 8x - 52 = 8x - 8x + 12$

$16x - 52 + 52 = 12 + 52$

$16x = 64$

$\dfrac{16x}{16} = \dfrac{64}{16}$

$x = 4$

3. $x - 2 = \dfrac{x + 2}{3}$

$3(x - 2) = 3 \cdot \dfrac{x + 2}{3}$

$3x - 6 = x + 2$

$3x - x - 6 = x - x + 2$

$2x - 6 + 6 = 2 + 6$

$2x = 8$

$\dfrac{2x}{2} = \dfrac{8}{2}$

$x = 4$

5.
$$7x - 9 \le 5$$
$$7x - 9 + 9 \le 5 + 9$$
$$7x \le 14$$
$$\frac{7x}{7} \le \frac{14}{7}$$
$$x \le 2$$

7. coefficient

9. general

11. 15

13.
$$\begin{array}{rcl} x + y &=& 5 \\ x - y &=& -3 \\ \hline 2x &=& 2 \\ x &=& 1 \end{array}$$
Substitute and solve for y:
$$x + y = 5$$
$$1 + y = 5$$
$$y = 4 \quad \boxed{(1, 4)}$$

15.
$$\begin{array}{rcl} x - y &=& -5 \\ x + y &=& 1 \\ \hline 2x &=& -4 \\ x &=& -2 \end{array}$$
Substitute and solve for y:
$$x + y = 1$$
$$-2 + y = 1$$
$$y = 3 \quad \boxed{(-2, 3)}$$

17.
$$\begin{array}{rcl} 2x + y &=& -1 \\ -2x + y &=& 3 \\ \hline 2y &=& 2 \\ y &=& 1 \end{array}$$
Substitute and solve for x:
$$2x + y = -1$$
$$2x + 1 = -1$$
$$2x = -2$$
$$x = -1 \quad \boxed{(-1, 1)}$$

19.
$$\begin{array}{rcl} 2x - 3y &=& -11 \\ 3x + 3y &=& 21 \\ \hline 5x &=& 10 \\ x &=& 2 \end{array}$$
Substitute and solve for y:
$$3x + 3y = 21$$
$$3(2) + 3y = 21$$
$$6 + 3y = 21$$
$$3y = 15$$
$$y = 5 \quad \boxed{(2, 5)}$$

21.
$$\begin{array}{rcl} 2x + y &=& -2 \\ -2x - 3y &=& -6 \\ \hline -2y &=& -8 \\ y &=& 4 \end{array}$$
Substitute and solve for x:
$$2x + y = -2$$
$$2x + 4 = -2$$
$$2x = -6$$
$$x = -3 \quad \boxed{(-3, 4)}$$

23.
$$\begin{array}{rcl} 4x + 3y &=& 24 \\ 4x - 3y &=& -24 \\ \hline 8x &=& 0 \\ x &=& 0 \end{array}$$
Substitute and solve for y:
$$4x + 3y = 24$$
$$4(0) + 3y = 24$$
$$0 + 3y = 24$$
$$3y = 24$$
$$y = 8 \quad \boxed{(0, 8)}$$

25.
$$\begin{array}{ll} x + y = 5 \Rightarrow \times (-1) & -x - y = -5 \\ x + 2y = 8 & x + 2y = 8 \\ \hline & y = 3 \end{array}$$
$$x + 2y = 8 \qquad \text{Solution:}$$
$$x + 2(3) = 8 \qquad \boxed{(2, 3)}$$
$$x + 6 = 8$$
$$x = 2$$

27. $2x + \ y = 4 \Rightarrow \times \ (-1)$ $\quad -2x - \ y = -4$ $\qquad 2x + y = 4$ $\qquad$ Solution:
$\quad \underline{2x + 3y = 0}$ $\qquad\qquad\qquad \underline{2x + 3y = \ \ 0}$ $\quad 2x + (-2) = 4$ $\qquad \boxed{(3, -2)}$
$\qquad\qquad\qquad\qquad\qquad\qquad\qquad\quad 2y = -4$ $\qquad\quad 2x = 6$
$\qquad\qquad\qquad\qquad\qquad\qquad\qquad\quad\ y = -2$ $\qquad\qquad x = 3$

29. $3x + 29 = \ \ 5y \ \Rightarrow \ 3x - 5y = -29 \Rightarrow \times \ (-1) \quad -3x + 5y = 29$
$\quad \underline{4y - 34 = -3x} \ \Rightarrow \ \underline{3x + 4y = \ \ 34}$ $\qquad\qquad\qquad\quad \underline{3x + 4y = 34}$
$\qquad\qquad\qquad\qquad\qquad\qquad\qquad\qquad\qquad\qquad\qquad\quad 9y = 63$
$\qquad\qquad\qquad\qquad\qquad\qquad\qquad\qquad\qquad\qquad\qquad\ \ y = \ \ 7$
$\qquad\qquad\qquad\qquad\qquad 3x + 4y = 34$ $\qquad$ Solution:
$\qquad\qquad\qquad\qquad\qquad 3x + 4(7) = 34$ $\qquad \boxed{(2, 7)}$
$\qquad\qquad\qquad\qquad\qquad 3x + 28 = 34$
$\qquad\qquad\qquad\qquad\qquad\qquad 3x = 6$
$\qquad\qquad\qquad\qquad\qquad\qquad\ x = 2$

31. $2x = 3(y - 2) \ \Rightarrow \ 2x = 3y - 6 \ \Rightarrow \ 2x - 3y = -6 \Rightarrow \times (-1) \ -2x + 3y = \ \ \ 6$ $\boxed{\text{inconsistent}}$
$\quad \underline{2(x + 4) = 3y} \Rightarrow \underline{2x + 8 = 3y} \ \Rightarrow \ \underline{2x - 3y = -8}$ $\qquad\qquad\qquad \underline{2x - 3y = -8}$ $\boxed{\text{system}}$
$\qquad\qquad\qquad\qquad\qquad\qquad\qquad\qquad\qquad\qquad\qquad\qquad\qquad 0 \neq -2$

33. $\quad -2(x + 1) = 3(y - 2) \ \Rightarrow \ \ -2x - 2 = 3y - 6 \ \Rightarrow \ -2x - 3y = -4$ $\boxed{\text{dependent}}$
$\quad \underline{3(y + 2) = 6 - 2(x - 2)} \Rightarrow \underline{3y + 6 = 6 - 2x + 4} \ \Rightarrow \ \underline{2x + 3y = 4}$ $\boxed{\text{equations}}$
$\qquad\qquad\qquad\qquad\qquad\qquad\qquad\qquad\qquad\qquad\qquad\qquad\quad 0 = 0$

35. $4(x + 1) = 17 - 3(y - 1) \ \Rightarrow \ 4x + 4 = 17 - 3y + 3 \ \Rightarrow \ 4x + 3y = 16$
$\quad \underline{2(x + 2) + 3(y - 1) = 9} \Rightarrow \underline{2x + 4 + 3y - 3 = 9} \ \Rightarrow \ \underline{2x + 3y = 8}$

$\qquad 4x + 3y = 16 \Rightarrow \times \ (-1)$ $\quad -4x - 3y = -16$ $\qquad 4x + 3y = 16$ $\qquad$ Solution:
$\qquad \underline{2x + 3y = \ \ 8}$ $\qquad\qquad\qquad \underline{2x + 3y = \ \ \ \ 8}$ $\quad 4(4) + 3y = 16$ $\qquad \boxed{(4, 0)}$
$\qquad\qquad\qquad\qquad\qquad\qquad\qquad -2x \qquad\ \ = -8$ $\quad 16 + 3y = 16$
$\qquad\qquad\qquad\qquad\qquad\qquad\qquad\quad x \qquad\ \ = \ \ \ 4$ $\qquad\qquad 3y = 0$
$\qquad\qquad\qquad\qquad\qquad\qquad\qquad\qquad\qquad\qquad\qquad\qquad\quad y = 0$

37. $2x + \ y = 10 \Rightarrow \times \ (-2)$ $\quad -4x - 2y = -20$ $\qquad 2x + y = 10$ $\qquad$ Solution:
$\quad \underline{x + 2y = 10}$ $\qquad\qquad\qquad\quad \underline{x + 2y = \ \ \ 10}$ $\quad 2\left(\frac{10}{3}\right) + y = 10$ $\qquad \boxed{\left(\dfrac{10}{3}, \dfrac{10}{3}\right)}$
$\qquad\qquad\qquad\qquad\qquad\qquad\quad -3x \qquad\ = -10$ $\qquad \frac{20}{3} + y = \frac{30}{3}$
$\qquad\qquad\qquad\qquad\qquad\qquad\qquad x \qquad\ = \ \ \frac{10}{3}$ $\qquad\qquad\quad y = \frac{10}{3}$

39. $2x - \ y = 16 \Rightarrow \times \ (2)$ $\quad 4x - 2y = 32$ $\qquad 3x + 2y = 3$ $\qquad$ Solution:
$\quad \underline{3x + 2y = \ \ 3}$ $\qquad\qquad\quad \underline{3x + 2y = \ \ 3}$ $\quad 3(5) + 2y = 3$ $\qquad \boxed{(5, -6)}$
$\qquad\qquad\qquad\qquad\qquad\qquad 7x \qquad = 35$ $\qquad 15 + 2y = 3$
$\qquad\qquad\qquad\qquad\qquad\qquad\ x \qquad = \ \ 5$ $\qquad\qquad 2y = -12$
$\qquad\qquad\qquad\qquad\qquad\qquad\qquad\qquad\qquad\qquad\qquad y = -6$

41. $2x + 3y = 2 \Rightarrow \times (-2)$ $\quad -4x - 6y = -4$ $\qquad 2x + 3y = 2$ $\qquad$ Solution:
$ 4x - 9y = -1$ $\qquad\qquad \underline{4x - 9y = -1}$ $\quad 2x + 3\left(\frac{1}{3}\right) = 2$ $\qquad \boxed{\left(\dfrac{1}{2}, \dfrac{1}{3}\right)}$
$\qquad\qquad\qquad\qquad\qquad\qquad -15y = -5$ $\qquad\quad 2x + 1 = 2$
$\qquad\qquad\qquad\qquad\qquad\qquad\quad y = \frac{-5}{-15} = \frac{1}{3}$ $\qquad\quad 2x = 1$
$\qquad\qquad\qquad\qquad\qquad\qquad\qquad\qquad\qquad\qquad x = \frac{1}{2}$

43. $4x + 5y = -20 \Rightarrow \times (4)$ $\quad 16x + 20y = -80$ $\qquad 4x + 5y = -20$ $\qquad$ Solution:
$ 5x - 4y = -25 \Rightarrow \times (5)$ $\quad \underline{25x - 20y = -125}$ $\quad 4(-5) + 5y = -20$ $\qquad \boxed{(-5, 0)}$
$\qquad\qquad\qquad\qquad\qquad\qquad 41x = -205$ $\quad\; -20 + 5y = -20$
$\qquad\qquad\qquad\qquad\qquad\qquad\quad x = -5$ $\qquad\qquad 5y = 0$
$\qquad\qquad\qquad\qquad\qquad\qquad\qquad\qquad\qquad\qquad\qquad y = 0$

45. $6x = -3y \qquad \Rightarrow \quad 6x + 3y = 0$ $\qquad 6x + 3y = 0$ $\quad 6x = -3y$ $\qquad$ Solution:
$ 5y = 2x + 12 \Rightarrow -2x + 5y = 12 \Rightarrow \times (3)$ $\quad \underline{-6x + 15y = 36}$ $\quad 6x = -3(2)$ $\qquad \boxed{(-1, 2)}$
$\qquad\qquad\qquad\qquad\qquad\qquad\qquad\qquad\qquad 18y = 36$ $\qquad 6x = -6$
$\qquad\qquad\qquad\qquad\qquad\qquad\qquad\qquad\quad\; y = 2$ $\qquad\; x = -1$

47. $4(2x - y) = 18 \qquad \Rightarrow 8x - 4y = 18 \qquad \Rightarrow 8x - 4y = 18$
$ 3(x - 3) = 2y - 1 \Rightarrow 3x - 9 = 2y - 1 \Rightarrow 3x - 2y = 8$

$8x - 4y = 18$ $\qquad\qquad 8x - 4y = 18$ $\quad 8x - 4y = 18$ $\qquad$ Solution:
$3x - 2y = 8 \Rightarrow \times (-2)$ $\quad \underline{-6x + 4y = -16}$ $\quad 8(1) - 4y = 18$ $\qquad \boxed{\left(1, -\dfrac{5}{2}\right)}$
$\qquad\qquad\qquad\qquad\qquad\qquad 2x = 2$ $\qquad 8 - 4y = 18$
$\qquad\qquad\qquad\qquad\qquad\qquad\; x = 1$ $\qquad\quad -4y = 10$
$\qquad\qquad\qquad\qquad\qquad\qquad\qquad\qquad\qquad\qquad\qquad y = -\frac{10}{4} = -\frac{5}{2}$

49. $3(x - 2y) = 12 \qquad \Rightarrow 3x - 6y = 12$ $\quad 3x - 6y = 12$ $\qquad\qquad 3x - 6y = 12$
$\qquad\quad x = 2(y + 2) \Rightarrow \qquad x = 2y + 4 \Rightarrow \quad x - 2y = 4 \Rightarrow \times (-3)$ $\quad \underline{-3x + 6y = -12}$
$\qquad\qquad\qquad\qquad\qquad\qquad\qquad\qquad\qquad\qquad\qquad\qquad\qquad\qquad\qquad 0 = 0$

dependent equations

51. $\dfrac{3}{5}x + \dfrac{4}{5}y = 1 \Rightarrow \times (5)$ $\quad 3x + 4y = 5 \Rightarrow \times (2)$ $\quad 6x + 8y = 10$ $\qquad 3x + 4y = 5$
$ -\dfrac{1}{4}x + \dfrac{3}{8}y = 1 \Rightarrow \times (8)$ $\quad -2x + 3y = 8 \Rightarrow \times (3)$ $\quad \underline{-6x + 9y = 24}$ $\qquad 3x + 4(2) = 5$
$\qquad\qquad\qquad\qquad\qquad\qquad\qquad\qquad\qquad\qquad\qquad 17y = 34$ $\qquad\quad 3x + 8 = 5$
$\qquad\qquad\qquad\qquad\qquad\qquad\qquad\qquad\qquad\qquad\quad\; y = 2$ $\qquad\qquad 3x = -3$
$\qquad\qquad\qquad\qquad\qquad\qquad\qquad\qquad\qquad\qquad\qquad\qquad\qquad\qquad x = -1$
$\qquad\qquad\qquad\qquad\qquad\qquad\qquad\qquad\qquad\qquad\qquad$ Solution: $\boxed{(-1, 2)}$

53. $\dfrac{3}{5}x + y = 1 \Rightarrow \times (5)$ $\quad 3x + 5y = 5$ $\qquad 3x + 5y = 5$ $\qquad$ Solution:
$ \dfrac{4}{5}x - y = -1 \Rightarrow \times (5)$ $\quad \underline{4x - 5y = -5}$ $\quad 3(0) + 5y = 5$ $\qquad \boxed{(0, 1)}$
$\qquad\qquad\qquad\qquad\qquad\qquad 7x = 0$ $\qquad\qquad 5y = 5$
$\qquad\qquad\qquad\qquad\qquad\qquad\; x = 0$ $\qquad\qquad\; y = 1$

55.

$$\frac{x}{2} - \frac{y}{3} = -2 \Rightarrow \times (6) \qquad 3x - 2y = -12 \Rightarrow \qquad 3x - 2y = -12$$

$$\frac{2x-3}{2} + \frac{6y+1}{3} = \frac{17}{6} \Rightarrow \times (6) \quad 3(2x-3) + 2(6y+1) = \quad 17 \Rightarrow 6x - 9 + 12y + 2 = \quad 17$$

$$\begin{array}{llll} 3x - \ 2y = -12 \Rightarrow \times (-2) & -6x + \ 4y = 24 & 3x - 2y = -12 & \text{Solution:} \\ 6x + 12y = \ \ 24 & \underline{6x + 12y = 24} & 3x - 2(3) = -12 & \boxed{(-2, 3)} \\ & 16y = 48 & 3x - 6 = -12 & \\ & y = \ \ 3 & 3x = -6 & \\ & & x = -2 & \end{array}$$

57.

$$\frac{x-3}{2} + \frac{y+5}{3} = \frac{11}{6} \Rightarrow \times (6) \quad 3(x-3) + 2(y+5) = 11 \Rightarrow 3x - 9 + 2y + 10 = 11$$

$$\frac{x+3}{3} - \frac{5}{12} = \frac{y+3}{4} \Rightarrow \times (12) \quad 4(x+3) - 5 = 3(y+3) \Rightarrow 4x + 12 - 5 = 3y + 9$$

$$\begin{array}{llll} 3x + 2y = 10 \Rightarrow \times (3) & 9x + 6y = 30 & 3x + 2y = 10 & \text{Solution:} \\ 4x - 3y = \ \ 2 \Rightarrow \times (2) & \underline{8x - 6y = \ \ 4} & 3(2) + 2y = 10 & \boxed{(2, 2)} \\ & 17x \quad\quad = 34 & 6 + 2y = 10 & \\ & x \quad\quad = 2 & 2y = 4 & \\ & & y = 2 & \end{array}$$

59. Answers may vary.   **61.** $(1, 4)$

Exercise 7.4 (page 452)

1. $x < 4$ **3.** $-1 < x \le 2$ **5.** $8 \cdot 8 \cdot 8 \cdot c = 8^3 c$

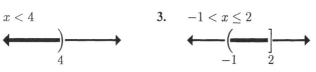

7. $a \cdot a \cdot b \cdot b = a^2 b^2$ **9.** variable   **11.** system

13. Let x and y represent the integers.   Substitute $x = 2y$ into (2):   Substitute; solve for x:

$$\begin{cases} (1) & x = 2y \\ (2) & x + y = 96 \end{cases}$$

$$\begin{aligned} x + y &= 96 \\ 2y + y &= 96 \\ 3y &= 96 \\ y &= 32 \end{aligned}$$

$$\begin{aligned} x &= 2y \\ x &= 2(32) = 64 \end{aligned}$$

The integers are 32 and 64.

15. Let x and y represent the integers.

$$\begin{array}{llll} 3x + \ y = 29 \Rightarrow \times (-2) & -6x - 2y = -58 & x + 2y = 18 & \text{The integers are 5 and 8.} \\ \underline{x + 2y = 18} & \underline{x + 2y = \ \ 18} & 8 + 2y = 18 & \\ & -5x \quad\quad = -40 & 2y = 10 & \\ & x \quad\quad = \ \ 8 & y = 5 & \end{array}$$

17. Let x = number of horses and y = number of cows.

$\begin{cases} (1) & y = 5x \\ (2) & x + y = 168 \end{cases}$

Substitute $y = 5x$ from (1) into (2):

$x + y = 168$
$x + 5x = 168$
$6x = 168$
$x = 28$

Substitute and solve for y:

$y = 5x$
$y = 5(28) = 140$
He has 140 cows.

19. Let x = cost of a gallon of paint and y = cost of a brush.

$8x + 3y = 135 \quad \Rightarrow \times (2) \qquad 16x + 6y = 270$
$6x + 2y = 100 \Rightarrow \times (-3) \quad -18x - 6y = -300$
$ \quad \overline{-2x = -30}$
$x = 15$

$6x + 2y = 100$
$6(15) + 2y = 100$
$90 + 2y = 100$
$2y = 10$
$y = 5$

The paint costs $15 per gallon, while each brush costs $5.

21. Let x = cost of contact cleaner and y = cost of soaking solution.

$2x + 3y = 29.40 \Rightarrow \times (-2) \quad -4x - 6y = -58.80$
$3x + 2y = 28.60 \Rightarrow \times (3) \quad 9x + 6y = 85.80$
$\overline{5x = 27}$
$\phantom{3x + 2y = 28.60 \Rightarrow \times (3) \quad }x = 5.40$

$2x + 3y = 29.40$
$2(5.40) + 3y = 29.40$
$10.80 + 3y = 29.40$
$3y = 18.60$
$y = 6.20$

The contact cleaner costs $5.40, while the soaking solution costs $6.20.

23. Let x = the length of one piece.
Let y = the other length.

$\begin{cases} (1) & x = y + 5 \\ (2) & x + y = 25 \end{cases}$

Substitute $x = y + 5$ into (2):

$x + y = 25$
$y + 5 + y = 25$
$2y = 20$
$y = 10$

Substitute; solve for x:

$x = y + 5$
$x = 10 + 5 = 15$
The lengths are 10 ft and 15 ft.

25. Let x = Lena's amount.
Let y = Chayla's amount.

$\begin{cases} (1) & x = y + 50000 \\ (2) & x + y = 250000 \end{cases}$

Substitute $x = y + 50000$ into (2):

$x + y = 250000$
$y + 50000 + y = 250000$
$2y = 200000$
$y = 100000$

Chayla will get $100,000.

27. Let x = time for causes.
Let y = time for outcome.

$\begin{cases} (1) & x = 4y \\ (2) & x + y = 30 \end{cases}$

Substitute $x = 4y$ into (2):

$x + y = 30$
$4y + y = 30$
$5y = 30$
$y = 6$

Substitute; solve for x:

$x = 4y$
$x = 4(6) = 24$
24 minutes will be devoted to the causes, and 6 minutes to the outcome.

29. Let x = deaths from cancer.   Substitute $x = 6y$ into (2):   Substitute; solve for x:
Let y = deaths from accidents. $x + y = 630000$ $x = 6y$
$\begin{cases} (1) & x = 6y \\ (2) & x + y = 630000 \end{cases}$ $6y + y = 630000$ $x = 6(90000) = 540000$
                                    $7y = 630000$ 90,000 die from accidents and
                                         $y = 90000$ 540,000 die from cancer.

31. Let w = the width.   Substitute $l = w + 5$ into (2):   Substitute; solve for l:
Let l = the length. $2l + 2w = 110$ $l = w + 5$
$\begin{cases} (1) & l = w + 5 \\ (2) & 2l + 2w = 110 \end{cases}$ $2(w + 5) + 2w = 110$ $l = 25 + 5 = 30$
                         $2w + 10 + 2w = 110$ The dimensions are 25 ft by 30 ft.
                                   $4w = 100$
                                       $w = 25$

33. Let x = the length.   Substitute $x = 2y + 2$ into (2):   Substitute; solve for x:
Let y = the width. $2x + 2y = 34$ $x = 2y + 2$
$\begin{cases} (1) & x = 2y + 2 \\ (2) & 2x + 2y = 34 \end{cases}$ $2(2y + 2) + 2y = 34$ $x = 2(5) + 2 = 12$
                         $4y + 4 + 2y = 34$ The area is $(5)(12) = 60$ ft^2.
                                      $6y = 30$
                                       $y = 5$

35. Let c = the cost.   Substitute: The break point is about 9.9 years.
Let n = number of years. $2250 + 412n = 1715 + 466n$
$\begin{cases} (1) & c = 2250 + 412n \\ (2) & c = 1715 + 466n \end{cases}$ $-54n = -535$
                                   $n \approx 9.9$

37. Since 7 years is less than the break point, choose the furnace with the lower up-front cost (80+).

39. Let x = Bill's principal and y = Janette's principal.
       $x + \quad y = 5000$ $x + \; y = \;\; 5000 \Rightarrow \times (-5)$ $-5x - 5y = -25000$
       $0.05x + 0.07y = \;\; 310 \Rightarrow \times (100)$ $5x + 7y = 31000$ $\underline{5x + 7y = \;\;\; 31000}$
                                                           $2y = \;\;\; 6000$
                                                             $y = \;\;\; 3000$

           $x + y = 5000$ Bill invested \$2000.
        $x + 3000 = 5000$
             $x = 2000$

41. Let x = number of student tickets and y = number of nonstudent tickets.
       $x + \; y = 350 \Rightarrow \times (-1)$ $-x - \; y = -350$ $x + y = 350$ There were 250 student
       $\underline{x + 2y = 450}$ $\underline{x + 2y = \;\;\; 450}$ $x + 100 = 350$ tickets sold.
                                         $y = \;\;\; 100$ $x = 250$

43. Let b = speed of boat in still water and c = speed of current.

	d	r	t	Equation $(d = r \cdot t)$
Downstream	24	$b + c$	2	$24 = (b + c) \cdot 2 \Rightarrow 2b + 2c = 24$
Upstream	24	$b - c$	3	$24 = (b - c) \cdot 3 \Rightarrow 3b - 3c = 24$

$$\begin{array}{l} 2b + 2c = 24 \Rightarrow \times (3) \\ 3b - 3c = 24 \Rightarrow \times (2) \end{array} \quad \begin{array}{rl} 6b + 6c = & 72 \\ \underline{6b - 6c = } & \underline{48} \\ 12b = & 120 \\ b = & 10 \end{array}$$

The speed of the boat is 10 mph in still water.

45. Let p = airspeed of plane and w = speed of wind.

	d	r	t	Equation $(d = r \cdot t)$
With wind	600	$p + w$	2	$600 = (p + w) \cdot 2 \Rightarrow 2p + 2w = 600$
Against wind	600	$p - w$	3	$600 = (p - w) \cdot 3 \Rightarrow 3p - 3w = 600$

$$\begin{array}{l} 2p + 2w = 600 \Rightarrow \times (3) \\ 3p - 3w = 600 \Rightarrow \times (2) \end{array} \quad \begin{array}{rl} 6p + 6w = & 1800 \\ \underline{6p - 6w = } & \underline{1200} \\ 12p = & 3000 \\ p = & 250 \end{array}$$

$$\begin{array}{rl} 2p + 2w = & 600 \\ 2(250) + 2w = & 600 \\ 500 + 2w = & 600 \\ 2w = & 100 \\ w = & 50 \end{array}$$

The speed of the wind is 50 mph.

47. Let x = liters of first solution and y = liters of second solution.

	Fractional part that is alcohol	Number of liters of solution	Number of liters of alcohol
First solution	0.40	x	$0.40x$
Second solution	0.55	y	$0.55y$
Final solution	0.50	15	$0.50(15) = 7.5$

$$\begin{array}{l} x + y = 15 \\ 0.40x + 0.55y = 7.5 \Rightarrow \times (100) \end{array} \quad \begin{array}{l} x + y = 15 \Rightarrow \times (-40) \\ 40x + 55y = 750 \end{array} \quad \begin{array}{rl} -40x - 40y = & -600 \\ \underline{40x + 55y = } & \underline{750} \\ 15y = & 150 \\ y = & 10 \end{array}$$

$$\begin{array}{l} x + y = 15 \\ x + 10 = 15 \\ x = 5 \end{array}$$

The chemist should use 5 L of the 40% solution and 10 L of the 55% solution.

49. Let x = pounds of peanuts and y = pounds of cashews.

	Cost per pound	Number of pounds	Total value
Peanuts	3	x	$3x$
Cashews	6	y	$6y$
Mixture	4	48	$4(48) = 192$

$$\begin{array}{l} x + y = 48 \Rightarrow \times (-3) \\ 3x + 6y = 192 \end{array} \quad \begin{array}{rl} -3x - 3y = & -144 \\ \underline{3x + 6y = } & \underline{192} \\ 3y = & 48 \\ y = & 16 \end{array} \quad \begin{array}{l} x + y = 48 \\ x + 16 = 48 \\ x = 32 \end{array}$$

The merchant should use 32 pounds of peanuts and 16 pounds of cashews.

51. Let x = number of inexpensive radios and y = number of expensive radios.

$$x + \quad y = \quad 25 \Rightarrow \times (-87) \quad -87x - \quad 87y = -2175 \quad x + y = 25 \quad \text{There were 15}$$
$$\underline{87x + 119y = 2495} \qquad\qquad \underline{87x + 119y = \quad 2495} \quad x + 10 = 25 \quad \text{inexpensive radios}$$
$$32y = \qquad 320 \qquad\quad x = 15 \quad \text{sold.}$$
$$y = \qquad 10$$

53. Let r = the lower rate. Then $r + 0.01$ = the higher rate.

$$\boxed{\text{Interest on \$950}} + \boxed{\text{Interest on \$1,200}} = \boxed{\text{Total interest}}$$
$$950r + 1200(r + 0.01) = 205.50$$
$$950r + 1200r + 12 = 205.50$$
$$2150r = 193.50$$
$$r = 0.09 \Rightarrow \text{The lower rate is 9\%.}$$

55. $\begin{cases} (1) \quad p = -\frac{1}{2}q + 1300 \\ (2) \quad p = \frac{1}{3}q + \frac{1400}{3} \end{cases}$ Substitute:

$$-\frac{1}{2}q + 1300 = \frac{1}{3}q + \frac{1400}{3}$$
$$6\left(-\frac{1}{2}q + 1300\right) = 6\left(\frac{1}{3}q + \frac{1400}{3}\right)$$
$$-3q + 7800 = 2q + 2800$$
$$-5q = -5000$$
$$q = 1000$$

Substitute $q = 1000$ and solve for p:

$$p = -\frac{1}{2}q + 1300$$
$$= -\frac{1}{2}(1000) + 1300$$
$$= -500 + 1300 = 800$$

The equilibrium price is \$800.

57. Answers may vary.

Exercise 7.5 (page 464)

1. $3x + 5 = 14$
$3x = 9$
$x = 3$

3. $A = P + Prt$
$A - P = Prt$
$\dfrac{A - P}{Pr} = \dfrac{Prt}{Pr}$
$\dfrac{A - P}{Pr} = t$, or $t = \dfrac{A - P}{Pr}$

5. $2a + 5(a - 3) = 2a + 5a - 15 = 7a - 15$

7. $4(b - a) + 3b + 2a = 4b - 4a + 3b + 2a$
$= -2a + 7b$

9. inequality $\qquad$ **11.** boundary $\qquad$ **13.** inequalities $\qquad$ **15.** doubly shaded

17. a. $5x - 3y \geq 0$
$5(1) - 3(1) \overset{?}{\geq} 0$
$5 - 3 \overset{?}{\geq} 0$
$2 \geq 0$
$(1, 1)$ is a solution.

b. $5x - 3y \geq 0$
$5(-2) - 3(-3) \overset{?}{\geq} 0$
$-10 + 9 \overset{?}{\geq} 0$
$-1 \not\geq 0$
$(-2, -3)$ is not a solution.

17. c.
$$5x - 3y \geq 0$$
$$5(0) - 3(0) \overset{?}{\geq} 0$$
$$0 - 0 \overset{?}{\geq} 0$$
$$0 \geq 0$$
$(0, 0)$ is a solution.

d.
$$5x - 3y \geq 0$$
$$5\left(\frac{1}{5}\right) - 3\left(\frac{4}{3}\right) \overset{?}{\geq} 0$$
$$1 - 4 \overset{?}{\geq} 0$$
$$-3 \not\geq 0$$
$\left(\dfrac{1}{5}, \dfrac{4}{3}\right)$ is not a solution.

19. a. does not include boundary

b. does include boundary

21.
Test point: $(0, 0)$
$$y \leq x + 2$$
$$0 \overset{?}{\leq} 0 + 2$$
$$0 \leq 2$$
Shade half-plane with test point.

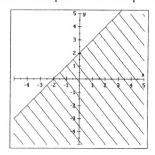

23.
Test point: $(0, 0)$
$$y > 2x - 4$$
$$0 \overset{?}{>} 2(0) - 4$$
$$0 \overset{?}{>} 0 - 4$$
$$0 > -4$$
Shade half-plane with test point.

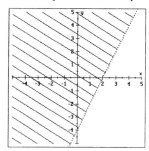

25.
Test point: $(0, 0)$
$$x - 2y \leq 4$$
$$0 - 2(0) \overset{?}{\leq} 4$$
$$0 - 0 \overset{?}{\leq} 4$$
$$0 \leq 4$$
Shade half-plane with test point.

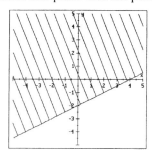

27.
Test point: $(1, 1)$
$$y \leq 4x$$
$$1 \overset{?}{\leq} 4(1)$$
$$1 \leq 4$$
Shade half-plane with test point.

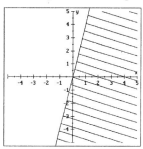

29.
Boundary (solid)

$y = 3 - x$

x	y
0	3
3	0

Test point: $(0, 0)$

$y \geq 3 - x$

$0 \overset{?}{\geq} 3 - 0$

$0 \not\geq 3$

opposite half-plane

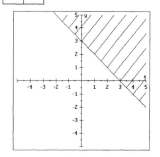

31.
Boundary (dotted)

$y = 2 - 3x$

x	y
0	2
1	-1

Test point: $(0, 0)$

$y < 2 - 3x$

$0 \overset{?}{<} 2 - 3(0)$

$0 \overset{?}{<} 2 - 0$

$0 < 2$

same half-plane

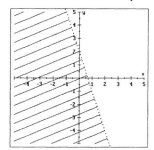

33.
Boundary (solid)

$y = 2x$

x	y
0	0
2	4

Test point: $(1, 1)$

$y \geq 2x$

$1 \overset{?}{\geq} 2(1)$

$1 \not\geq 2$

opposite half-plane

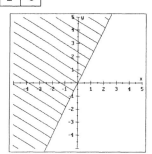

35.
Boundary (dotted)

$2y - x = 8$

x	y
0	4
-8	0

Test point: $(0, 0)$

$2y - x < 8$

$2(0) - 0 \overset{?}{<} 8$

$0 - 0 \overset{?}{<} 8$

$0 < 8$

same half-plane

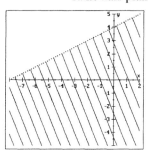

37.   Boundary   Test point: $(0,0)$
(dotted) $3x - 4y > 12$
$3x - 4y = 12$ $3(0) - 4(0) \overset{?}{>} 12$

x	y
0	-3
4	0

$0 - 0 \overset{?}{>} 12$
$0 \not> 12$
opposite half-plane

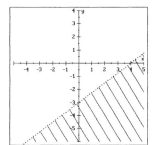

39.   Boundary   Test point: $(0,0)$
(solid) $5x + 4y \geq 20$
$5x + 4y = 20$ $5(0) + 4(0) \overset{?}{\geq} 20$

x	y
0	5
4	0

$0 + 0 \overset{?}{\geq} 20$
$0 \not\geq 20$
opposite half-plane

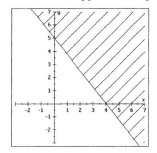

41.   Boundary   Test point: $(0,0)$
(dotted) $x < 2$
$x = 2$ $0 < 2$

x	y
2	0
2	4

same half-plane

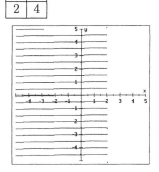

43.   Boundary   Test point: $(0,0)$
(solid) $y \leq 1$
$y = 1$ $0 \leq 1$

x	y
0	1
4	1

same half-plane

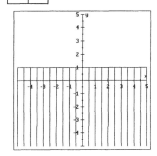

45. $x + 2y \leq 3$ $2x - y \geq 1$

x	y
1	1
3	0

x	y
0	-1
2	3

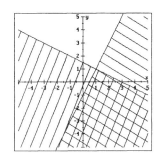

47. $x + y < -1$ $x - y > -1$

x	y
0	-1
-1	0

x	y
0	1
-1	0

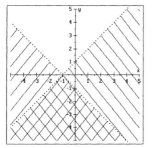

49. $2x - y < 4$ $x + y \geq -1$

x	y
0	-4
2	0

x	y
0	-1
-1	0

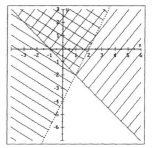

51. $x > 2$ $y \leq 3$

x	y
2	0
2	2

x	y
0	3
1	3

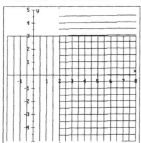

53. $x + y < 1$ $x + y > 3$

x	y
0	1
1	0

x	y
0	3
3	0

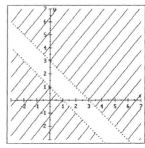

(no solutions)

55. $3x + 4y > -7$ $2x - 3y \geq 1$

x	y
-1	-1
-5	2

x	y
2	1
-1	-1

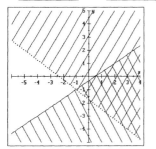

57. $2x - 4y > -6$ $3x + y \geq 5$

x	y
1	2
-3	0

x	y
0	5
1	2

59. $3x - y \leq -4$ $3y > -2(x + 5)$

x	y
0	4
-1	1

x	y
1	-4
-5	0

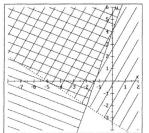

61. $\dfrac{x}{2} + \dfrac{y}{3} \geq 2$ $\dfrac{x}{2} - \dfrac{y}{2} < -1$

$3x + 2y \geq 12$ $x - y < -2$

x	y
0	6
4	0

x	y
0	2
-2	0

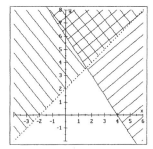

63. $\boxed{\text{Cake cost}} + \boxed{\text{Pie cost}} \leq 120$

$3x + 4y \leq 120$

Boundary   Test point: $(0,0)$
(solid) $3x + 4y \leq 120$

$3x + 4y = 120$ $3(0) + 4(0) \overset{?}{\leq} 120$

x	y
0	30
40	0

$0 + 0 \overset{?}{\leq} 120$

$0 \leq 120$

same half-plane

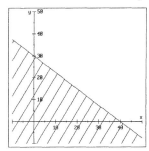

Solutions: $(10, 10), (20, 10), (10, 20)$

65. $\boxed{\text{Leather}} + \boxed{\text{Nylon}} \geq 4400$

$100x + 88y \geq 4400$

Boundary   Test point: $(0,0)$
(solid) $100x + 88y \geq 4400$

x	y
0	50
44	0

$100(0) + 88(0) \overset{?}{\geq} 4400$

$0 + 0 \overset{?}{\geq} 4400$

$0 \not\geq 4400$

opposite half-plane

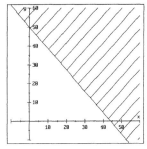

Solutions: $(50, 50), (30, 40), (40, 40)$

67. $\boxed{\text{R. stock}} + \boxed{\text{M. stock}} \le 8000$

$40x + 50y \le 8000$

Boundary   Test point: $(0, 0)$

(solid) $40x + 50y \le 8000$

$40x + 50y = 8000$ $40(0) + 50(0) \overset{?}{\le} 8000$

x	y
0	160
200	0

$0 + 0 \overset{?}{\le} 8000$

$0 \le 8000$

same half-plane

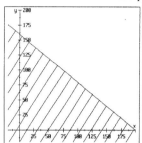

Solutions: $(80, 40), (80, 80), (120, 40)$

69. $10x + 15y \ge 30 \qquad 10x + 15y \le 60$

x	y
0	2
3	0

x	y
0	4
6	0

Solutions: $(1, 2), (4, 1)$

71. $150x + 100y \le 900 \qquad y > x$

x	y
0	9
6	0

x	y
0	0
1	1

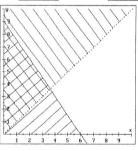

Solutions: $(2, 4), (1, 5)$

73-79. **Answers may vary.**

Chapter 7 Summary (page 470)

1. $3x - y = -2 \qquad 2x + 3y = 17$

$3(1) - 5 \overset{?}{=} -2 \quad 2(1) + 3(5) \overset{?}{=} 17$

$3 - 5 \overset{?}{=} -2 \qquad 2 + 15 \overset{?}{=} 17$

$-2 = -2 \qquad\qquad 17 = 17$

$(1, 5)$ is a solution to the system.

2. $5x + 3y = 2 \qquad -3x + 2y = 16$

$5(-2) + 3(4) \overset{?}{=} 2 \quad -3(-2) + 2(4) \overset{?}{=} 16$

$-10 + 12 \overset{?}{=} 2 \qquad 6 + 8 \overset{?}{=} 16$

$2 = 2 \qquad\qquad 14 \ne 16$

$(-2, 4)$ is not a solution to the system.

3.
$$2x + 4y = 30 \qquad \frac{x}{4} - y = 3$$
$$2(14) + 4\left(\frac{1}{2}\right) \overset{?}{=} 30 \qquad \frac{14}{4} - \frac{1}{2} \overset{?}{=} 3$$
$$28 + 2 \overset{?}{=} 30 \qquad \frac{7}{2} - \frac{1}{2} \overset{?}{=} 3$$
$$30 = 30 \qquad \frac{6}{2} = 3$$

$\left(14, \frac{1}{2}\right)$ is a solution to the system.

4.
$$4x - 6y = 18 \qquad \frac{x}{3} + \frac{y}{2} = \frac{5}{6}$$
$$4\left(\frac{7}{2}\right) - 6\left(-\frac{2}{3}\right) \overset{?}{=} 18 \qquad \frac{\frac{7}{2}}{3} + \frac{-\frac{2}{3}}{2} \overset{?}{=} \frac{5}{6}$$
$$14 + 4 \overset{?}{=} 18 \qquad \frac{7}{6} - \frac{2}{6} \overset{?}{=} \frac{5}{6}$$
$$18 = 18 \qquad \frac{5}{6} = \frac{5}{6}$$

$\left(\frac{7}{2}, -\frac{2}{3}\right)$ is a solution to the system.

5. $x + y = 7 \qquad 2x - y = 5$

x	y
0	7
7	0

x	y
2	-1
3	1

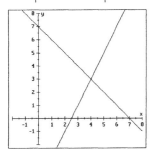

$\boxed{(4,3)}$

6. $\frac{x}{3} + \frac{y}{5} = -1 \qquad x - 3y = -3$

x	y
0	-5
-3	0

x	y
0	1
-3	0

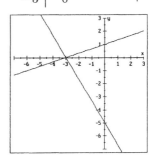

$\boxed{(-3,0)}$

7. $3x + 6y = 6 \qquad x + 2y = 2$

x	y
0	1
2	0

x	y
0	1
2	0

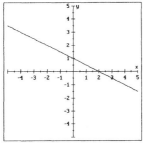

(dependent equations)

8. $6x + 3y = 12 \qquad 2x + y = 2$

x	y
0	4
2	0

x	y
0	2
1	0

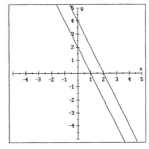

(inconsistent system)

9. $\begin{cases} (1) \quad x = 3y + 5 \\ (2) \quad 5x - 4y = 3 \end{cases}$

Substitute $x = 3y + 5$ from (1): Substitute and solve for x: Solution:

$$5x - 4y = 3$$
$$5(3y + 5) - 4y = 3$$
$$15y + 25 - 4y = 3$$
$$11y = -22$$
$$y = -2$$

$$x = 3y + 5$$
$$x = 3(-2) + 5$$
$$x = -6 + 5 = -1$$

$\boxed{x = -1, y = -2}$

10. $\begin{cases} 3x - \dfrac{2y}{5} = 2(x - 2) \;\Rightarrow\; 15x - 2y = 10(x - 2) \;\Rightarrow\; 5x - 2y = -20 \;\;(1) \\ 2x - 3 = 3 - 2y \qquad\quad \Rightarrow 2x + 2y = 6 \qquad\qquad\quad \Rightarrow 2x + 2y = 6 \;\;(2) \end{cases}$

Substitute $y = \dfrac{6 - 2x}{2}$ from (2): Substitute and solve for y: Solution:

$$5x - 2y = -20$$
$$5x - 2\left(\frac{6 - 2x}{2}\right) = -20$$
$$5x - (6 - 2x) = -20$$
$$5x - 6 + 2x = -20$$
$$7x = -14$$
$$x = -2$$

$$y = \frac{6 - 2x}{2}$$
$$y = \frac{6 - 2(-2)}{2}$$
$$y = \frac{6 + 4}{2} = \frac{10}{2} = 5$$

$\boxed{x = -2, y = 5}$

11. $\begin{cases} 8x + 5y = 3 \;\;(1) \\ 5x - 8y = 13 \;\;(2) \end{cases}$

Substitute $y = \dfrac{3 - 8x}{5}$ from (1): Substitute and solve for y: Solution:

$$5x - 8y = 13$$
$$5x - 8\left(\frac{3 - 8x}{5}\right) = 13$$
$$25x - 8(3 - 8x) = 65$$
$$25x - 24 + 64x = 65$$
$$89x = 89$$
$$x = 1$$

$$y = \frac{3 - 8x}{5}$$
$$y = \frac{3 - 8(1)}{5}$$
$$y = \frac{3 - 8}{5} = \frac{-5}{5} = -1$$

$\boxed{x = 1, y = -1}$

12. $\begin{cases} 6(x + 2) = y - 1 \;\Rightarrow\; 6x + 12 = y - 1 \;\Rightarrow\; 6x - y = -13 \;\;(1) \\ 5(y - 1) = x + 2 \;\Rightarrow\; 5y - 5 = x + 2 \;\;\;\Rightarrow\; -x + 5y = 7 \;\;(2) \end{cases}$

Substitute $y = 6x + 13$ from (1): Substitute and solve for y: Solution:

$$-x + 5y = 7$$
$$-x + 5(6x + 13) = 7$$
$$-x + 30x + 65 = 7$$
$$29x = -58$$
$$x = -2$$

$$y = 6x + 13$$
$$y = 6(-2) + 13$$
$$y = -12 + 13 = 1$$

$\boxed{x = -2, y = 1}$

13.
$$\begin{aligned} 2x + y &= 1 \\ 5x - y &= 20 \\ \hline 7x &= 21 \\ x &= 3 \end{aligned}$$

$$\begin{aligned} 2x + y &= 1 \\ 2(3) + y &= 1 \\ 6 + y &= 1 \\ y &= -5 \end{aligned}$$

Solution:

$\boxed{(3, -5)}$

14.
$$\begin{aligned} x + 8y &= 7 \Rightarrow \times (-1) \\ x - 4y &= 1 \end{aligned}$$

$$\begin{aligned} -x - 8y &= -7 \\ x - 4y &= 1 \\ \hline -12y &= -6 \\ y &= \tfrac{1}{2} \end{aligned}$$

$$\begin{aligned} x + 8y &= 7 \\ x + 8\left(\tfrac{1}{2}\right) &= 7 \\ x + 4 &= 7 \\ x &= 3 \end{aligned}$$

Solution:

$\boxed{\left(3, \tfrac{1}{2}\right)}$

15.
$$\begin{aligned} 5x + y &= 2 \Rightarrow \times (-2) \\ 3x + 2y &= 11 \end{aligned}$$

$$\begin{aligned} -10x - 2y &= -4 \\ 3x + 2y &= 11 \\ \hline -7x &= 7 \\ x &= -1 \end{aligned}$$

$$\begin{aligned} 5x + y &= 2 \\ 5(-1) + y &= 2 \\ -5 + y &= 2 \\ y &= 7 \end{aligned}$$

Solution:

$\boxed{(-1, 7)}$

16.
$$\begin{aligned} x + y &= 3 \Rightarrow \\ 3x &= 2 - y \Rightarrow \end{aligned}$$
$$\begin{aligned} x + y &= 3 \Rightarrow \times (-1) \\ 3x + y &= 2 \end{aligned}$$

$$\begin{aligned} -x - y &= -3 \\ 3x + y &= 2 \\ \hline 2x &= -1 \\ x &= -\tfrac{1}{2} \end{aligned}$$

$$\begin{aligned} x + y &= 3 \\ -\tfrac{1}{2} + y &= \tfrac{6}{2} \\ y &= \tfrac{7}{2} \end{aligned}$$

Solution:

$\boxed{\left(-\tfrac{1}{2}, \tfrac{7}{2}\right)}$

17.
$$\begin{aligned} 11x + 3y &= 27 \Rightarrow \times (4) \\ 8x + 4y &= 36 \Rightarrow \times (-3) \end{aligned}$$

$$\begin{aligned} 44x + 12y &= 108 \\ -24x - 12y &= -108 \\ \hline 20x &= 0 \\ x &= 0 \end{aligned}$$

$$\begin{aligned} 11x + 3y &= 27 \\ 11(0) + 3y &= 27 \\ 0 + 3y &= 27 \\ 3y &= 27 \\ y &= 9 \end{aligned}$$

Solution:

$\boxed{(0, 9)}$

18.
$$\begin{aligned} 9x + 3y &= 5 \Rightarrow \\ 3x &= 4 - y \Rightarrow \end{aligned}$$
$$\begin{aligned} 9x + 3y &= 5 \\ 3x + y &= 4 \Rightarrow \times (-3) \end{aligned}$$

$$\begin{aligned} 9x + 3y &= 5 \\ -9x - 3y &= -12 \\ \hline 0 &\neq -7 \end{aligned}$$

$\boxed{\text{inconsistent system}}$

19.
$$\begin{aligned} 9x + 3y &= 5 \\ 3x + y &= \tfrac{5}{3} \Rightarrow \times (-3) \end{aligned}$$

$$\begin{aligned} 9x + 3y &= 5 \\ -9x - 3y &= -5 \\ \hline 0 &= 0 \end{aligned}$$

$\boxed{\text{dependent equations}}$

20.
$$\begin{aligned} \frac{x}{3} + \frac{y+2}{2} &= 1 \Rightarrow \times (6) \\ \frac{x+8}{8} + \frac{y-3}{3} &= 0 \Rightarrow \times (24) \end{aligned}$$

$$\begin{aligned} 2x + 3(y+2) &= 6 \Rightarrow \\ 3(x+8) + 8(y-3) &= 0 \Rightarrow \end{aligned}$$

$$\begin{aligned} 2x + 3y + 6 &= 6 \\ 3x + 24 + 8y - 24 &= 0 \end{aligned}$$

$$\begin{aligned} 2x + 3y &= 0 \Rightarrow \times (3) \\ 3x + 8y &= 0 \Rightarrow \times (-2) \end{aligned}$$

$$\begin{aligned} 6x + 9y &= 0 \\ -6x - 16y &= 0 \\ \hline -7y &= 0 \\ y &= 0 \end{aligned}$$

$$\begin{aligned} 2x + 3y &= 0 \\ 2x + 3(0) &= 0 \\ 2x &= 0 \\ x &= 0 \end{aligned}$$

Solution:

$\boxed{(0, 0)}$

21. Let x and y represent the integers. Substitute $x = 5y$ into (2): Substitute; solve for x:

$$\begin{cases} (1) & x = 5y \\ (2) & x + y = 18 \end{cases}$$

$$\begin{aligned} x + y &= 18 \\ 5y + y &= 18 \\ 6y &= 18 \\ y &= 3 \end{aligned}$$

$$\begin{aligned} x &= 5y \\ x &= 5(3) = 15 \end{aligned}$$

The integers are 3 and 15.

22. Let $w =$ the width.   Substitute $l = 3w$ into (2): Substitute; solve for l:

Let $l =$ the length.

$$\begin{cases} (1) & l = 3w \\ (2) & 2l + 2w = 24 \end{cases}$$

$$\begin{aligned} 2l + 2w &= 24 \\ 2(3w) + 2w &= 24 \\ 6w + 2w &= 24 \\ 8w &= 24 \\ w &= 3 \end{aligned}$$

$$\begin{aligned} l &= 3w \\ l &= 3(3) = 9 \end{aligned}$$

The dimensions are 3 ft by 9 ft.

23. Let $x =$ the cost of a grapefruit.   Substitute $x = y + 15$ into (2): Substitute; solve for x:

Let $y =$ the cost of an orange.

$$\begin{cases} (1) & x = y + 15 \\ (2) & x + y = 85 \end{cases}$$

$$\begin{aligned} x + y &= 85 \\ y + 15 + y &= 85 \\ 2y &= 70 \\ y &= 35 \end{aligned}$$

$$\begin{aligned} x &= y + 15 \\ x &= 35 + 15 = 50 \end{aligned}$$

A grapefruit costs 50¢.

24. Let $x =$ the electric bill. Substitute $x = y - 23$ into (2): The gas bill was \$66.

Let $y =$ the gas bill.

$$\begin{cases} (1) & x = y - 23 \\ (2) & x + y = 109 \end{cases}$$

$$\begin{aligned} x + y &= 109 \\ y - 23 + y &= 109 \\ 2y &= 132 \\ y &= 66 \end{aligned}$$

25. Let $x =$ cost of a gallon of milk and $y =$ cost of one dozen eggs.

$$\begin{aligned} 2x + 3y = 6.80 &\Rightarrow \times(-2) & -4x - 6y &= -13.60 \\ 3x + 2y = 7.35 &\Rightarrow \times(3) & 9x + 6y &= 22.05 \\ \hline & & 5x &= 8.45 \\ & & x &= 1.69 \end{aligned}$$

A gallon of milk costs \$1.69.

26. Let $x =$ principal at 10% and $y =$ principal at 6%.

$$\begin{aligned} x + y &= 3000 \\ 0.10x + 0.06y &= 270 \end{aligned}$$

$$\begin{aligned} x + y &= 3000 \Rightarrow \times(-6) \\ 10x + 6y &= 27000 \end{aligned}$$

$$\begin{aligned} -6x - 6y &= -18000 \\ 10x + 6y &= 27000 \\ \hline 4x &= 9000 \\ x &= 2250 \end{aligned}$$

$$\begin{aligned} x + y &= 3000 \\ 2250 + y &= 3000 \\ y &= 750 \end{aligned}$$

He invested \$750 at the 6% rate.

27. Boundary   Test point: $(0, 0)$

(solid) $y \geq x + 2$

$y = x + 2$ $\overset{?}{0 \geq 0 + 2}$

x	y
0	2
-2	0

$0 \not\geq 2$

opposite half-plane

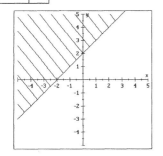

28. Boundary   Test point: $(0, 0)$

(dotted) $x < 3$

$x = 3$ $0 < 3$

x	y
3	0
3	4

same half-plane

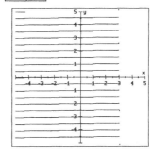

29. $5x + 3y < 15$ $3x - y > 3$

x	y
0	5
3	0

x	y
0	-3
1	0

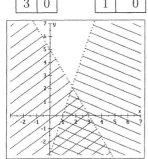

30. $5x - 3y \geq 5$ $3x + 2y \geq 3$

x	y
1	0
4	5

x	y
1	0
3	-3

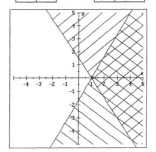

31. $x \geq 3y$ $y < 3x$

x	y
0	0
3	1

x	y
0	0
1	3

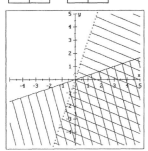

32. $x \geq 0$ $x \leq 3$

x	y
0	0
0	4

x	y
3	1
3	0

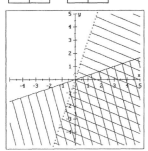

Chapter 7 Test (page 473)

1.
$$3x - 2y = 12 \qquad 2x + 3y = -5$$
$$3(2) - 2(-3) \stackrel{?}{=} 12 \quad 2(2) + 3(-3) \stackrel{?}{=} -5$$
$$6 + 6 \stackrel{?}{=} 12 \qquad 4 - 9 \stackrel{?}{=} -5$$
$$12 = 12 \qquad\qquad -5 = -5$$
$(2, -3)$ is a solution to the system.

2.
$$4x + y = -9 \qquad 2x - 3y = -7$$
$$4(-2) + (-1) \stackrel{?}{=} -9 \quad 2(-2) - 3(-1) \stackrel{?}{=} -7$$
$$-8 - 1 \stackrel{?}{=} -9 \qquad -4 + 3 \stackrel{?}{=} -7$$
$$-9 = -9 \qquad\qquad -1 \neq -7$$
$(-2, -1)$ is not a solution to the system.

3. $\quad 3x + y = 7 \quad x - 2y = 0$

x	y
0	7
2	1

x	y
0	0
4	2

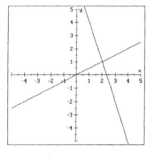

$\boxed{(2, 1)}$

4. $\quad x + \dfrac{y}{2} = 1 \quad y = 1 - 3x$

x	y
0	2
1	0

x	y
0	1
1	-2

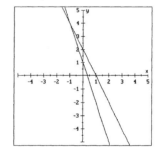

$\boxed{(-1, 4)}$

5.
$$\begin{cases} (1) \quad y = x - 1 \\ (2) \quad x + y = -7 \end{cases}$$

Substitute $y = x - 1$ from (1):
$$x + y = -7$$
$$x + x - 1 = -7$$
$$2x = -6$$
$$x = -3$$

Substitute and solve for y:
$$y = x - 1$$
$$y = -3 - 1$$
$$y = -4$$

Solution:
$$\boxed{x = -3, y = -4}$$

6. $\begin{cases} (1) & \dfrac{x}{6} + \dfrac{y}{10} = 3 \quad \Rightarrow 5x + 3y = 90 \quad \Rightarrow x = \dfrac{90 - 3y}{5} \\ (2) & \dfrac{5x}{16} - \dfrac{3y}{16} = \dfrac{15}{8} \quad \Rightarrow 5x - 3y = 30 \end{cases}$

Substitute $x = \dfrac{90 - 3y}{5}$ from (1) into (2):   Substitute and solve for x: Solution:

$$5x - 3y = 30$$
$$5\left(\dfrac{90 - 3y}{5}\right) - 3y = 30$$
$$90 - 3y - 3y = 30$$
$$-6y = -60$$
$$y = 10$$

$x = \dfrac{90 - 3y}{5}$

$x = \dfrac{90 - 3(10)}{5}$

$x = \dfrac{90 - 30}{5} = \dfrac{60}{5} = 12$

$\boxed{x = 12, y = 10}$

7. $\begin{array}{l} 3x - y = 2 \\ 2x + y = 8 \\ \hline 5x \quad\;\; = 10 \\ \;x \quad\;\; = 2 \end{array}$ $\begin{array}{l} 2x + y = 8 \\ 2(2) + y = 8 \\ 4 + y = 8 \\ y = 4 \end{array}$ Solution: $\boxed{(2, 4)}$

8. $4x + \;\;3 = -3y$ $4x + 3y = -3 \Rightarrow \times (3)$ $\begin{array}{l} 12x + \;\,9y = -9 \\ -12x + 16y = \;\,84 \\ \hline 25y = \;\,75 \\ y = \;\;\,3 \end{array}$ $\begin{array}{l} 4x + 3y = -3 \\ 4x + 3(3) = -3 \\ 4x + 9 = -3 \\ 4x = -12 \\ x = -3 \end{array}$

$\dfrac{-x}{7} + \dfrac{4y}{21} = \;\;1 \Rightarrow \times (21)$ $-3x + 4y = \;21 \Rightarrow \times (4)$

$\boxed{(-3, 3)}$

9. $\begin{array}{l} 2x + 3(y - 2) = 0 \Rightarrow 2x + 3y - 6 = 0 \Rightarrow \;\;\;2x + 3y = 6 \\ -3y = 2(x - 4) \Rightarrow \;\;\;\;-3y = 2x - 8 \Rightarrow -2x - 3y = -8 \\ \hline \hspace{9cm} 0 \neq -2 \quad \Rightarrow \text{inconsistent} \end{array}$

10. $\begin{array}{l} \dfrac{x}{3} + y - 4 = 0 \Rightarrow x + 3y - 12 = 0 \Rightarrow \;\;\;\;x + 3y = 12 \\ -3y = x - 12 \Rightarrow -x - 3y = -12 \Rightarrow -x - 3y = -12 \\ \hline \hspace{9cm} 0 = 0 \quad \Rightarrow \text{dependent (and consistent)} \end{array}$

11. Let x and y represent the numbers. Substitute $x = 3y + 2$. Substitute; solve for x:

$\begin{cases} (1) & x = 3y + 2 \\ (2) & x + y = -18 \end{cases}$

$\begin{array}{l} x + y = -18 \\ 3y + 2 + y = -18 \\ 4y = -20 \\ y = -5 \end{array}$

$\begin{array}{l} x = 3y + 2 \\ x = 3(-5) + 2 \\ x = -15 + 2 = -13 \\ \text{The numbers are } -5 \text{ and } -13. \\ \text{The product is } (-5)(-13) = 65. \end{array}$

12. Let x = principal at 8% and y = principal at 9%.

$$x + y = 10000 \qquad\qquad x + y = 10000 \Rightarrow \times(-9) \quad -9x - 9y = -90000$$
$$\underline{0.08x + 0.09y = 840} \Rightarrow \times(100) \quad \underline{8x + 9y = 84000} \qquad\qquad \underline{8x + 9y = 84000}$$
$$\begin{aligned}-x &= -6000\\ x &= 6000\end{aligned}$$

$$x + y = 10000 \qquad \text{She invested \$4000 at the 9\% rate.}$$
$$6000 + y = 10000$$
$$\phantom{6000 + {}}y = 4000$$

13. $\quad x + y < 3 \quad x - y < 1$

x	y
0	3
3	0

x	y
0	-1
1	0

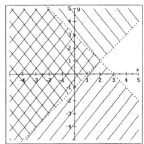

14. $\quad 2x + 3y \le 6 \quad x \ge 2$

x	y
0	2
3	0

x	y
2	0
2	3

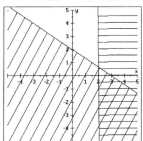

Exercise 8.1 (page 482)

1. $\qquad x = 3$

x	y
3	0
3	2

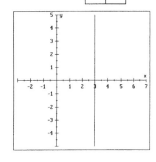

3. $\quad -2x + y = 4$

x	y
0	4
-2	0

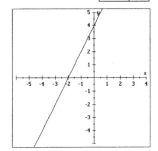

5. $\quad b^2 = a$

7. positive

9. right

11. two; 5; -5

13. square

15. hypotenuse; $a^2 + b^2$

17. $\sqrt{9} = \sqrt{3^2} = 3$

19. $\sqrt{49} = \sqrt{7^2} = 7$

21. $\sqrt{36} = \sqrt{6^2} = 6$

23. $\sqrt{\dfrac{1}{81}} = \sqrt{\left(\dfrac{1}{9}\right)^2} = \dfrac{1}{9}$

25. $-\sqrt{25} = -\sqrt{5^2} = -5$

27. $-\sqrt{81} = -\sqrt{9^2} = -9$

29. $\sqrt{196} = \sqrt{14^2} = 14$

31. $\sqrt{\dfrac{9}{256}} = \sqrt{\left(\dfrac{3}{16}\right)^2} = \dfrac{3}{16}$

33. $-\sqrt{289} = -\sqrt{17^2} = -17$

35. $\sqrt{10,000} = \sqrt{100^2} = 100$

37. $\sqrt{324} = \sqrt{18^2} = 18$

39. $-\sqrt{3,600} = -\sqrt{60^2} = -60$

41. $\sqrt{2} \approx 1.414$

43. $\sqrt{5} \approx 2.236$

45. $\sqrt{6} \approx 2.449$

47. $\sqrt{11} \approx 3.317$

49. $\sqrt{23} \approx 4.796$

51. $\sqrt{95} \approx 9.747$

53. $\sqrt{6,428} \approx 80.175$

55. $-\sqrt{9,876} \approx -99.378$

57. $\sqrt{21.35} \approx 4.621$

59. $\sqrt{0.3588} \approx 0.599$

61. $\sqrt{0.9925} \approx 0.996$

63. $-\sqrt{0.8372} \approx -0.915$

65. $\sqrt{9} = 3$
rational

67. $\sqrt{49} = 7$
rational

69. $-\sqrt{5} \approx -2.236$
irrational

71. $\sqrt{-100}$
imaginary

73. $f(x) = 1 + \sqrt{x}$

x	$f(x)$
0	1
1	2
4	3

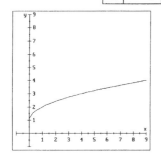

75. $f(x) = -\sqrt{x}$

x	$f(x)$
0	0
1	-1
4	-2

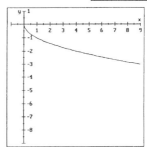

77.
$a^2 + b^2 = c^2$
$4^2 + 3^2 = c^2$
$16 + 9 = c^2$
$25 = c^2$
$\sqrt{25} = c$
$5 = c$

79.
$a^2 + b^2 = c^2$
$5^2 + 12^2 = c^2$
$25 + 144 = c^2$
$169 = c^2$
$\sqrt{169} = c$
$13 = c$

81.
$a^2 + b^2 = c^2$
$21^2 + b^2 = 29^2$
$441 + b^2 = 841$
$b^2 = 400$
$b = \sqrt{400}$
$b = 20$

83.
$a^2 + b^2 = c^2$
$a^2 + 45^2 = 53^2$
$a^2 + 2025 = 2809$
$a^2 = 784$
$a = \sqrt{784}$
$a = 28$

85. See the figure to the right.

$$b^2 + 16^2 = 20^2$$
$$b^2 + 256 = 400$$
$$b^2 = 144$$
$$b = \sqrt{144} = 12$$

The base of the ladder is 12 feet from the wall.

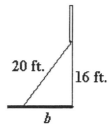

20 ft. 16 ft. b

87. See the figure to the right.

$$16^2 + x^2 = 34^2$$
$$256 + x^2 = 1156$$
$$x^2 = 900$$
$$x = \sqrt{900} = 30$$

The pole is 30 feet tall.

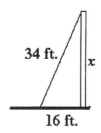

34 ft. x 16 ft.

89.
$$90^2 + 90^2 = x^2$$
$$8100 + 8100 = x^2$$
$$16200 = x^2$$
$$\sqrt{16200} = x$$
$$127.3 \approx x$$

The distance is about 127.3 feet.

91. See the figure to the right.

$$(4.2)^2 + (4.0)^2 = x^2$$
$$17.64 + 16 = x^2$$
$$33.64 = x^2$$
$$\sqrt{33.64} = x$$
$$5.8 = x$$

She is 5.8 miles from her starting place.

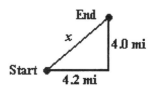

End x 4.0 mi Start 4.2 mi

93. See the figure to the right.

$$(0.6)^2 + (0.6)^2 = x^2$$
$$0.36 + 0.36 = x^2$$
$$0.72 = x^2$$
$$\sqrt{0.72} = x$$
$$0.8485 = x$$

The diameter is $2(0.375 + 0.8485)$, or about 2.4 inches.

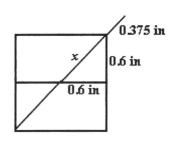

0.375 in x 0.6 in 0.6 in

95. See the figure to the right.

$$x^2 + x^2 = 5^2$$
$$2x^2 = 25$$
$$x^2 = 12.5$$
$$x = \sqrt{12.5}$$
$$x \approx 3.5$$

The play will only gain about 9.5 yards, not enough for a first down.

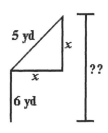

97. See the figure.

$$3^2 + 3^2 = x^2$$
$$18 = x^2$$
$$\sqrt{18} = x$$
$$4.2 \text{ ft} \approx x$$

99. See the figure.

$$10^2 + x^2 = 26^2$$
$$100 + x^2 = 676$$
$$x^2 = 576$$
$$x = \sqrt{576}$$
$$x = 24 \text{ in.}$$

The altitude is 24 in.

101. See the figure and find x.

$$4^2 + 14^2 = x^2$$
$$212 = x^2$$
$$\sqrt{212} = x$$
$$14.56 \approx x$$

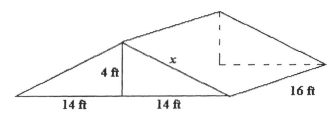

The ceiling is actually two rectangles, each with area $A = lw = (16)(14.56) = 232.92$ ft^2. The total area of the ceiling is then $2(232.96) = 465.92$ ft^2. Each sheet of plaster board has an area of $A = lw = (4)(8) = 32$ ft^2. Thus, the number of sheets needed is $465.92 \div 32 = 14.56$. Since a partial sheet cannot be purchased, 15 sheets will be needed.

103. Answers may vary.

105. Answers may vary.

Exercise 8.2 (page 491)

1. $f(0) = 2(0)^2 - 0 - 1 = 2(0) - 0 - 1$
$$= 0 - 0 - 1 = -1$$

3. $f(-2) = 2(-2)^2 - (-2) - 1$
$$= 2(4) + 2 - 1 = 8 + 2 - 1 = 9$$

5. $x^2 - 16y^2 = x^2 - (4y)^2 = (x + 4y)(x - 4y)$

7. $ax + ay + bx + by = a(x + y) + b(x + y)$
$$= (x + y)(a + b)$$

9. cube root

11. index; radicand

13. $V = s^3$

15. $(-6)^3 = -216$

17. $\sqrt[3]{1} = \sqrt[3]{1^3} = 1$

19. $\sqrt[3]{27} = \sqrt[3]{3^3} = 3$

21. $\sqrt[3]{-8} = \sqrt[3]{(-2)^3} = -2$ **23.** $\sqrt[3]{-64} = \sqrt[3]{(-4)^3} = -4$ **25.** $\sqrt[3]{125} = \sqrt[3]{5^3} = 5$

27. $-\sqrt[3]{-1} = -\sqrt[3]{(-1)^3}$ **29.** $-\sqrt[3]{64} = -\sqrt[3]{4^3} = -4$ **31.** $\sqrt[3]{729} = \sqrt[3]{9^3} = 9$
$\qquad = -(-1) = 1$

33. $\sqrt[3]{32,100} \approx 31.78$ **35.** $\sqrt[3]{-0.11324} \approx -0.48$

37.
$$f(x) = -\sqrt[3]{x}$$

x	-8	-1	0	1	8
$f(x)$	2	1	0	-1	-2

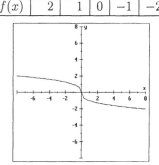

39.
$$f(x) = \sqrt[3]{x} - 2$$

x	-8	-1	0	1	8
$f(x)$	-4	-3	-2	-1	0

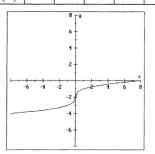

41. $\sqrt[4]{16} = \sqrt[4]{2^4} = 2$ **43.** $-\sqrt[5]{32} = -\sqrt[5]{2^5} = -2$ **45.** $\sqrt[6]{1} = \sqrt[6]{1^6} = 1$

47. $\sqrt[5]{-32} = \sqrt[5]{(-2)^5} = -2$ **49.** $\sqrt[4]{125} \approx 3.34$ **51.** $\sqrt[5]{-6,000} \approx -5.70$

53. $\sqrt{x^2y^2} = xy$ **55.** $\sqrt{x^4z^4} = x^2z^2$ **57.** $-\sqrt{x^4y^2} = -x^2y$

59. $\sqrt{4z^2} = 2z$ **61.** $-\sqrt{9x^4y^2} = -3x^2y$ **63.** $\sqrt{x^2y^2z^2} = xyz$

65. $-\sqrt{x^2y^2z^4} = -xyz^2$ **67.** $-\sqrt{25x^4z^{12}} = -5x^2z^6$ **69.** $\sqrt{36z^{36}} = 6z^{18}$

71. $-\sqrt{16z^2} = -4z$ **73.** $\sqrt[3]{27y^3z^6} = 3yz^2$ **75.** $\sqrt[3]{-8p^6q^3} = -2p^2q$

77. $V = 2$
$\qquad s^3 = 2$
$\qquad s = \sqrt[3]{2} \approx 1.26$ ft

79. $S = \sqrt[3]{\dfrac{P}{0.02}} = \sqrt[3]{\dfrac{400}{0.02}} \approx 27.14$ mph

81. $r = \sqrt[3]{\dfrac{l}{a}} = \sqrt[3]{\dfrac{192}{3}} = 4$ **83.** **Answers may vary.**

85. $\sqrt{x^2 - 4x + 4} = \sqrt{(x-2)^2} = x - 2$ if $x - 2 \ge 0$, that is, if $x \ge 2$.

Exercise 8.3 (page 497)

1. $\dfrac{5xy^2z^3}{10x^2y^2z^4} = \dfrac{5}{10}x^{1-2}y^{2-2}z^{3-4}$

$\quad = \dfrac{1}{2}x^{-1}y^0z^{-1} = \dfrac{1}{2xz}$

3. $\dfrac{a^2 - a - 2}{a^2 + a - 6} = \dfrac{(a-2)(a+1)}{(a+3)(a-2)} = \dfrac{a+1}{a+3}$

5. perfect

7. $\sqrt{a}\sqrt{b}$

9. $\sqrt{9+4} \neq \sqrt{9} + \sqrt{4}$

11. $\sqrt{20} = \sqrt{4}\sqrt{5} = 2\sqrt{5}$

13. $\sqrt{50} = \sqrt{25}\sqrt{2} = 5\sqrt{2}$

15. $\sqrt{45} = \sqrt{9}\sqrt{5} = 3\sqrt{5}$

17. $\sqrt{98} = \sqrt{49}\sqrt{2} = 7\sqrt{2}$

19. $\sqrt{48} = \sqrt{16}\sqrt{3} = 4\sqrt{3}$

21. $\sqrt{200} = \sqrt{100}\sqrt{2} = 10\sqrt{2}$

23. $\sqrt{192} = \sqrt{64}\sqrt{3} = 8\sqrt{3}$

25. $\sqrt{88} = \sqrt{4}\sqrt{22} = 2\sqrt{22}$

27. $\sqrt{324} = 18$

29. $\sqrt{147} = \sqrt{49}\sqrt{3} = 7\sqrt{3}$

31. $\sqrt{180} = \sqrt{36}\sqrt{5} = 6\sqrt{5}$

33. $\sqrt{432} = \sqrt{144}\sqrt{3} = 12\sqrt{3}$

35. $4\sqrt{288} = 4\sqrt{144}\sqrt{2} = 4(12)\sqrt{2} = 48\sqrt{2}$

37. $-7\sqrt{1,000} = -7\sqrt{100}\sqrt{10} = -7(10)\sqrt{10}$
$\quad = -70\sqrt{10}$

39. $2\sqrt{245} = 2\sqrt{49}\sqrt{5} = 2(7)\sqrt{5} = 14\sqrt{5}$

41. $-5\sqrt{162} = -5\sqrt{81}\sqrt{2} = -5(9)\sqrt{2}$
$\quad = -45\sqrt{2}$

43. $\sqrt{25x} = \sqrt{25}\sqrt{x} = 5\sqrt{x}$

45. $\sqrt{a^2b} = \sqrt{a^2}\sqrt{b} = a\sqrt{b}$

47. $\sqrt{9x^2y} = \sqrt{9x^2}\sqrt{y} = 3x\sqrt{y}$

49. $\dfrac{1}{5}x^2y\sqrt{50x^2y^2} = \dfrac{1}{5}x^2y\sqrt{25x^2y^2}\sqrt{2} = \dfrac{1}{5}x^2y \cdot 5xy\sqrt{2} = x^3y^2\sqrt{2}$

51. $12x\sqrt{16x^2y^3} = 12x\sqrt{16x^2y^2}\sqrt{y} = 12x \cdot 4xy\sqrt{y} = 48x^2y\sqrt{y}$

53. $-3xyz\sqrt{18x^3y^5} = -3xyz\sqrt{9x^2y^4}\sqrt{2xy} = -3xyz \cdot 3xy^2\sqrt{2xy} = -9x^2y^3z\sqrt{2xy}$

55. $\dfrac{3}{4}\sqrt{192a^3b^5} = \dfrac{3}{4}\sqrt{64a^2b^4}\sqrt{3ab} = \dfrac{3}{4} \cdot 8ab^2\sqrt{3ab} = 6ab^2\sqrt{3ab}$

57. $-\dfrac{2}{5}\sqrt{80mn^2} = -\dfrac{2}{5}\sqrt{16n^2}\sqrt{5m} = -\dfrac{2}{5} \cdot 4n\sqrt{5m} = -\dfrac{8n\sqrt{5m}}{5}$

59. $\sqrt{\dfrac{25}{9}} = \dfrac{\sqrt{25}}{\sqrt{9}} = \dfrac{5}{3}$

61. $\sqrt{\dfrac{81}{64}} = \dfrac{\sqrt{81}}{\sqrt{64}} = \dfrac{9}{8}$

63. $\sqrt{\dfrac{26}{25}} = \dfrac{\sqrt{26}}{\sqrt{25}} = \dfrac{\sqrt{26}}{5}$

65. $\sqrt{\dfrac{20}{49}} = \dfrac{\sqrt{20}}{\sqrt{49}} = \dfrac{\sqrt{4}\sqrt{5}}{7} = \dfrac{2\sqrt{5}}{7}$

67. $\sqrt{\dfrac{48}{81}} = \dfrac{\sqrt{48}}{\sqrt{81}} = \dfrac{\sqrt{16}\sqrt{3}}{9} = \dfrac{4\sqrt{3}}{9}$

69. $\sqrt{\dfrac{32}{25}} = \dfrac{\sqrt{16}\sqrt{2}}{5} = \dfrac{4\sqrt{2}}{5}$

71. $\sqrt{\dfrac{125}{121}} = \dfrac{\sqrt{125}}{\sqrt{121}} = \dfrac{\sqrt{25}\sqrt{5}}{11} = \dfrac{5\sqrt{5}}{11}$

73. $\sqrt{\dfrac{245}{36}} = \dfrac{\sqrt{245}}{\sqrt{36}} = \dfrac{\sqrt{49}\sqrt{5}}{6} = \dfrac{7\sqrt{5}}{6}$

75. $\sqrt{\dfrac{72x^3}{y^2}} = \dfrac{\sqrt{72x^3}}{\sqrt{y^2}} = \dfrac{\sqrt{36x^2}\sqrt{2x}}{y}$
$= \dfrac{6x\sqrt{2x}}{y}$

77. $\sqrt{\dfrac{125m^2n^5}{64n}} = \sqrt{\dfrac{125m^2n^4}{64}} = \dfrac{\sqrt{125m^2n^4}}{\sqrt{64}} = \dfrac{\sqrt{25m^2n^4}\sqrt{5}}{8} = \dfrac{5mn^2\sqrt{5}}{8}$

79. $\sqrt{\dfrac{128m^3n^5}{36mn^7}} = \sqrt{\dfrac{32m^2}{9n^2}} = \dfrac{\sqrt{32m^2}}{\sqrt{9n^2}} = \dfrac{\sqrt{16m^2}\sqrt{2}}{3n} = \dfrac{4m\sqrt{2}}{3n}$

81. $\sqrt{\dfrac{12r^7s^6t}{81r^5s^2t}} = \sqrt{\dfrac{12r^2s^4}{81}} = \dfrac{\sqrt{12r^2s^4}}{\sqrt{81}} = \dfrac{\sqrt{4r^2s^4}\sqrt{3}}{9} = \dfrac{2rs^2\sqrt{3}}{9}$

83. $\sqrt[3]{8x^3} = 2x$

85. $\sqrt[3]{-64x^5} = \sqrt[3]{-64x^3}\sqrt[3]{x^2} = -4x\sqrt[3]{x^2}$

87. $\sqrt[3]{54x^3y^4z^6} = \sqrt[3]{27x^3y^3z^6}\sqrt[3]{2y}$
$= 3xyz^2\sqrt[3]{2y}$

89. $\sqrt[3]{-81x^2y^3z^4} = \sqrt[3]{-27y^3z^3}\sqrt[3]{3x^2z}$
$= -3yz\sqrt[3]{3x^2z}$

91. $\sqrt[3]{\dfrac{27m^3}{8n^6}} = \dfrac{\sqrt[3]{27m^3}}{\sqrt[3]{8n^6}} = \dfrac{3m}{2n^2}$

93. $\sqrt[3]{\dfrac{16r^4s^5}{1000t^3}} = \dfrac{\sqrt[3]{16r^4s^5}}{\sqrt[3]{1000t^3}} = \dfrac{\sqrt[3]{8r^3s^3}\sqrt[3]{2rs^2}}{10t} = \dfrac{2rs\sqrt[3]{2rs^2}}{10t} = \dfrac{rs\sqrt[3]{2rs^2}}{5t}$

95. $\sqrt[3]{\dfrac{250a^3b^4}{16b}} = \sqrt[3]{\dfrac{125a^3b^3}{8}} = \dfrac{\sqrt[3]{125a^3b^3}}{\sqrt[3]{8}} = \dfrac{5ab}{2}$

97. Answers may vary.

99. $\sqrt{a+b} \neq \sqrt{a} + \sqrt{b}$
$\left(\sqrt{a} + \sqrt{b}\right)^2 \neq \left(\sqrt{a}\right)^2 + \left(\sqrt{b}\right)^2$
The conclusion is correct despite the errors made in the problem

Exercise 8.4 (page 502)

1.
$$\frac{a-2}{8} = \frac{a+10}{24}$$
$$24(a-2) = 8(a+10)$$
$$24a - 48 = 8a + 80$$
$$16a = 128$$
$$a = 8$$

3.
$$\frac{-2}{x+14} = \frac{6}{x-6}$$
$$-2(x-6) = 6(x+14)$$
$$-2x + 12 = 6x + 84$$
$$-8x = 72$$
$$x = -9$$

5. Like terms

7. not like radicals

9. like radicals

11. The radicals are not like radicals, so they may not be combined.

13. $\sqrt{12} + \sqrt{27} = \sqrt{4}\sqrt{3} + \sqrt{9}\sqrt{3}$
$$= 2\sqrt{3} + 3\sqrt{3} = 5\sqrt{3}$$

15. $\sqrt{48} + \sqrt{75} = \sqrt{16}\sqrt{3} + \sqrt{25}\sqrt{3}$
$$= 4\sqrt{3} + 5\sqrt{3} = 9\sqrt{3}$$

17. $\sqrt{45} + \sqrt{80} = \sqrt{9}\sqrt{5} + \sqrt{16}\sqrt{5}$
$$= 3\sqrt{5} + 4\sqrt{5} = 7\sqrt{5}$$

19. $\sqrt{125} + \sqrt{245} = \sqrt{25}\sqrt{5} + \sqrt{49}\sqrt{5}$
$$= 5\sqrt{5} + 7\sqrt{5} = 12\sqrt{5}$$

21. $\sqrt{20} + \sqrt{180} = \sqrt{4}\sqrt{5} + \sqrt{36}\sqrt{5}$
$$= 2\sqrt{5} + 6\sqrt{5} = 8\sqrt{5}$$

23. $\sqrt{160} + \sqrt{360} = \sqrt{16}\sqrt{10} + \sqrt{36}\sqrt{10}$
$$= 4\sqrt{10} + 6\sqrt{10} = 10\sqrt{10}$$

25. $3\sqrt{45} + 4\sqrt{245} = 3\sqrt{9}\sqrt{5} + 4\sqrt{49}\sqrt{5} = 3(3)\sqrt{5} + 4(7)\sqrt{5} = 9\sqrt{5} + 28\sqrt{5} = 37\sqrt{5}$

27. $2\sqrt{28} + 2\sqrt{112} = 2\sqrt{4}\sqrt{7} + 2\sqrt{16}\sqrt{7} = 2(2)\sqrt{7} + 2(4)\sqrt{7} = 4\sqrt{7} + 8\sqrt{7} = 12\sqrt{7}$

29. $5\sqrt{32} + 3\sqrt{72} = 5\sqrt{16}\sqrt{2} + 3\sqrt{36}\sqrt{2} = 5(4)\sqrt{2} + 3(6)\sqrt{2} = 20\sqrt{2} + 18\sqrt{2} = 38\sqrt{2}$

31. $3\sqrt{98} + 8\sqrt{128} = 3\sqrt{49}\sqrt{2} + 8\sqrt{64}\sqrt{2} = 3(7)\sqrt{2} + 8(8)\sqrt{2} = 21\sqrt{2} + 64\sqrt{2} = 85\sqrt{2}$

33. $\sqrt{20} + \sqrt{45} + \sqrt{80} = \sqrt{4}\sqrt{5} + \sqrt{9}\sqrt{5} + \sqrt{16}\sqrt{5} = 2\sqrt{5} + 3\sqrt{5} + 4\sqrt{5} = 9\sqrt{5}$

35. $\sqrt{24} + \sqrt{150} + \sqrt{240} = \sqrt{4}\sqrt{6} + \sqrt{25}\sqrt{6} + \sqrt{16}\sqrt{15} = 2\sqrt{6} + 5\sqrt{6} + 4\sqrt{15}$
$$= 7\sqrt{6} + 4\sqrt{15}$$

37. $\sqrt{18} - \sqrt{8} = \sqrt{9}\sqrt{2} - \sqrt{4}\sqrt{2}$
$$= 3\sqrt{2} - 2\sqrt{2} = \sqrt{2}$$

39. $\sqrt{9} - \sqrt{50} = 3 - \sqrt{25}\sqrt{2} = 3 - 5\sqrt{2}$

41. $\sqrt{72} - \sqrt{32} = \sqrt{36}\sqrt{2} - \sqrt{16}\sqrt{2}$
$$= 6\sqrt{2} - 4\sqrt{2} = 2\sqrt{2}$$

43. $\sqrt{12} - \sqrt{48} = \sqrt{4}\sqrt{3} - \sqrt{16}\sqrt{3}$
$$= 2\sqrt{3} - 4\sqrt{3} = -2\sqrt{3}$$

45. $\sqrt{108} - \sqrt{75} = \sqrt{36}\sqrt{3} - \sqrt{25}\sqrt{3} = 6\sqrt{3} - 5\sqrt{3} = \sqrt{3}$

47. $\sqrt{1,000} - \sqrt{360} = \sqrt{100}\sqrt{10} - \sqrt{36}\sqrt{10} = 10\sqrt{10} - 6\sqrt{10} = 4\sqrt{10}$

49. $2\sqrt{80} - 3\sqrt{125} = 2\sqrt{16}\sqrt{5} - 3\sqrt{25}\sqrt{5} = 2(4)\sqrt{5} - 3(5)\sqrt{5} = 8\sqrt{5} - 15\sqrt{5} = -7\sqrt{5}$

51. $8\sqrt{96} - 5\sqrt{24} = 8\sqrt{16}\sqrt{6} - 5\sqrt{4}\sqrt{6} = 8(4)\sqrt{6} - 5(2)\sqrt{6} = 32\sqrt{6} - 10\sqrt{6} = 22\sqrt{6}$

53. $\sqrt{288} - 3\sqrt{200} = \sqrt{144}\sqrt{2} - 3\sqrt{100}\sqrt{2} = 12\sqrt{2} - 3(10)\sqrt{2} = 12\sqrt{2} - 30\sqrt{2} = -18\sqrt{2}$

55. $5\sqrt{250} - 3\sqrt{160} = 5\sqrt{25}\sqrt{10} - 3\sqrt{16}\sqrt{10} = 5(5)\sqrt{10} - 3(4)\sqrt{10} = 25\sqrt{10} - 12\sqrt{10}$
$$= 13\sqrt{10}$$

57. $\sqrt{12} + \sqrt{18} - \sqrt{27} = \sqrt{4}\sqrt{3} + \sqrt{9}\sqrt{2} - \sqrt{9}\sqrt{3} = 2\sqrt{3} + 3\sqrt{2} - 3\sqrt{3} = 3\sqrt{2} - \sqrt{3}$

59. $\sqrt{200} - \sqrt{75} + \sqrt{48} = \sqrt{100}\sqrt{2} - \sqrt{25}\sqrt{3} + \sqrt{16}\sqrt{3} = 10\sqrt{2} - 5\sqrt{3} + 4\sqrt{3} = 10\sqrt{2} - \sqrt{3}$

61. $\sqrt{24} - \sqrt{150} - \sqrt{54} = \sqrt{4}\sqrt{6} - \sqrt{25}\sqrt{6} - \sqrt{9}\sqrt{6} = 2\sqrt{6} - 5\sqrt{6} - 3\sqrt{6} = -6\sqrt{6}$

63. $\sqrt{200} + \sqrt{300} - \sqrt{75} = \sqrt{100}\sqrt{2} + \sqrt{100}\sqrt{3} - \sqrt{25}\sqrt{3} = 10\sqrt{2} + 10\sqrt{3} - 5\sqrt{3}$
$$= 10\sqrt{2} + 5\sqrt{3}$$

65. $\sqrt{48} - \sqrt{8} + \sqrt{27} - \sqrt{32} = \sqrt{16}\sqrt{3} - \sqrt{4}\sqrt{2} + \sqrt{9}\sqrt{3} - \sqrt{16}\sqrt{2}$
$$= 4\sqrt{3} - 2\sqrt{2} + 3\sqrt{3} - 4\sqrt{2} = 7\sqrt{3} - 6\sqrt{2}$$

67. $\sqrt{147} + \sqrt{216} - \sqrt{108} - \sqrt{27} = \sqrt{49}\sqrt{3} + \sqrt{36}\sqrt{6} - \sqrt{36}\sqrt{3} - \sqrt{9}\sqrt{3}$
$$= 7\sqrt{3} + 6\sqrt{6} - 6\sqrt{3} - 3\sqrt{3} = 6\sqrt{6} - 2\sqrt{3}$$

69. $\sqrt{2x^2} + \sqrt{8x^2} = \sqrt{x^2}\sqrt{2} + \sqrt{4x^2}\sqrt{2} = x\sqrt{2} + 2x\sqrt{2} = 3x\sqrt{2}$

71. $\sqrt{2x^3} + \sqrt{8x^3} = \sqrt{x^2}\sqrt{2x} + \sqrt{4x^2}\sqrt{2x} = x\sqrt{2x} + 2x\sqrt{2x} = 3x\sqrt{2x}$

73. $\sqrt{18x^2y} - \sqrt{27x^2y} = \sqrt{9x^2}\sqrt{2y} - \sqrt{9x^2}\sqrt{3y} = 3x\sqrt{2y} - 3x\sqrt{3y}$

75. $\sqrt{32x^5} - \sqrt{18x^5} = \sqrt{16x^4}\sqrt{2x} - \sqrt{9x^4}\sqrt{2x} = 4x^2\sqrt{2x} - 3x^2\sqrt{2x} = x^2\sqrt{2x}$

77. $3\sqrt{54x^2} + 5\sqrt{24x^2} = 3\sqrt{9x^2}\sqrt{6} + 5\sqrt{4x^2}\sqrt{6} = 3(3x)\sqrt{6} + 5(2x)\sqrt{6}$
$$= 9x\sqrt{6} + 10x\sqrt{6} = 19x\sqrt{6}$$

79. $y\sqrt{490y} - 2\sqrt{360y^3} = y\sqrt{49}\sqrt{10y} - 2\sqrt{36y^2}\sqrt{10y} = y(7)\sqrt{10y} - 2(6y)\sqrt{10y}$
$$= 7y\sqrt{10y} - 12y\sqrt{10y} = -5y\sqrt{10y}$$

81. $\sqrt{20x^3y} + \sqrt{45x^5y^3} - \sqrt{80x^7y^5} = \sqrt{4x^2}\sqrt{5xy} + \sqrt{9x^4y^2}\sqrt{5xy} - \sqrt{16x^6y^4}\sqrt{5xy}$
$$= 2x\sqrt{5xy} + 3x^2y\sqrt{5xy} - 4x^3y^2\sqrt{5xy}$$

83. $\sqrt[3]{16} + \sqrt[3]{54} = \sqrt[3]{8}\sqrt[3]{2} + \sqrt[3]{27}\sqrt[3]{2}$ **85.** $\sqrt[3]{81} - \sqrt[3]{24} = \sqrt[3]{27}\sqrt[3]{3} - \sqrt[3]{8}\sqrt[3]{3}$
$$= 2\sqrt[3]{2} + 3\sqrt[3]{2} = 5\sqrt[3]{2} \qquad\qquad\qquad = 3\sqrt[3]{3} - 2\sqrt[3]{3} = \sqrt[3]{3}$$

87. $\sqrt[3]{40} + \sqrt[3]{125} = \sqrt[3]{8}\sqrt[3]{5} + 5 = 2\sqrt[3]{5} + 5$ **89.** $\sqrt[3]{x^4} - \sqrt[3]{x^7} = \sqrt[3]{x^3}\sqrt[3]{x} - \sqrt[3]{x^6}\sqrt[3]{x}$
$$= x\sqrt[3]{x} - x^2\sqrt[3]{x}$$

91. $\sqrt[3]{192x^4y^5} - \sqrt[3]{24x^4y^5} = \sqrt[3]{64x^3y^3}\sqrt[3]{3xy^2} - \sqrt[3]{8x^3y^3}\sqrt[3]{3xy^2} = 4xy\sqrt[3]{3xy^2} - 2xy\sqrt[3]{3xy^2}$
$$= 2xy\sqrt[3]{3xy^2}$$

93. $\sqrt[3]{135x^7y^4} - \sqrt[3]{40x^7y^4} = \sqrt[3]{27x^6y^3}\sqrt[3]{5xy} - \sqrt[3]{8x^6y^3}\sqrt[3]{5xy} = 3x^2y\sqrt[3]{5xy} - 2x^2y\sqrt[3]{5xy}$
$$= x^2y\sqrt[3]{5xy}$$

95. Answers may vary.   **97.** The answer should be $-2\sqrt{3}$.

Exercise 8.5 (page 510)

1. $x^2 - 4x - 21 = (x - 7)(x + 3)$ **3.** $6x^2y - 15xy = 3xy(2x - 5)$

5. radical   **7.** $\sqrt{11}$ **9.** $\sqrt{7}$ **11.** $\sqrt{3}\sqrt{3} = \sqrt{9} = 3$

13. $\sqrt{2}\sqrt{8} = \sqrt{16} = 4$ **15.** $\sqrt{16}\sqrt{4} = \sqrt{64} = 8$ **17.** $\sqrt[3]{8}\sqrt[3]{8} = \sqrt[3]{64} = 4$

19. $\sqrt{x^3}\sqrt{x^3} = \sqrt{x^6} = x^3$ **21.** $\sqrt{b^8}\sqrt{b^6} = \sqrt{b^{14}} = b^7$ **23.** $\left(2\sqrt{5}\right)\left(2\sqrt{3}\right) = 4\sqrt{15}$

25. $\left(-5\sqrt{6}\right)\left(4\sqrt{3}\right) = -20\sqrt{18} = -20\sqrt{9}\sqrt{2} = -20(3)\sqrt{2} = -60\sqrt{2}$

27. $\left(2\sqrt[3]{4}\right)\left(3\sqrt[3]{16}\right) = 6\sqrt[3]{64} = 6(4) = 24$ **29.** $\left(4\sqrt{x}\right)\left(-2\sqrt{x}\right) = -8\sqrt{x^2} = -8x$

31. $\left(-14\sqrt{50x}\right)\left(-5\sqrt{20x}\right) = 70\sqrt{1,000x^2} = 70\sqrt{100x^2}\sqrt{10} = 70(10x)\sqrt{10} = 700x\sqrt{10}$

33. $\sqrt{8x}\sqrt{2x^3y} = \sqrt{16x^4y} = \sqrt{16x^4}\sqrt{y}$ **35.** $\sqrt{2}\left(\sqrt{2} + 1\right) = \sqrt{2}\sqrt{2} + \sqrt{2}(1)$
$$= 4x^2\sqrt{y} \qquad\qquad\qquad\qquad = \sqrt{4} + \sqrt{2} = 2 + \sqrt{2}$$

37. $\sqrt{3}\left(\sqrt{27} - 1\right) = \sqrt{3}\sqrt{27} - \sqrt{3}(1)$ **39.** $\sqrt{7}\left(\sqrt{7} - 3\right) = \sqrt{7}\sqrt{7} - \sqrt{7}(3)$
$$= \sqrt{81} - \sqrt{3} = 9 - \sqrt{3} \qquad\qquad = \sqrt{49} - 3\sqrt{7} = 7 - 3\sqrt{7}$$

41. $\sqrt{5}\left(3 - \sqrt{5}\right) = \sqrt{5}(3) + \sqrt{5}\left(-\sqrt{5}\right) = 3\sqrt{5} - 5$

43. $\sqrt{3}\left(\sqrt{6} + 1\right) = \sqrt{3}\sqrt{6} + \sqrt{3}(1) = \sqrt{18} + \sqrt{3} = \sqrt{9}\sqrt{2} + \sqrt{3} = 3\sqrt{2} + \sqrt{3}$

45. $\sqrt[3]{7}\left(\sqrt[3]{49} - 2\right) = \sqrt[3]{7}\sqrt[3]{49} - \sqrt[3]{7}(2) = \sqrt[3]{343} - 2\sqrt[3]{7} = 7 - 2\sqrt[3]{7}$

47. $\sqrt{x}\left(\sqrt{3x} - 2\right) = \sqrt{x}\sqrt{3x} - \sqrt{x}(2) = \sqrt{3x^2} - 2\sqrt{x} = \sqrt{x^2}\sqrt{3} - 2\sqrt{x} = x\sqrt{3} - 2\sqrt{x}$

49. $2\sqrt{x}\left(\sqrt{9x} + 3\right) = 2\sqrt{x}\sqrt{9x} + 2\sqrt{x}(3) = 2\sqrt{9x^2} + 6\sqrt{x} = 2(3x) + 6\sqrt{x} = 6x + 6\sqrt{x}$

51. $3\sqrt{x}\left(2 + \sqrt{x}\right) = 3\sqrt{x}(2) + 3\sqrt{x}\sqrt{x} = 6\sqrt{x} + 3\sqrt{x^2} = 6\sqrt{x} + 3x$

53. $\left(\sqrt{2} + 1\right)\left(\sqrt{2} - 1\right) = \sqrt{2}\sqrt{2} + \sqrt{2}(-1) + 1\sqrt{2} - 1 = \sqrt{4} - \sqrt{2} + \sqrt{2} - 1 = 2 - 1 = 1$

55. $\left(\sqrt{5} + 2\right)\left(\sqrt{5} - 2\right) = \sqrt{5}\sqrt{5} + \sqrt{5}(-2) + 2\sqrt{5} - 4 = \sqrt{25} - 2\sqrt{5} + 2\sqrt{5} - 4 = 5 - 4 = 1$

57. $\left(\sqrt[3]{2} + 1\right)\left(\sqrt[3]{2} + 1\right) = \sqrt[3]{2}\sqrt[3]{2} + \sqrt[3]{2}(1) + 1\sqrt[3]{2} + 1 = \sqrt[3]{4} + 2\sqrt[3]{2} + 1$

59. $\left(\sqrt{7} - x\right)\left(\sqrt{7} + x\right) = \sqrt{7}\sqrt{7} + \sqrt{7}(x) - x\sqrt{7} - x^2 = 7 - x^2$

61. $\left(\sqrt{6x} + \sqrt{7}\right)\left(\sqrt{6x} - \sqrt{7}\right) = \sqrt{6x}\sqrt{6x} + \sqrt{6x}\left(-\sqrt{7}\right) + \sqrt{7}\sqrt{6x} - \sqrt{7}\sqrt{7}$

$$= \sqrt{36x^2} - \sqrt{42x} + \sqrt{42x} - \sqrt{49} = 6x - 7$$

63. $\left(\sqrt{2x} + 3\right)\left(\sqrt{8x} - 6\right) = \sqrt{2x}\sqrt{8x} + \sqrt{2x}(-6) + 3\sqrt{8x} - 18$

$$= \sqrt{16x^2} - 6\sqrt{2x} + 3\sqrt{4}\sqrt{2x} - 18$$
$$= 4x - 6\sqrt{2x} + 3(2)\sqrt{2x} - 18$$
$$= 4x - 6\sqrt{2x} + 6\sqrt{2x} - 18 = 4x - 18$$

65. $\left(\sqrt{8xy} + 1\right)\left(\sqrt{8xy} + 1\right) = \sqrt{8xy}\sqrt{8xy} + \sqrt{8xy}(1) + 1\sqrt{8xy} + 1$

$$= \sqrt{64x^2y^2} + \sqrt{8xy} + \sqrt{8xy} + 1$$
$$= 8xy + 2\sqrt{8xy} + 1$$
$$= 8xy + 2\sqrt{4}\sqrt{2xy} + 1$$
$$= 8xy + 2(2)\sqrt{2xy} + 1 = 8xy + 4\sqrt{2xy} + 1$$

67. $\dfrac{\sqrt{12x^3}}{\sqrt{27x}} = \sqrt{\dfrac{12x^3}{27x}} = \sqrt{\dfrac{4x^2}{9}} = \dfrac{\sqrt{4x^2}}{\sqrt{9}} = \dfrac{2x}{3}$

69. $\dfrac{\sqrt{18xy^2}}{\sqrt{25x}} = \sqrt{\dfrac{18xy^2}{25x}} = \sqrt{\dfrac{18y^2}{25}} = \dfrac{\sqrt{9y^2}\sqrt{2}}{\sqrt{25}} = \dfrac{3y\sqrt{2}}{5}$

71. $\dfrac{\sqrt{196xy^3}}{\sqrt{49x^3y}} = \sqrt{\dfrac{196xy^3}{49x^3y}} = \sqrt{\dfrac{196y^2}{49x^2}} = \dfrac{\sqrt{196y^2}}{\sqrt{49x^2}} = \dfrac{14y}{7x} = \dfrac{2y}{x}$

73. $\dfrac{\sqrt[3]{16x^6}}{\sqrt[3]{54x^3}} = \sqrt[3]{\dfrac{16x^6}{54x^3}} = \sqrt[3]{\dfrac{8x^3}{27}} = \dfrac{\sqrt[3]{8x^3}}{\sqrt[3]{27}}$
$= \dfrac{2x}{3}$

75. $\dfrac{\sqrt{3x^2y^3}}{\sqrt{27x}} = \sqrt{\dfrac{3x^2y^3}{27x}} = \sqrt{\dfrac{xy^3}{9}} = \dfrac{\sqrt{y^2}\sqrt{xy}}{\sqrt{9}}$
$= \dfrac{y\sqrt{xy}}{3}$

77. $\dfrac{\sqrt{5x}\sqrt{10y^2}}{\sqrt{x^3y}} = \dfrac{\sqrt{50xy^2}}{\sqrt{x^3y}} = \sqrt{\dfrac{50xy^2}{x^3y}} = \sqrt{\dfrac{50y}{x^2}} = \dfrac{\sqrt{50y}}{\sqrt{x^2}} = \dfrac{\sqrt{25}\sqrt{2y}}{x} = \dfrac{5\sqrt{2y}}{x}$

79. $\dfrac{1}{\sqrt{3}} = \dfrac{1 \cdot \sqrt{3}}{\sqrt{3} \cdot \sqrt{3}} = \dfrac{\sqrt{3}}{3}$

81. $\dfrac{2}{\sqrt{7}} = \dfrac{2 \cdot \sqrt{7}}{\sqrt{7} \cdot \sqrt{7}} = \dfrac{2\sqrt{7}}{7}$

83. $\dfrac{5}{\sqrt[3]{5}} = \dfrac{5 \cdot \sqrt[3]{25}}{\sqrt[3]{5} \cdot \sqrt[3]{25}} = \dfrac{5\sqrt[3]{25}}{\sqrt[3]{125}} = \dfrac{5\sqrt[3]{25}}{5}$
$= \sqrt[3]{25}$

85. $\dfrac{9}{\sqrt{27}} = \dfrac{9 \cdot \sqrt{3}}{\sqrt{27} \cdot \sqrt{3}} = \dfrac{9\sqrt{3}}{\sqrt{81}} = \dfrac{9\sqrt{3}}{9} = \sqrt{3}$

87. $\dfrac{3}{\sqrt{32}} = \dfrac{3 \cdot \sqrt{2}}{\sqrt{32} \cdot \sqrt{2}} = \dfrac{3\sqrt{2}}{\sqrt{64}} = \dfrac{3\sqrt{2}}{8}$

89. $\dfrac{4}{\sqrt[3]{4}} = \dfrac{4 \cdot \sqrt[3]{2}}{\sqrt[3]{4} \cdot \sqrt[3]{2}} = \dfrac{4\sqrt[3]{2}}{\sqrt[3]{8}} = \dfrac{4\sqrt[3]{2}}{2} = 2\sqrt[3]{2}$

91. $\dfrac{\sqrt{5}}{\sqrt{3}} = \dfrac{\sqrt{5} \cdot \sqrt{3}}{\sqrt{3} \cdot \sqrt{3}} = \dfrac{\sqrt{15}}{3}$

93. $\dfrac{10}{\sqrt{x}} = \dfrac{10\sqrt{x}}{\sqrt{x}\sqrt{x}} = \dfrac{10\sqrt{x}}{x}$

95. $\dfrac{\sqrt{9}}{\sqrt{2x}} = \dfrac{3 \cdot \sqrt{2x}}{\sqrt{2x} \cdot \sqrt{2x}} = \dfrac{3\sqrt{2x}}{2x}$

97. $\dfrac{\sqrt{2x}}{\sqrt{9y}} = \dfrac{\sqrt{2x} \cdot \sqrt{y}}{\sqrt{9y} \cdot \sqrt{y}} = \dfrac{\sqrt{2xy}}{\sqrt{9y^2}} = \dfrac{\sqrt{2xy}}{3y}$

99. $\dfrac{\sqrt[3]{5}}{\sqrt[3]{2}} = \dfrac{\sqrt[3]{5}\sqrt[3]{4}}{\sqrt[3]{2}\sqrt[3]{4}} = \dfrac{\sqrt[3]{20}}{\sqrt[3]{8}} = \dfrac{\sqrt[3]{20}}{2}$

101. $\dfrac{\sqrt[3]{2x^2}}{\sqrt[3]{2x}} = \dfrac{\sqrt[3]{2x^2}\sqrt[3]{4x^2}}{\sqrt[3]{2x}\sqrt[3]{4x^2}} = \dfrac{\sqrt[3]{8x^4}}{\sqrt[3]{8x^3}} = \dfrac{\sqrt[3]{8x^3}\sqrt[3]{x}}{2x} = \dfrac{2x\sqrt[3]{x}}{2x} = \sqrt[3]{x}$

103. $\dfrac{2}{\sqrt[3]{4x^2y}} = \dfrac{2\sqrt[3]{2xy^2}}{\sqrt[3]{4x^2y}\sqrt[3]{2xy^2}} = \dfrac{2\sqrt[3]{2xy^2}}{\sqrt[3]{8x^3y^3}} = \dfrac{2\sqrt[3]{2xy^2}}{2xy} = \dfrac{\sqrt[3]{2xy^2}}{xy}$

105. $\dfrac{-5}{\sqrt[3]{25a^2b^2}} = \dfrac{-5\sqrt[3]{5ab}}{\sqrt[3]{25a^2b^2}\sqrt[3]{5ab}} = \dfrac{-5\sqrt[3]{5ab}}{\sqrt[3]{125a^3b^3}} = \dfrac{-5\sqrt[3]{5ab}}{5ab} = -\dfrac{\sqrt[3]{5ab}}{ab}$

107. $\dfrac{3}{\sqrt{3}-1} = \dfrac{3\left(\sqrt{3}+1\right)}{\left(\sqrt{3}-1\right)\left(\sqrt{3}+1\right)} = \dfrac{3\left(\sqrt{3}+1\right)}{3+\sqrt{3}-\sqrt{3}-1} = \dfrac{3\left(\sqrt{3}+1\right)}{2}$

109. $\dfrac{3}{\sqrt{7}+2} = \dfrac{3\left(\sqrt{7}-2\right)}{\left(\sqrt{7}+2\right)\left(\sqrt{7}-2\right)} = \dfrac{3\left(\sqrt{7}-2\right)}{7-2\sqrt{7}+2\sqrt{7}-4} = \dfrac{3\left(\sqrt{7}-2\right)}{3} = \sqrt{7}-2$

111. $\dfrac{12}{3-\sqrt{3}} = \dfrac{12\left(3+\sqrt{3}\right)}{\left(3-\sqrt{3}\right)\left(3+\sqrt{3}\right)} = \dfrac{12\left(3+\sqrt{3}\right)}{9+3\sqrt{3}-3\sqrt{3}-3} = \dfrac{12\left(3+\sqrt{3}\right)}{6} = 2\left(3+\sqrt{3}\right)$

$$= 6+2\sqrt{3}$$

113. $\dfrac{\sqrt{2}}{\sqrt{2}+1} = \dfrac{\sqrt{2}\left(\sqrt{2}-1\right)}{\left(\sqrt{2}+1\right)\left(\sqrt{2}-1\right)} = \dfrac{2-\sqrt{2}}{2-\sqrt{2}+\sqrt{2}-1} = \dfrac{2-\sqrt{2}}{1} = 2-\sqrt{2}$

115. $\dfrac{-\sqrt{3}}{\sqrt{3}+1} = \dfrac{-\sqrt{3}\left(\sqrt{3}-1\right)}{\left(\sqrt{3}+1\right)\left(\sqrt{3}-1\right)} = \dfrac{-3+\sqrt{3}}{3-\sqrt{3}+\sqrt{3}-1} = \dfrac{\sqrt{3}-3}{2}$

117. $\dfrac{5}{\sqrt{3}+\sqrt{2}} = \dfrac{5\left(\sqrt{3}-\sqrt{2}\right)}{\left(\sqrt{3}+\sqrt{2}\right)\left(\sqrt{3}-\sqrt{2}\right)} = \dfrac{5\left(\sqrt{3}-\sqrt{2}\right)}{3-\sqrt{6}+\sqrt{6}-2} = \dfrac{5\left(\sqrt{3}-\sqrt{2}\right)}{1} = 5\sqrt{3}-5\sqrt{2}$

119. $\dfrac{\sqrt{x}+2}{\sqrt{x}-2} = \dfrac{\left(\sqrt{x}+2\right)\left(\sqrt{x}+2\right)}{\left(\sqrt{x}-2\right)\left(\sqrt{x}+2\right)} = \dfrac{x+2\sqrt{x}+2\sqrt{x}+4}{x+2\sqrt{x}-2\sqrt{x}-4} = \dfrac{x+4\sqrt{x}+4}{x-4}$

121. $\dfrac{\sqrt{3x}-2}{\sqrt{3x}+2} = \dfrac{\left(\sqrt{3x}-2\right)\left(\sqrt{3x}-2\right)}{\left(\sqrt{3x}+2\right)\left(\sqrt{3x}-2\right)} = \dfrac{3x-2\sqrt{3x}-2\sqrt{3x}+4}{3x-2\sqrt{3x}+2\sqrt{3x}-4} = \dfrac{3x-4\sqrt{3x}+4}{3x-4}$

123. $\dfrac{\sqrt{3x}-1}{\sqrt{3x}+1} = \dfrac{\left(\sqrt{3x}-1\right)\left(\sqrt{3x}-1\right)}{\left(\sqrt{3x}+1\right)\left(\sqrt{3x}-1\right)} = \dfrac{3x-\sqrt{3x}-\sqrt{3x}+1}{3x-\sqrt{3x}+\sqrt{3x}-1} = \dfrac{3x-2\sqrt{3x}+1}{3x-1}$

125. $\dfrac{\sqrt{3y}+3}{\sqrt{3y}-2} = \dfrac{\left(\sqrt{3y}+3\right)\left(\sqrt{3y}+2\right)}{\left(\sqrt{3y}-2\right)\left(\sqrt{3y}+2\right)} = \dfrac{3y+2\sqrt{3y}+3\sqrt{3y}+6}{3y+2\sqrt{3y}-2\sqrt{3y}-4} = \dfrac{3y+5\sqrt{3y}+6}{3y-4}$

127. Answers may vary.

129. $\dfrac{\sqrt{3}}{2} = \dfrac{\sqrt{3}\sqrt{3}}{2\sqrt{3}} = \dfrac{3}{2\sqrt{3}}$

Exercise 8.6 (page 519)

1.
$$\begin{aligned} x + y &= 5 \\ x - y &= -1 \\ \hline 2x &= 4 \\ x &= 2 \end{aligned}$$
$$\begin{aligned} x + y &= 5 \\ 2 + y &= 5 \\ y &= 3 \end{aligned}$$
Solution:
$\boxed{(2, 3)}$

3.
$$\begin{aligned} 2x + 3y &= 0 \Rightarrow \times (2) \\ 3x - 2y &= 13 \Rightarrow \times (3) \end{aligned}$$
$$\begin{aligned} 4x + 6y &= 0 \\ 9x - 6y &= 39 \\ \hline 13x &= 39 \\ x &= 3 \end{aligned}$$
$$\begin{aligned} 2x + 3y &= 0 \\ 2(3) + 3y &= 0 \\ 6 + 3y &= 0 \\ 3y &= -6 \\ y &= -2 \end{aligned}$$
Solution:
$\boxed{(3, -2)}$

5. extraneous

7. b^2

9. The left side of the equation was squared, but the right side was not.

11.
$$\begin{aligned} \sqrt{x} &= 3 \\ \left(\sqrt{x}\right)^2 &= 3^2 \\ x &= 9 \end{aligned}$$

13.
$$\begin{aligned} \sqrt{x} &= 7 \\ \left(\sqrt{x}\right)^2 &= 7^2 \\ x &= 49 \end{aligned}$$

15.
$$\begin{aligned} \sqrt{x} &= -4 \\ \left(\sqrt{x}\right)^2 &= (-4)^2 \\ \boxed{x = 16} \end{aligned}$$
does not check, no solutions

17.
$$\begin{aligned} \sqrt{x+3} &= 2 \\ \left(\sqrt{x+3}\right)^2 &= 2^2 \\ x + 3 &= 4 \\ x &= 1 \end{aligned}$$

19.
$$\begin{aligned} \sqrt{x-5} &= 5 \\ \left(\sqrt{x-5}\right)^2 &= 5^2 \\ x - 5 &= 25 \\ x &= 30 \end{aligned}$$

21.
$$\begin{aligned} \sqrt{3-x} &= -2 \\ \left(\sqrt{3-x}\right)^2 &= (-2)^2 \\ 3 - x &= 4 \\ \boxed{-1 = x} \end{aligned}$$
does not check, no solutions

23.
$$\begin{aligned} \sqrt{6+2x} &= 4 \\ \left(\sqrt{6+2x}\right)^2 &= 4^2 \\ 6 + 2x &= 16 \\ 2x &= 10 \\ x &= 5 \end{aligned}$$

25.
$$\begin{aligned} \sqrt{5x-5} &= 5 \\ \left(\sqrt{5x-5}\right)^2 &= 5^2 \\ 5x - 5 &= 25 \\ 5x &= 30 \\ x &= 6 \end{aligned}$$

27.
$$\begin{aligned} \sqrt{4x-3} &= 3 \\ \left(\sqrt{4x-3}\right)^2 &= 3^2 \\ 4x - 3 &= 9 \\ 4x &= 12 \\ x &= 3 \end{aligned}$$

29.
$$\begin{aligned} \sqrt{13x+14} &= 1 \\ \left(\sqrt{13x+14}\right)^2 &= 1^2 \\ 13x + 14 &= 1 \\ 13x &= -13 \\ x &= -1 \end{aligned}$$

31.
$$\begin{aligned} \sqrt{x+3} + 5 &= 12 \\ \sqrt{x+3} &= 7 \\ \left(\sqrt{x+3}\right)^2 &= 7^2 \\ x + 3 &= 49 \\ x &= 46 \end{aligned}$$

33. $\sqrt{2x+10}+3=5$

$\sqrt{2x+10}=2$

$\left(\sqrt{2x+10}\right)^2=2^2$

$2x+10=4$

$2x=-6$

$x=-3$

35. $\sqrt{5x+9}+4=7$

$\sqrt{5x+9}=3$

$\left(\sqrt{5x+9}\right)^2=3^2$

$5x+9=9$

$5x=0$

$x=0$

37. $\sqrt{7-5x}+4=3$

$\sqrt{7-5x}=-1$

$\left(\sqrt{7-5x}\right)^2=(-1)^2$

$7-5x=1$

$-5x=-6$

$\boxed{x=\frac{-6}{-5}=\frac{6}{5}}$

does not check, no solutions

39. $\sqrt{x+1}=x-1$

$\left(\sqrt{x+1}\right)^2=(x-1)^2$

$x+1=(x-1)(x-1)$

$x+1=x^2-2x+1$

$0=x^2-3x$

$0=x(x-3)$

$x=0$ **or** $x-3=0$

$\boxed{x=0}$ $x=3$

$x=0$ does not check.

41. $\sqrt{x+1}=x+1$

$\left(\sqrt{x+1}\right)^2=(x+1)^2$

$x+1=(x+1)(x+1)$

$x+1=x^2+2x+1$

$0=x^2+x$

$0=x(x+1)$

$x=0$ **or** $x+1=0$

$x=0$ $x=-1$

43. $\sqrt{7x+2}-2x=0$

$\sqrt{7x+2}=2x$

$\left(\sqrt{7x+2}\right)^2=(2x)^2$

$7x+2=4x^2$

$0=4x^2-7x-2$

$0=(4x+1)(x-2)$

$4x+1=0$ **or** $x-2=0$

$4x=-1$ $x=2$

$\boxed{x=-\frac{1}{4}}$ $x=2$

$x=-\frac{1}{4}$ does not check.

45. $x-1=\sqrt{x-1}$

$(x-1)^2=\left(\sqrt{x-1}\right)^2$

$(x-1)(x-1)=x-1$

$x^2-2x+1=x-1$

$x^2-3x+2=0$

$(x-2)(x-1)=0$

$x-2=0$ **or** $x-1=0$

$x=2$ $x=1$

47.
$$x = \sqrt{3 - x} + 3$$
$$x - 3 = \sqrt{3 - x}$$
$$(x - 3)^2 = \left(\sqrt{3 - x}\right)^2$$
$$(x - 3)(x - 3) = 3 - x$$
$$x^2 - 6x + 9 = 3 - x$$
$$x^2 - 5x + 6 = 0$$
$$(x - 2)(x - 3) = 0$$
$$x - 2 = 0 \quad \textbf{or} \quad x - 3 = 0$$
$$\boxed{x = 2} \qquad x = 3$$
$x = 2$ does not check.

49.
$$\sqrt{3x + 3} = 3\sqrt{x - 1}$$
$$\left(\sqrt{3x + 3}\right)^2 = \left(3\sqrt{x - 1}\right)^2$$
$$3x + 3 = 3^2\left(\sqrt{x - 1}\right)^2$$
$$3x + 3 = 9(x - 1)$$
$$3x + 3 = 9x - 9$$
$$-6x = -12$$
$$x = 2$$

51.
$$2\sqrt{3x + 4} = \sqrt{5x + 9}$$
$$\left(2\sqrt{3x + 4}\right)^2 = \left(\sqrt{5x + 9}\right)^2$$
$$2^2\left(\sqrt{3x + 4}\right)^2 = 5x + 9$$
$$4(3x + 4) = 5x + 9$$
$$12x + 16 = 5x + 9$$
$$7x = -7$$
$$x = -1$$

53.
$$\sqrt{3x + 6} = 2\sqrt{2x - 11}$$
$$\left(\sqrt{3x + 6}\right)^2 = \left(2\sqrt{2x - 11}\right)^2$$
$$3x + 6 = 2^2\left(\sqrt{2x - 11}\right)^2$$
$$3x + 6 = 4(2x - 11)$$
$$3x + 6 = 8x - 44$$
$$-5x = -50$$
$$x = 10$$

55.
$$\sqrt[3]{x - 1} = 4$$
$$\left(\sqrt[3]{x - 1}\right)^3 = 4^3$$
$$x - 1 = 64$$
$$x = 65$$

57.
$$\sqrt[3]{\frac{1}{2}x - 3} = 2$$
$$\left(\sqrt[3]{\frac{1}{2}x - 3}\right)^3 = 2^3$$
$$\tfrac{1}{2}x - 3 = 8$$
$$\tfrac{1}{2}x = 11$$
$$x = 22$$

59.
$$d = \sqrt{(x_2 - x_1)^2 + (y_2 - y_1)^2}$$
$$= \sqrt{(0 - 3)^2 + (0 - (-4))^2}$$
$$= \sqrt{(-3)^2 + (4)^2}$$
$$= \sqrt{9 + 16} = \sqrt{25} = 5$$

61.
$$d = \sqrt{(x_2 - x_1)^2 + (y_2 - y_1)^2}$$
$$= \sqrt{(5 - 2)^2 + (8 - 4)^2}$$
$$= \sqrt{(3)^2 + (4)^2}$$
$$= \sqrt{9 + 16} = \sqrt{25} = 5$$

63.
$$d = \sqrt{(x_2 - x_1)^2 + (y_2 - y_1)^2}$$
$$= \sqrt{(3 - (-2))^2 + (4 - (-8))^2}$$
$$= \sqrt{(5)^2 + (12)^2}$$
$$= \sqrt{25 + 144} = \sqrt{169} = 13$$

65.
$$d = \sqrt{(x_2 - x_1)^2 + (y_2 - y_1)^2}$$
$$= \sqrt{(12 - 6)^2 + (16 - 8)^2}$$
$$= \sqrt{(6)^2 + (8)^2}$$
$$= \sqrt{36 + 64} = \sqrt{100} = 10$$

67. Let $t = 4$.

$$t = \frac{\sqrt{s}}{4}$$
$$4 = \frac{\sqrt{s}}{4}$$
$$16 = \sqrt{s}$$
$$16^2 = \left(\sqrt{s}\right)^2$$
$$256 = s$$

The shaft is 256 ft deep.

69. Let $t = \frac{3}{2} = 1.5$.

$$t = 1.11\sqrt{L}$$
$$1.5 = 1.11\sqrt{L}$$
$$\frac{1.5}{1.11} = \sqrt{L}$$
$$\left(\frac{1.5}{1.11}\right)^2 = \left(\sqrt{L}\right)^2$$
$$1.826 \approx L$$

The length is about 1.8 ft.

71. Let $I = 7$ and $R = 20$.

$$I = \sqrt{\frac{P}{R}}$$
$$7 = \sqrt{\frac{P}{20}}$$
$$7^2 = \left(\sqrt{\frac{P}{20}}\right)^2$$
$$49 = \frac{P}{20}$$
$$980 = P$$

The power used is 980 watts.

73. Let $k = 3.24$ and $s = 56$.

$$s = k\sqrt{d}$$
$$56 = 3.24\sqrt{d}$$
$$(56)^2 = \left(3.24\sqrt{d}\right)^2$$
$$3136 = 10.4976d$$
$$299 \text{ ft} \approx d$$

75. Let $k = 5.34$ and $s = 50$.

$$s = k\sqrt{d}$$
$$50 = 5.34\sqrt{d}$$
$$(50)^2 = \left(5.34\sqrt{d}\right)^2$$
$$2500 = 28.5156d$$
$$88 \text{ ft} \approx d$$

77.

$$s = \frac{2.029 \times 10^7}{\sqrt{r}}$$
$$7 \times 10^3 = \frac{2.029 \times 10^7}{\sqrt{r}}$$
$$\sqrt{r} = \frac{2.029 \times 10^7}{7 \times 10^3}$$
$$\sqrt{r} = \frac{2.029}{7} \times 10^4$$
$$\left(\sqrt{r}\right)^2 = \left(\frac{2.029}{7} \times 10^4\right)^2$$
$$r \approx 0.0840 \times 10^8$$
$$r \approx 8.4 \times 10^6$$
$$a \approx 2 \times 10^6 \text{ m}$$

79.

$$f = \sqrt{1 - \frac{v^2}{c^2}}$$
$$f^2 = \left(\sqrt{1 - \frac{v^2}{c^2}}\right)^2$$
$$f^2 = 1 - \frac{v^2}{c^2}$$
$$c^2 \cdot f^2 = c^2\left(1 - \frac{v^2}{c^2}\right)$$
$$c^2 f^2 = c^2 - v^2$$
$$v^2 = c^2 - c^2 f^2$$

81. Let $s = 30$.

$$s = \sqrt[3]{\frac{P}{0.02}}$$
$$30 = \sqrt[3]{\frac{P}{0.02}}$$
$$30^3 = \left(\sqrt[3]{\frac{P}{0.02}}\right)^3$$
$$27{,}000 = \frac{P}{0.02}$$
$$0.02(27{,}000) = P$$
$$540 \text{ watts} = P$$

83. From $(21, 4)$ to $(120, 70)$:

$$d = \sqrt{(x_2 - x_1)^2 + (y_2 - y_1)^2}$$
$$= \sqrt{(120 - 21)^2 + (70 - 4)^2}$$
$$= \sqrt{(99)^2 + (66)^2}$$
$$= \sqrt{9801 + 4356}$$
$$= \sqrt{14{,}157} \approx 119 \text{ blocks}$$

$119(750 \text{ ft}) = 89{,}250 \text{ ft}$

$$\frac{89{,}250 \text{ ft}}{5{,}280 \text{ ft/mi}} \approx 16.9 \text{ miles}$$

85. Answers may vary.

87.
$$\sqrt[4]{3x + 4} = 5$$
$$\left(\sqrt[4]{3x + 4}\right)^4 = 5^4$$
$$3x + 4 = 625$$
$$3x = 621$$
$$x = 207$$

Exercise 8.7 (page 525)

1. $3z^2 - 15tz + 12t^2 = 3(z^2 - 5tz + 4t^2)$
$$= 3(z - 4t)(z - t)$$

3.
$$\frac{x - 5}{7} + \frac{2}{5} = \frac{7 - x}{5}$$
$$35\left(\frac{x - 5}{7} + \frac{2}{5}\right) = 35\left(\frac{7 - x}{5}\right)$$
$$5(x - 5) + 7(2) = 7(7 - x)$$
$$5x - 25 + 14 = 49 - 7x$$
$$12x = 60$$
$$x = 5$$

5. rational

7. x^{m+n}

9. $\dfrac{x^n}{y^n}$

11. $\dfrac{1}{x}$

13. $81^{1/2} = 9$

15. $-144^{1/2} = -1 \cdot 12 = -12$

17. $\left(\dfrac{1}{4}\right)^{1/2} = \dfrac{1}{2}$

19. $\left(\dfrac{4}{49}\right)^{1/2} = \dfrac{2}{7}$

21. $27^{1/3} = 3$

23. $-125^{1/3} = -1 \cdot 5 = -5$

25. $(-8)^{1/3} = -2$

27. $\left(\dfrac{1}{64}\right)^{1/3} = \dfrac{1}{4}$

29. $\left(\dfrac{27}{64}\right)^{1/3} = \dfrac{3}{4}$

31. $16^{1/4} = 2$

33. $32^{1/5} = 2$

35. $-243^{1/5} = -1 \cdot 3 = -3$

37. $81^{3/2} = \left(81^{1/2}\right)^3 = 9^3 = 729$

39. $25^{3/2} = \left(25^{1/2}\right)^3 = 5^3 = 125$

41. $125^{2/3} = \left(125^{1/3}\right)^2 = 5^2 = 25$

43. $1{,}000^{2/3} = \left(1{,}000^{1/3}\right)^2 = 10^2 = 100$

45. $(-8)^{2/3} = \left((-8)^{1/3}\right)^2 = (-2)^2 = 4$

47. $32^{3/5} - \left(32^{1/5}\right)^3 = 2^3 = 8$

49. $81^{3/4} = \left(81^{1/4}\right)^3 = 3^3 = 27$

51. $(-32)^{3/5} = \left((-32)^{1/5}\right)^3 = (-2)^3 = -8$

53. $\left(\dfrac{8}{27}\right)^{2/3} = \left(\left(\dfrac{8}{27}\right)^{1/3}\right)^2 = \left(\dfrac{2}{3}\right)^2 = \dfrac{4}{9}$

55. $\left(\dfrac{16}{625}\right)^{3/4} = \left(\left(\dfrac{16}{625}\right)^{1/4}\right)^3 = \left(\dfrac{2}{5}\right)^3 = \dfrac{8}{125}$

57. $6^{3/5}6^{2/5} = 6^{3/5+2/5} = 6^{5/5} = 6^1 = 6$

59. $5^{2/3}5^{4/3} = 5^{2/3+4/3} = 5^{6/3} = 5^2 = 25$

61. $\left(7^{2/5}\right)^{5/2} = 7^1 = 7$

63. $\left(5^{2/7}\right)^7 = 5^2 = 25$

65. $\dfrac{8^{3/2}}{8^{1/2}} = 8^{2/2} = 8^1 = 8$

67. $\dfrac{5^{11/3}}{5^{2/3}} = 5^{9/3} = 5^3 = 125$

69. $\left(2^{1/2}3^{1/2}\right)^2 = \left(2^{1/2}\right)^2\left(3^{1/2}\right)^2 = (2)(3) = 6$

71. $\left(4^{3/4}3^{1/4}\right)^4 = \left(4^{3/4}\right)^4\left(3^{1/4}\right)^4$
$\qquad = \left(4^3\right)(3) = 64(3) = 192$

73. $4^{-1/2} = \dfrac{1}{4^{1/2}} = \dfrac{1}{2}$

75. $27^{-2/3} = \dfrac{1}{27^{2/3}} = \dfrac{1}{\left(27^{1/3}\right)^2} = \dfrac{1}{3^2} = \dfrac{1}{9}$

77. $16^{-3/2} = \dfrac{1}{16^{3/2}} = \dfrac{1}{\left(16^{1/2}\right)^3} = \dfrac{1}{4^3} = \dfrac{1}{64}$

79. $(-27)^{-4/3} = \dfrac{1}{(-27)^{4/3}} = \dfrac{1}{\left((-27)^{1/3}\right)^4} = \dfrac{1}{(-3)^4} = \dfrac{1}{81}$

81. $\left(x^{1/2}\right)^2 = x^{2/2} = x^1 = x$

83. $\left(x^{12}\right)^{1/6} = x^{12/6} = x^2$

85. $\left(x^{18}\right)^{2/9} = x^{36/9} = x^4$

87. $x^{5/6}x^{7/6} = x^{12/6} = x^2$

89. $y^{4/7}y^{10/7} = y^{14/7} = y^2$

91. $\dfrac{x^{3/5}}{x^{1/5}} = x^{2/5}$

93. $\dfrac{x^{1/7}x^{3/7}}{x^{2/7}} = \dfrac{x^{4/7}}{x^{2/7}} = x^{2/7}$

95. $\left(\dfrac{x^{3/5}}{x^{2/5}}\right)^5 = \left(x^{1/5}\right)^5 = x$

97. $\left(\dfrac{y^{2/7}y^{3/7}}{y^{4/7}}\right)^{49} = \left(\dfrac{y^{5/7}}{y^{4/7}}\right)^{49} = \left(y^{1/7}\right)^{49} = y^{49/7} = y^7$

99. $\left(\dfrac{y^{5/6}y^{7/6}}{y^{1/3}y}\right)^3 = \left(\dfrac{y^{12/6}}{y^{1/3}y^{3/3}}\right)^3 = \left(\dfrac{y^2}{y^{4/3}}\right)^3 = \left(\dfrac{y^{6/3}}{y^{4/3}}\right)^3 = \left(y^{2/3}\right)^3 = y^2$

101. $x^{2/3}x^{3/4} = x^{8/12}x^{9/12} = x^{17/12}$

103. $\left(b^{1/2}\right)^{3/5} = b^{3/10}$

105. $\dfrac{t^{2/3}}{t^{2/5}} = \dfrac{t^{10/15}}{t^{6/15}} = t^{4/15}$

107. $\dfrac{x^{4/5}x^{1/3}}{x^{2/15}} = \dfrac{x^{12/15}x^{5/15}}{x^{2/15}} = \dfrac{x^{17/15}}{x^{2/15}}$
$\qquad\qquad = x^{15/15} = x$

109. $\dfrac{a^{2/5}a^{1/5}}{a^{-1/3}} = \dfrac{a^{3/5}}{a^{-1/3}} = \dfrac{a^{9/15}}{a^{-5/15}} = a^{14/15}$ **111. Answers may vary.**

113. 2^x is larger.

Chapter 8 Summary (page 527)

1. $\sqrt{25} = \sqrt{5^2} = 5$ **2.** $\sqrt{64} = \sqrt{8^2} = 8$ **3.** $-\sqrt{144} = -\sqrt{12}^2 = -12$

4. $-\sqrt{289} = -\sqrt{17^2} = -17$ **5.** $\sqrt{256} = \sqrt{16}^2 = 16$ **6.** $-\sqrt{64} = -\sqrt{8^2} = -8$

7. $\sqrt{169} = \sqrt{13^2} = 13$ **8.** $-\sqrt{225} = -\sqrt{15^2} = -15$ **9.** $\sqrt{21} \approx 4.583$

10. $-\sqrt{15} \approx -3.873$ **11.** $-\sqrt{57.3} \approx -7.570$ **12.** $\sqrt{751.9} \approx 27.421$

13. $f(x) = \sqrt{x}$

x	$f(x)$
0	0
1	1
4	2

14. $f(x) = 2 - \sqrt{x}$

x	$f(x)$
0	2
1	1
4	0

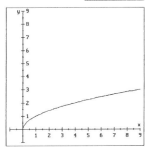

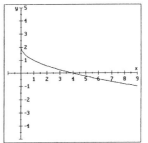

15.
$$a^2 + b^2 = c^2$$
$$21^2 + 28^2 = c^2$$
$$441 + 784 = c^2$$
$$1225 = c^2$$
$$\sqrt{1225} = c$$
$$35 = c$$

16.
$$a^2 + b^2 = c^2$$
$$25^2 + b^2 = 65^2$$
$$625 + b^2 = 4225$$
$$b^2 = 3600$$
$$b = \sqrt{3600}$$
$$b = 60$$

17.
$$a^2 + b^2 = c^2$$
$$1^2 + b^2 = \left(\sqrt{2}\right)^2$$
$$1 + b^2 = 2$$
$$b^2 = 1$$
$$b = \sqrt{1}$$
$$b = 1$$

18.
$$a^2 + b^2 = c^2$$
$$a^2 + 6^2 = 7^2$$
$$a^2 + 36 = 49$$
$$a^2 = 13$$
$$a = \sqrt{13}$$

19.
$$32^2 + 60^2 = x^2$$
$$1024 + 3600 = x^2$$
$$4624 = x^2$$
$$\sqrt{4624} = x$$
$$68 = x$$
The brace is 68 in. long.

20.
$$x^2 + 28^2 = 53^2$$
$$x^2 + 784 = 2809$$
$$x^2 = 2025$$
$$x = \sqrt{2025}$$
$$x - 45$$
The mast is 45 ft tall.

21. $\sqrt[3]{-27} = -3$

22. $-\sqrt[3]{125} = -1 \cdot 5 = -5$

23. $\sqrt[4]{81} = 3$

24. $\sqrt[5]{32} = 2$

25. $\sqrt[3]{54.3} \approx 3.787$

26. $\sqrt[3]{0.003} \approx 0.144$

27. $\sqrt[3]{-0.055} \approx -0.380$

28. $\sqrt[3]{-63{,}777} \approx -39.953$

29. $\sqrt{x^6} = x^3$

30. $\sqrt{16x^4y^2} = 4x^2y$

31. $\sqrt[3]{27x^3} = 3x$

32. $\sqrt[3]{1{,}000a^6b^3} = 10a^2b$

33. $\sqrt{32} = \sqrt{16}\sqrt{2} = 4\sqrt{2}$

34. $\sqrt{50} = \sqrt{25}\sqrt{2} = 5\sqrt{2}$

35. $\sqrt{500} = \sqrt{100}\sqrt{5} = 10\sqrt{5}$

36. $\sqrt{112} = \sqrt{16}\sqrt{7} = 4\sqrt{7}$

37. $\sqrt{80x^2} = \sqrt{16x^2}\sqrt{5} = 4x\sqrt{5}$

38. $\sqrt{63y^2} = \sqrt{9y^2}\sqrt{7} = 3y\sqrt{7}$

39. $-\sqrt{250t^3} = -\sqrt{25t^2}\sqrt{10t} = -5t\sqrt{10t}$

40. $-\sqrt{700z^5} = -\sqrt{100z^4}\sqrt{7z} = -10z^2\sqrt{7z}$

41. $\sqrt{200x^2y} = \sqrt{100x^2}\sqrt{2y} = 10x\sqrt{2y}$

42. $\sqrt{75y^2z} = \sqrt{25y^2}\sqrt{3z} = 5y\sqrt{3z}$

43. $\sqrt[3]{8x^2y^3} = \sqrt[3]{8y^3}\sqrt[3]{x^2} = 2y\sqrt[3]{x^2}$

44. $\sqrt[3]{250x^4y^3} = \sqrt[3]{125x^3y^3}\sqrt[3]{2x} = 5xy\sqrt[3]{2x}$

45. $\sqrt{\dfrac{16}{25}} = \dfrac{\sqrt{16}}{\sqrt{25}} = \dfrac{4}{5}$

46. $\sqrt{\dfrac{100}{49}} = \dfrac{\sqrt{100}}{\sqrt{49}} = \dfrac{10}{7}$

47. $\sqrt[3]{\dfrac{1{,}000}{27}} = \dfrac{\sqrt[3]{1{,}000}}{\sqrt[3]{27}} = \dfrac{10}{3}$

48. $\sqrt[3]{\dfrac{16}{64}} = \sqrt[3]{\dfrac{2}{8}} = \dfrac{\sqrt[3]{2}}{\sqrt[3]{8}} = \dfrac{\sqrt[3]{2}}{2}$

49. $\sqrt{\dfrac{60}{49}} = \dfrac{\sqrt{60}}{\sqrt{49}} = \dfrac{\sqrt{4}\sqrt{15}}{7} = \dfrac{2\sqrt{15}}{7}$

50. $\sqrt{\dfrac{80}{225}} = \dfrac{\sqrt{80}}{\sqrt{225}} = \dfrac{\sqrt{16}\sqrt{5}}{15} = \dfrac{4\sqrt{5}}{15}$

51. $\sqrt{\dfrac{242x^4}{169x^2}} = \sqrt{\dfrac{242x^2}{169}} = \dfrac{\sqrt{242x^2}}{\sqrt{169}} = \dfrac{\sqrt{121x^2}\sqrt{2}}{13} = \dfrac{11x\sqrt{2}}{13}$

52. $\sqrt{\dfrac{450a^6}{196a^2}} = \sqrt{\dfrac{450a^4}{196}} = \dfrac{\sqrt{450a^4}}{\sqrt{196}} = \dfrac{\sqrt{225a^4}\sqrt{2}}{14} = \dfrac{15a^2\sqrt{2}}{14}$

53. $\sqrt{2} + \sqrt{8} - \sqrt{18} = \sqrt{2} + \sqrt{4}\sqrt{2} - \sqrt{9}\sqrt{2} = \sqrt{2} + 2\sqrt{2} - 3\sqrt{2} = 0$

54. $\sqrt{3} + \sqrt{27} - \sqrt{12} = \sqrt{3} + \sqrt{9}\sqrt{3} - \sqrt{4}\sqrt{3} = \sqrt{3} + 3\sqrt{3} - 2\sqrt{3} = 2\sqrt{3}$

55. $3\sqrt{5} + 5\sqrt{45} = 3\sqrt{5} + 5\sqrt{9}\sqrt{5} = 3\sqrt{5} + 5(3)\sqrt{5} = 3\sqrt{5} + 15\sqrt{5} = 18\sqrt{5}$

56. $5\sqrt{28} - 3\sqrt{63} = 5\sqrt{4}\sqrt{7} - 3\sqrt{9}\sqrt{7} = 5(2)\sqrt{7} - 3(3)\sqrt{3} = 10\sqrt{7} - 9\sqrt{7} = \sqrt{7}$

57. $3\sqrt{2x^2y} + 2x\sqrt{2y} = 3\sqrt{x^2}\sqrt{2y} + 2x\sqrt{2y} = 3x\sqrt{2y} + 2x\sqrt{2y} = 5x\sqrt{2y}$

58. $3y\sqrt{5xy^3} - y^2\sqrt{20xy} = 3y\sqrt{y^2}\sqrt{5xy} - y^2\sqrt{4}\sqrt{5xy} = 3y(y)\sqrt{5xy} - y^2(2)\sqrt{5xy}$
$$= 3y^2\sqrt{5xy} - 2y^2\sqrt{5xy} = y^2\sqrt{5xy}$$

59. $\sqrt[3]{16} + \sqrt[3]{54} = \sqrt[3]{8}\sqrt[3]{2} + \sqrt[3]{27}\sqrt[3]{2} = 2\sqrt[3]{2} + 3\sqrt[3]{2} = 5\sqrt[3]{2}$

60. $\sqrt[3]{2,000x^3} - \sqrt[3]{128x^3} = \sqrt[3]{1,000x^3}\sqrt[3]{2} - \sqrt[3]{64x^3}\sqrt[3]{2} = 10x\sqrt[3]{2} - 4x\sqrt[3]{2} = 6x\sqrt[3]{2}$

61. $\left(3\sqrt{2}\right)\left(-2\sqrt{3}\right) = -6\sqrt{6}$ 　　　　　**62.** $\left(-5\sqrt{x}\right)\left(-2\sqrt{x}\right) = 10\sqrt{x^2} = 10x$

63. $\left(3\sqrt{3x}\right)\left(4\sqrt{6x}\right) = 12\sqrt{18x^2} = 12\sqrt{9x^2}\sqrt{2} = 12(3x)\sqrt{2} = 36x\sqrt{2}$

64. $\left(-2\sqrt{27y^3}\right)\left(y\sqrt{2y}\right) = -2y\sqrt{54y^4} = -2y\sqrt{9y^4}\sqrt{6} = -2y(3y^2)\sqrt{6} = -6y^3\sqrt{6}$

65. $\left(\sqrt[3]{4}\right)\left(2\sqrt[3]{4}\right) = 2\sqrt[3]{16} = 2\sqrt[3]{8}\sqrt[3]{2} = 2(2)\sqrt[3]{2} = 4\sqrt[3]{2}$

66. $\left(-2\sqrt[3]{32x^2}\right)\left(3\sqrt[3]{2x^2}\right) = -6\sqrt[3]{64x^4} = -6\sqrt[3]{64x^3}\sqrt[3]{x} = -6(4x)\sqrt[3]{x} = -24x\sqrt[3]{x}$

67. $\sqrt{2}\left(\sqrt{8} - \sqrt{18}\right) = \sqrt{16} - \sqrt{36} = 4 - 6 = -2$

68. $\sqrt{6y}\left(\sqrt{2y} + \sqrt{75}\right) = \sqrt{12y^2} + \sqrt{450y} = \sqrt{4y^2}\sqrt{3} + \sqrt{225}\sqrt{2y} = 2y\sqrt{3} + 15\sqrt{2y}$

69. $\left(\sqrt{3} + \sqrt{5}\right)\left(\sqrt{3} - \sqrt{5}\right) = \sqrt{3}\sqrt{3} + \sqrt{3}\left(-\sqrt{5}\right) + \sqrt{5}\sqrt{3} + \sqrt{5}\left(-\sqrt{5}\right)$
$$= 3 - \sqrt{15} + \sqrt{15} - 5 = 3 - 5 = -2$$

70. $\left(\sqrt{15} + 3x\right)\left(\sqrt{15} + 3x\right) = \sqrt{15}\sqrt{15} + \sqrt{15}(3x) + 3x\sqrt{15} + 9x^2 = 15 + 6x\sqrt{15} + 9x^2$

71. $\left(\sqrt[3]{3} + 2\right)\left(\sqrt[3]{3} - 1\right) = \sqrt[3]{3}\sqrt[3]{3} + \sqrt[3]{3}(-1) + 2\sqrt[3]{3} + 2(-1) = \sqrt[3]{9} - \sqrt[3]{3} + 2\sqrt[3]{3} - 2$
$$= \sqrt[3]{9} + \sqrt[3]{3} - 2$$

72. $\left(\sqrt[3]{5} - 1\right)\left(\sqrt[3]{5} + 1\right) = \sqrt[3]{5}\sqrt[3]{5} + \sqrt[3]{5}(1) - \sqrt[3]{5}(1) - 1(1) = \sqrt[3]{25} + \sqrt[3]{5} - \sqrt[3]{5} - 1 = \sqrt[3]{25} - 1$

73. $\dfrac{1}{\sqrt{7}} = \dfrac{1\sqrt{7}}{\sqrt{7}\sqrt{7}} = \dfrac{\sqrt{7}}{7}$ 　　　　**74.** $\dfrac{3}{\sqrt{18}} = \dfrac{3\sqrt{2}}{\sqrt{18}\sqrt{2}} = \dfrac{3\sqrt{2}}{\sqrt{36}} = \dfrac{3\sqrt{2}}{6} = \dfrac{\sqrt{2}}{2}$

75. $\dfrac{8}{\sqrt[3]{16}} = \dfrac{8\sqrt[3]{4}}{\sqrt[3]{16}\sqrt[3]{4}} = \dfrac{8\sqrt[3]{4}}{\sqrt[3]{64}} = \dfrac{8\sqrt[3]{4}}{4} = 2\sqrt[3]{4}$ 　　**76.** $\dfrac{10}{\sqrt[3]{32}} = \dfrac{10\sqrt[3]{2}}{\sqrt[3]{32}\sqrt[3]{2}} = \dfrac{10\sqrt[3]{2}}{\sqrt[3]{64}} = \dfrac{10\sqrt[3]{2}}{4} = \dfrac{5\sqrt[3]{2}}{2}$

77. $\dfrac{7}{\sqrt{2}+1} = \dfrac{7\left(\sqrt{2}-1\right)}{\left(\sqrt{2}+1\right)\left(\sqrt{2}-1\right)} = \dfrac{7\left(\sqrt{2}-1\right)}{2-\sqrt{2}+\sqrt{2}-1} = \dfrac{7\left(\sqrt{2}-1\right)}{1} = 7\left(\sqrt{2}-1\right)$

78. $\dfrac{3}{\sqrt{3}-1} = \dfrac{3\left(\sqrt{3}+1\right)}{\left(\sqrt{3}-1\right)\left(\sqrt{3}+1\right)} = \dfrac{3\left(\sqrt{3}+1\right)}{3+\sqrt{3}-\sqrt{3}-1} = \dfrac{3\left(\sqrt{3}+1\right)}{2} = \dfrac{3\sqrt{3}+3}{2}$

79. $\dfrac{2\sqrt{5}}{\sqrt{5}+\sqrt{3}} = \dfrac{2\sqrt{5}\left(\sqrt{5}-\sqrt{3}\right)}{\left(\sqrt{5}+\sqrt{3}\right)\left(\sqrt{5}-\sqrt{3}\right)} = \dfrac{2\sqrt{25}-2\sqrt{15}}{\sqrt{25}-\sqrt{15}+\sqrt{15}-\sqrt{9}} = \dfrac{2(5)-2\sqrt{15}}{5-3}$

$$= \dfrac{10-2\sqrt{15}}{2}$$

$$= \dfrac{2\left(5-\sqrt{15}\right)}{2}$$

$$= 5-\sqrt{15}$$

80. $\dfrac{\sqrt{7x}+\sqrt{x}}{\sqrt{7x}-\sqrt{x}} = \dfrac{\left(\sqrt{7x}+\sqrt{x}\right)\left(\sqrt{7x}+\sqrt{x}\right)}{\left(\sqrt{7x}-\sqrt{x}\right)\left(\sqrt{7x}+\sqrt{x}\right)} = \dfrac{\sqrt{49x^2}+\sqrt{7x^2}+\sqrt{7x^2}+\sqrt{x^2}}{\sqrt{49x^2}+\sqrt{7x^2}-\sqrt{7x^2}-\sqrt{x^2}}$

$$= \dfrac{7x+x\sqrt{7}+x\sqrt{7}+x}{7x-x}$$

$$= \dfrac{8x+2x\sqrt{7}}{6x} = \dfrac{2x\left(4+\sqrt{7}\right)}{6x} = \dfrac{4+\sqrt{7}}{3}$$

81. $\sqrt{x+3} = 3$
$\left(\sqrt{x+3}\right)^2 = 3^2$
$x+3 = 9$
$x = 6$

82. $\sqrt{2x+10} = 2$
$\left(\sqrt{2x+10}\right)^2 = 2^2$
$2x+10 = 4$
$2x = -6$
$x = -3$

83. $\sqrt{3x+4} = -2\sqrt{x}$
$\left(\sqrt{3x+4}\right)^2 = \left(-2\sqrt{x}\right)^2$
$3x+4 = 4x$
$\boxed{4 = x}$
does not check, no solution

84. $\sqrt{2(x+4)} - \sqrt{4x} = 0$
$\sqrt{2x+8} = \sqrt{4x}$
$\left(\sqrt{2x+8}\right)^2 = \left(\sqrt{4x}\right)^2$
$2x+8 = 4x$
$8 = 2x$
$4 = x$

85.
$$\sqrt{x+5} = x-1$$
$$\left(\sqrt{x+5}\right)^2 = (x-1)^2$$
$$x+5 = (x-1)(x-1)$$
$$x+5 = x^2 - 2x + 1$$
$$0 = x^2 - 3x - 4$$
$$0 = (x-4)(x+1)$$
$$x-4 = 0 \quad \textbf{or} \quad x+1 = 0$$
$$x = 4 \qquad \boxed{x = -1}$$
$x = -1$ does not check.

86.
$$\sqrt{2x+9} = x-3$$
$$\left(\sqrt{2x+9}\right)^2 = (x-3)^2$$
$$2x+9 = (x-3)(x-3)$$
$$2x+9 = x^2 - 6x + 9$$
$$0 = x^2 - 8x$$
$$0 = x(x-8)$$
$$x-8 = 0 \quad \textbf{or} \quad x = 0$$
$$x = 8 \qquad \boxed{x = 0}$$
$x = 0$ does not check.

87.
$$\sqrt{2x+5} - 1 = x$$
$$\sqrt{2x+5} = x+1$$
$$\left(\sqrt{2x+5}\right)^2 = (x+1)^2$$
$$2x+5 = (x+1)(x+1)$$
$$2x+5 = x^2 + 2x + 1$$
$$0 = x^2 - 4$$
$$0 = (x-2)(x+2)$$
$$x-2 = 0 \quad \textbf{or} \quad x+2 = 0$$
$$x = 2 \qquad \boxed{x = -2}$$
$x = -2$ does not check.

88.
$$\sqrt{4a+13} + 2 = a$$
$$\sqrt{4a+13} = a-2$$
$$\left(\sqrt{4a+13}\right)^2 = (a-2)^2$$
$$4a+13 = (a-2)(a-2)$$
$$4a+13 = a^2 - 4a + 4$$
$$0 = a^2 - 8a - 9$$
$$0 = (a-9)(a+1)$$
$$a-9 = 0 \quad \textbf{or} \quad a+1 = 0$$
$$a = 9 \qquad \boxed{a = -1}$$
$a = -1$ does not check.

89. $d = \sqrt{(x_2 - x_1)^2 + (y_2 - y_1)^2} = \sqrt{(-4 - (-7))^2 + (8 - 12)^2} = \sqrt{(3)^2 + (-4)^2}$
$$= \sqrt{9+16} = \sqrt{25} = 5$$

90. $d = \sqrt{(x_2 - x_1)^2 + (y_2 - y_1)^2} = \sqrt{(-10 - (-15))^2 + (-15 - (-3))^2} = \sqrt{(5)^2 + (-12)^2}$
$$= \sqrt{25 + 144}$$
$$= \sqrt{169} = 13$$

91. $d = \sqrt{(x_2 - x_1)^2 + (y_2 - y_1)^2}$
$$= \sqrt{(-1 - 1)^2 + (1 - 1)^2}$$
$$= \sqrt{(-2)^2 + (0)^2}$$
$$= \sqrt{4 + 0} = \sqrt{4} = 2$$

92. $d = \sqrt{(x_2 - x_1)^2 + (y_2 - y_1)^2}$
$$= \sqrt{(10 - (-10))^2 + (-10 - 11)^2}$$
$$= \sqrt{(20)^2 + (-21)^2}$$
$$= \sqrt{400 + 441} = \sqrt{841} = 29$$

93. $49^{1/2} = 7$

94. $(-1{,}000)^{1/3} = -10$

95. $36^{3/2} = \left(36^{1/2}\right)^3 = 6^3 = 216$

96. $\left(\dfrac{4}{9}\right)^{5/2} = \left(\left(\dfrac{4}{9}\right)^{1/2}\right)^5 = \left(\dfrac{2}{3}\right)^5 = \dfrac{32}{243}$

97. $8^{2/3}8^{4/3} = 8^{6/3} = 8^2 = 64$

98. $\dfrac{5^{17/7}}{5^{3/7}} = 5^{14/7} = 5^2 = 25$

99. $\dfrac{x^{4/5}x^{3/5}}{\left(x^{2/5}\right)^3} = \dfrac{x^{7/5}}{x^{6/5}} = x^{1/5}$

100. $\left(\dfrac{r^{1/3}r^{2/3}}{r^{4/3}}\right)^3 = \left(\dfrac{r^{3/3}}{r^{4/3}}\right)^3 = \left(r^{-1/3}\right)^3$
$$= r^{-1} = \tfrac{1}{r}$$

101. $6^{5/3}6^{-2/3} = 6^{5/3+(-2/3)} = 6^{3/3} = 6^1 = 6$

102. $\dfrac{5^{2/3}}{5^{-1/3}} = 5^{2/3-(-1/3)} = 5^{3/3} = 5^1 = 5$

103. $\dfrac{x^{2/5}x^{1/5}}{x^{-2/5}} = \dfrac{x^{3/5}}{x^{-2/5}} = x^{5/5} = x^1 = x$

104. $\left(a^4b^8\right)^{-1/2} = \left(a^4\right)^{-1/2}\left(b^8\right)^{-1/2} = a^{-4/2}b^{-8/2} = a^{-2}b^{-4} = \dfrac{1}{a^2b^4}$

105. $x^{1/3}x^{2/5} = x^{5/15}x^{6/15} = x^{11/15}$

106. $\dfrac{t^{3/4}}{t^{2/3}} = \dfrac{t^{9/12}}{t^{8/12}} = t^{1/12}$

107. $\dfrac{x^{-4/5}x^{1/3}}{x^{1/3}} = x^{-4/5} \cdot \dfrac{x^{1/3}}{x^{1/3}} = x^{-4/5} = \dfrac{1}{x^{4/5}}$

108. $\dfrac{r^{1/4}r^{1/3}}{r^{5/6}} = \dfrac{r^{3/12}r^{4/12}}{r^{10/12}} = \dfrac{r^{7/12}}{r^{10/12}} = r^{-3/12} = r^{-1/4} = \dfrac{1}{r^{1/4}}$

Chapter 8 Test (page 532)

1. $\sqrt{100} = 10$

2. $-\sqrt{400} = -20$

3. $\sqrt[3]{-27} = -3$

4. $\sqrt{3x}\sqrt{27x} = \sqrt{81x^2} = 9x$

5. Let $x =$ the hypotenuse length.
$$5^2 + 12^2 = x^2$$
$$25 + 144 = x^2$$
$$169 = x^2$$
13 inches $= x$

6. Let $x =$ distance from the wall to the base of the ladder.
$$x^2 + 24^2 = 26^2$$
$$x^2 + 576 = 676$$
$$x^2 = 100$$
$$x = 10 \text{ ft from the wall}$$

7. $\sqrt{8x^2} = \sqrt{4x^2}\sqrt{2} = 2x\sqrt{2}$

8. $\sqrt{54x^3y} = \sqrt{9x^2}\sqrt{6xy} = 3x\sqrt{6xy}$

9. $\sqrt{\dfrac{320}{10}} = \sqrt{32} = \sqrt{16}\sqrt{2} = 4\sqrt{2}$

10. $\sqrt{\dfrac{18x^2y^3}{2xy}} = \sqrt{9xy^2} = \sqrt{9y^2}\sqrt{x} = 3y\sqrt{x}$

11. $\sqrt[3]{x^6 y^6} = x^2 y^2$

12. $\sqrt[4]{\dfrac{16x^8}{y^4}} = \dfrac{\sqrt[4]{16x^8}}{\sqrt[4]{y^4}} = \dfrac{2x^2}{y}$

13. $\sqrt{12} + \sqrt{27} = \sqrt{4}\sqrt{3} + \sqrt{9}\sqrt{3}$
$\qquad = 2\sqrt{3} + 3\sqrt{3} = 5\sqrt{3}$

14. $\sqrt{8x^3} - x\sqrt{18x} = \sqrt{4x^2}\sqrt{2x} - x\sqrt{9}\sqrt{2x}$
$\qquad = 2x\sqrt{2x} - 3x\sqrt{2x}$
$\qquad = -x\sqrt{2x}$

15. $\left(-2\sqrt{8x}\right)\left(3\sqrt{12x}\right) = -6\sqrt{96x^2} = -6\sqrt{16x^2}\sqrt{6} = -24x\sqrt{6}$

16. $\sqrt{3}\left(\sqrt{8} + \sqrt{6}\right) = \sqrt{24} + \sqrt{18} = \sqrt{4}\sqrt{6} + \sqrt{9}\sqrt{2} = 2\sqrt{6} + 3\sqrt{2}$

17. $\left(\sqrt{2} + \sqrt{3}\right)\left(\sqrt{2} - \sqrt{3}\right) = \sqrt{4} - \sqrt{6} + \sqrt{6} - \sqrt{9} = 2 - 3 = -1$

18. $\left(2\sqrt{x} + 2\right)\left(\sqrt{x} - 3\right) = 2\sqrt{x^2} + 2\sqrt{x}(-3) + 2\sqrt{x} - 6 = 2x - 6\sqrt{x} + 2\sqrt{x} - 6$
$\qquad = 2x - 4\sqrt{x} - 6$

19. $\dfrac{2}{\sqrt{2}} = \dfrac{2\sqrt{2}}{\sqrt{2}\sqrt{2}} = \dfrac{2\sqrt{2}}{2} = \sqrt{2}$

20. $\sqrt{\dfrac{3xy^3}{48x^2}} = \sqrt{\dfrac{y^3}{16x}} = \dfrac{\sqrt{y^3}}{\sqrt{16x}} = \dfrac{\sqrt{y^2}\sqrt{y}}{\sqrt{16}\sqrt{x}} = \dfrac{y\sqrt{y}}{4\sqrt{x}} = \dfrac{y\sqrt{y}\sqrt{x}}{4\sqrt{x}\sqrt{x}} = \dfrac{y\sqrt{xy}}{4x}$

21. $\dfrac{2}{\sqrt{5} - 2} = \dfrac{2\left(\sqrt{5} + 2\right)}{\left(\sqrt{5} - 2\right)\left(\sqrt{5} + 2\right)} = \dfrac{2\left(\sqrt{5} + 2\right)}{\sqrt{25} + 2\sqrt{5} - 2\sqrt{5} - 4} = \dfrac{2\left(\sqrt{5} + 2\right)}{5 - 4} = \dfrac{2\left(\sqrt{5} + 2\right)}{1}$
$\qquad = 2\sqrt{5} + 4$

22. $\dfrac{\sqrt{3x}}{\sqrt{x} + 2} = \dfrac{\sqrt{3x}\left(\sqrt{x} - 2\right)}{\left(\sqrt{x} + 2\right)\left(\sqrt{x} - 2\right)} = \dfrac{\sqrt{3x}\sqrt{x} - 2\sqrt{3x}}{x - 2\sqrt{x} + 2\sqrt{x} - 4} = \dfrac{x\sqrt{3} - 2\sqrt{3x}}{x - 4}$

23. $\sqrt{x} + 3 = 9$
$\qquad \sqrt{x} = 6$
$\qquad \left(\sqrt{x}\right)^2 = 6^2$
$\qquad x = 36$

24. $\sqrt{x - 2} - 2 = 6$
$\qquad \sqrt{x - 2} = 8$
$\qquad \left(\sqrt{x - 2}\right)^2 = 8^2$
$\qquad x - 2 = 64$
$\qquad x = 66$

25.
$$\sqrt{3x+9} = 2\sqrt{x+1}$$
$$\left(\sqrt{3x+9}\right)^2 = \left(2\sqrt{x+1}\right)^2$$
$$3x+9 = 2^2\left(\sqrt{x+1}\right)^2$$
$$3x+9 = 4(x+1)$$
$$3x+9 = 4x+4$$
$$-x = -5$$
$$x = 5$$

26.
$$3\sqrt{x-3} = \sqrt{2x+8}$$
$$\left(3\sqrt{x-3}\right)^2 = \left(\sqrt{2x+8}\right)^2$$
$$3^2\left(\sqrt{x-3}\right)^2 = 2x+8$$
$$9(x-3) = 2x+8$$
$$9x-27 = 2x+8$$
$$7x = 35$$
$$x = 5$$

27.
$$\sqrt{3x+1} = x-1$$
$$\left(\sqrt{3x+1}\right)^2 = (x-1)^2$$
$$3x+1 = (x-1)(x-1)$$
$$3x+1 = x^2-2x+1$$
$$0 = x^2-5x$$
$$0 = x(x-5)$$
$$x-5 = 0 \quad \textbf{or} \quad x = 0$$
$$x = 5 \qquad \boxed{x = 0}$$
$x = 0$ does not check.

28.
$$\sqrt[3]{x-2} = 3$$
$$\left(\sqrt[3]{x-2}\right)^3 = 3^3$$
$$x-2 = 27$$
$$x = 29$$

29.
$$d = \sqrt{(x_2-x_1)^2 + (y_2-y_1)^2}$$
$$= \sqrt{(7-1)^2 + (12-4)^2}$$
$$= \sqrt{(6)^2 + (8)^2}$$
$$= \sqrt{36+64} = \sqrt{100} = 10$$

30.
$$d = \sqrt{(x_2-x_1)^2 + (y_2-y_1)^2}$$
$$= \sqrt{(-5-(-2))^2 + (1-(-3))^2}$$
$$= \sqrt{(-3)^2 + (4)^2}$$
$$= \sqrt{9+16} = \sqrt{25} = 5$$

31. $121^{1/2} = 11$

32. $27^{-4/3} = \dfrac{1}{27^{4/3}} = \dfrac{1}{\left(27^{1/3}\right)^4} = \dfrac{1}{3^4} = \dfrac{1}{81}$

33. $\left(y^{15}\right)^{2/5} = y^{30/5} = y^6$

34. $\left(\dfrac{a^{5/3}a^{4/3}}{\left(a^{1/3}\right)^2 a^{2/3}}\right)^6 = \left(\dfrac{a^{9/3}}{a^{2/3}a^{2/3}}\right)^6 = \left(\dfrac{a^{9/3}}{a^{4/3}}\right)^6 = \left(a^{5/3}\right)^6 = a^{30/3} = a^{10}$

35. $p^{2/3}p^{3/4} = p^{8/12}p^{9/12} = p^{17/12}$

36. $\dfrac{x^{2/3}x^{-4/5}}{x^{2/15}} = \dfrac{x^{10/15}x^{-12/15}}{x^{2/15}} = \dfrac{x^{-2/15}}{x^{2/15}} = x^{-4/15} = \dfrac{1}{x^{4/15}}$

Cumulative Review Exercises (page 532)

1. $(3x^2 + 2x) + (6x^3 - 3x^2 + 1) = 3x^2 + 2x + 6x^3 - 3x^2 + 1 = 6x^3 + 2x + 1$

2. $(5x^3 - 2x) - (2x^3 - 3x^2 - 3x - 1) = 5x^3 - 2x - 2x^3 + 3x^2 + 3x + 1 = 3x^3 + 3x^2 + x + 1$

3. $3(6x^2 - 3x + 3) + 2(-x^2 + 2x - 5) = 18x^2 - 9x + 9 - 2x^2 + 4x - 10 = 16x^2 - 5x - 1$

4. $5(3x^2 - 4x - 1) - 2(-2x^2 + 4x + 3) = 15x^2 - 20x - 5 + 4x^2 - 8x - 6 = 19x^2 - 28x - 11$

5. $(5x^4y^3)(-3x^2y^3) = (5)(-3)x^4x^2y^3y^3 = -15x^6y^6$

6. $-3x^2(-5x^3 - 3x^2 + 2) = (-3x^2)(-5x^3) + (-3x^2)(-3x^2) + (-3x^2)(2)$
$$= 15x^5 + 9x^4 - 6x^2$$

7. $(2x + 3)(3x + 4) = (2x)(3x) + (2x)(4) + (3)(3x) + (3)(4) = 6x^2 + 8x + 9x + 12$
$$= 6x^2 + 17x + 12$$

8. $(4x - 3y)(3x - 2y) = (4x)(3x) + (4x)(-2y) + (-3y)(3x) + (-3y)(-2y)$
$$= 12x^2 - 8xy - 9xy + 6y^2 = 12x^2 - 17xy + 6y^2$$

9.
$$
\begin{array}{r}
x - 4 \\
x + 3 \overline{\smash{\big)}\, x^2 - x - 12} \\
\underline{x^2 + 3x} \\
-4x - 12 \\
\underline{-4x - 12} \\
0
\end{array}
$$

10.
$$
\begin{array}{r}
x^2 + x + \frac{-3}{2x-1} \\
2x - 1 \overline{\smash{\big)}\, 2x^3 + x^2 - x - 3} \\
\underline{2x^3 - x^2} \\
2x^2 - x \\
\underline{2x^2 - x} \\
-3
\end{array}
$$

11. $2x^2y - 4xy^2 = 2xy(x - 2y)$

12. $5(x + y) + a(x + y) = (x + y)(5 + a)$

13. $3a + 3b + ab + b^2 = 3(a + b) + b(a + b)$
$$= (a + b)(3 + b)$$

14. $49p^4 - 16q^2 = (7p^2)^2 - (4q)^2$
$$= (7p^2 + 4q)(7p^2 - 4q)$$

15. $x^2 - 9x - 36 = (x - 12)(x + 3)$

16. $x^2 - 3xy - 10y^2 = (x - 5y)(x + 2y)$

17. $12a^2 + a - 20 = (4a - 5)(3a + 4)$

18. $10m^2 - 13mn - 3n^2 = (5m + n)(2m - 3n)$

19. $p^3 - 64q^3 = p^3 - (4q)^3 = (p - 4q)(p^2 + p(4q) + (4q)^2) = (p - 4q)(p^2 + 4pq + 16q^2)$

20. $2r^3 + 54s^3 = 2(r^3 + 27s^3) = 2(r^3 + (3s)^3) = 2(r + 3s)(r^2 - r(3s) + (3s)^2)$
$$= 2(r + 3s)(r^2 - 3rs + 9s^2)$$

21.
$$a^2 + 3a = -2$$
$$a^2 + 3a + 2 = 0$$
$$(a+1)(a+2) = 0$$
$$a + 1 = 0 \quad \textbf{or} \quad a + 2 = 0$$
$$a = -1 \qquad\qquad a = -2$$

22.
$$2b^2 - 12 = -5b$$
$$2b^2 + 5b - 12 = 0$$
$$(2b-3)(b+4) = 0$$
$$2b - 3 = 0 \quad \textbf{or} \quad b + 4 = 0$$
$$2b = 3 \qquad\qquad b = -4$$
$$b = \tfrac{3}{2} \qquad\qquad b = -4$$

23.
$$\frac{4}{a} = \frac{6}{a} - 1$$
$$a \cdot \frac{4}{a} = a\left(\frac{6}{a} - 1\right)$$
$$4 = 6 - a$$
$$a = 2$$

24.
$$\frac{a+2}{a+3} - 1 = \frac{-1}{a^2 + 2a - 3}$$
$$\frac{a+2}{a+3} - 1 = \frac{-1}{(a+3)(a-1)}$$
$$(a+3)(a-1)\left(\frac{a+2}{a+3} - 1\right) = (a+3)(a-1) \cdot \frac{-1}{(a+3)(a-1)}$$
$$(a-1)(a+2) - (a+3)(a-1) = -1$$
$$a^2 + a - 2 - \left(a^2 + 2a - 3\right) = -1$$
$$a^2 + a - 2 - a^2 - 2a + 3 = -1$$
$$-a + 1 = -1$$
$$2 = a$$

25.
$$\frac{4-a}{13} = \frac{11}{26}$$
$$26(4-a) = 13(11)$$
$$104 - 26a = 143$$
$$-26a = 39$$
$$a = -\frac{39}{26} = -\frac{3}{2}$$

26.
$$\frac{3a-2}{7} = \frac{a}{28}$$
$$28(3a-2) = 7a$$
$$84a - 56 = 7a$$
$$77a = 56$$
$$a = \frac{56}{77} = \frac{8}{11}$$

27.

x	0	4
y	-5	0

28.

x	-2	-1	0	1	2
y	-4	-1	0	-1	-4

29. $m = \dfrac{y_2 - y_1}{x_2 - x_1} = \dfrac{8 - 4}{6 - (-2)} = \dfrac{4}{8} = \dfrac{1}{2}$

30. $3x + 6y = 13$

$\qquad 6y = -3x + 13$

$\qquad y = -\dfrac{1}{2}x + \dfrac{13}{6} \Rightarrow m = -\dfrac{1}{2}$

31. $y - y_1 = m(x - x_1)$

$\qquad y - 5 = \dfrac{2}{3}(x - 0)$

$\qquad y - 5 = \dfrac{2}{3}x$

$\qquad y = \dfrac{2}{3}x + 5$

32. $m = \dfrac{y_2 - y_1}{x_2 - x_1} = \dfrac{10 - 4}{6 - (-2)} = \dfrac{6}{8} = \dfrac{3}{4}$

$\qquad y - y_1 = m(x - x_1)$

$\qquad y - 10 = \dfrac{3}{4}(x - 6)$

$\qquad 4(y - 10) = 4 \cdot \dfrac{3}{4}(x - 6)$

$\qquad 4y - 40 = 3(x - 6)$

$\qquad 4y - 40 = 3x - 18$

$\qquad -22 = 3x - 4y, \text{ or } 3x - 4y = -22$

33. $\quad 3x + 4y = 15 \qquad\qquad 4x - 3y = 25$

$\qquad\quad 4y = -3x + 15 \qquad\quad -3y = -4x + 25$

$\qquad\qquad y = -\dfrac{3}{4}x + \dfrac{15}{4} \qquad\qquad y = \dfrac{4}{3}x - \dfrac{25}{3}$

$\qquad\qquad m = -\dfrac{3}{4} \qquad\qquad\qquad m = \dfrac{4}{3}$

$\qquad\qquad\qquad\text{perpendicular}$

34. $\quad 3x + 4y = 15 \qquad\qquad 6x = 15 - 8y$

$\qquad\quad 4y = -3x + 15 \qquad\quad 8y = -6x + 15$

$\qquad\qquad y = -\dfrac{3}{4}x + \dfrac{15}{4} \qquad\qquad y = -\dfrac{6}{8}x + \dfrac{15}{8}$

$\qquad\qquad m = -\dfrac{3}{4} \qquad\qquad\qquad m = -\dfrac{3}{4}$

$\qquad\qquad\qquad\text{parallel}$

35. $y \leq -2x + 4$

x	y
0	4
1	2

36. $3x + 4y \leq 12$

x	y
0	3
4	0

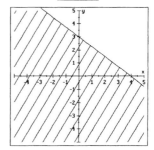

37. $f(0) = 2(0)^2 - 3 = 2(0) - 3 = 0 - 3$
$$= -3$$

38. $f(3) = 2(3)^2 - 3 = 2(9) - 3 = 18 - 3$
$$= 15$$

39. $f(-2) = 2(-2)^2 - 3 = 2(4) - 3 = 8 - 3$
$$= 5$$

40. $f(2x) = 2(2x)^2 - 3 = 2(4x^2) - 3 = 8x^2 - 3$

41.
$y = kx \qquad y = \dfrac{2}{5}x$
$4 = k(10) \qquad$
$\dfrac{4}{10} = k \qquad y = \dfrac{2}{5}(30)$
$\qquad\qquad y = 12$
$\dfrac{2}{5} = k$

42.
$y = \dfrac{k}{x} \qquad y = \dfrac{16}{x}$
$8 = \dfrac{k}{2} \qquad y = \dfrac{16}{8}$
$16 = k \qquad y = 2$

Exercise 9.1 (page 542)

1. $(y - 1)^2 = (y - 1)(y - 1) = y^2 - y - y + 1 = y^2 - 2y + 1$

3. $(x + y)^2 = (x + y)(x + y) = x^2 + xy + xy + y^2 = x^2 + 2xy + y^2$

5. $(2r - s)^2 = (2r - s)(2r - s) = 4r^2 - 2rs - 2rs + s^2 = 4r^2 - 4rs + s^2$

7. quadratic $\qquad\qquad$ **9.** 0; 0 $\qquad\qquad\qquad$ **11.** two

13.
$(x - 2)(x + 3) = 0$
$x - 2 = 0 \quad \textbf{or} \quad x + 3 = 0$
$x = 2 \qquad\qquad x = -3$

15.
$(x - 4)(x + 1) = 0$
$x - 4 = 0 \quad \textbf{or} \quad x + 1 = 0$
$x = 4 \qquad\qquad x = -1$

17.
$$(2x - 5)(3x + 6) = 0$$
$2x - 5 = 0$ **or** $3x + 6 = 0$
$\quad 2x = 5 \qquad\qquad 3x = -6$
$\quad x = \frac{5}{2} \qquad\qquad x = -\frac{6}{3} = -2$

19.
$$(x - 1)(x + 2)(x - 3) = 0$$
$x - 1 = 0$ **or** $x + 2 = 0$ **or** $x - 3 = 0$
$\quad x = 1 \qquad\qquad x = -2 \qquad\qquad x = 3$

21.
$$x^2 - 9 = 0$$
$(x + 3)(x - 3) = 0$
$x + 3 = 0$ **or** $x - 3 = 0$
$\quad x = -3 \qquad\qquad x = 3$

23.
$$3x^2 + 9x = 0$$
$3x(x + 3) = 0$
$3x = 0$ **or** $x + 3 = 0$
$\quad x = 0 \qquad\qquad x = -3$

25.
$$x^2 - 5x + 6 = 0$$
$(x - 2)(x - 3) = 0$
$x - 2 = 0$ **or** $x - 3 = 0$
$\quad x = 2 \qquad\qquad x = 3$

27.
$$3x^2 + x - 2 = 0$$
$(x + 1)(3x - 2) = 0$
$x + 1 = 0$ **or** $3x - 2 = 0$
$\quad x = -1 \qquad\qquad 3x = 2$
$\quad x = -1 \qquad\qquad x = \frac{2}{3}$

29.
$$6x^2 + 11x + 3 = 0$$
$(2x + 3)(3x + 1) = 0$
$2x + 3 = 0$ **or** $3x + 1 = 0$
$\quad 2x = -3 \qquad\qquad 3x = -1$
$\quad x = -\frac{3}{2} \qquad\qquad x = -\frac{1}{3}$

31.
$$10x^2 + x - 2 = 0$$
$(2x + 1)(5x - 2) = 0$
$2x + 1 = 0$ **or** $5x - 2 = 0$
$\quad 2x = -1 \qquad\qquad 5x = 2$
$\quad x = -\frac{1}{2} \qquad\qquad x = \frac{2}{5}$

33.
$$x^2 = 1$$
$\sqrt{x^2} = \pm\sqrt{1}$
$x = \pm 1$

35.
$$x^2 = 9$$
$\sqrt{x^2} = \pm\sqrt{9}$
$x = \pm 3$

37.
$$x^2 = 20$$
$\sqrt{x^2} = \pm\sqrt{20}$
$x = \pm\sqrt{4}\sqrt{5}$
$x = \pm 2\sqrt{5}$

39.
$$3x^2 = 27$$
$x^2 = 9$
$\sqrt{x^2} = \pm\sqrt{9}$
$x = \pm 3$

41.
$$4x^2 = 16$$
$x^2 = 4$
$\sqrt{x^2} = \pm\sqrt{4}$
$x = \pm 2$

43.
$$x^2 = a$$
$\sqrt{x^2} = \pm\sqrt{a}$
$x = \pm\sqrt{a}$

45.
$$(x + 1)^2 = 25$$
$\sqrt{(x + 1)^2} = \pm\sqrt{25}$
$x + 1 = \pm 5$
$x = -1 \pm 5$
$x = 4$ or $x = -6$

47.
$$(x + 2)^2 = 81$$
$\sqrt{(x + 2)^2} = \pm\sqrt{81}$
$x + 2 = \pm 9$
$x = -2 \pm 9$
$x = 7$ or $x = -11$

49.
$$(x - 2)^2 = 8$$
$\sqrt{(x - 2)^2} = \pm\sqrt{8}$
$x - 2 = \pm\sqrt{4}\sqrt{2}$
$x = 2 \pm 2\sqrt{2}$

51. $(x-a)^2 = 4a^2$

$\sqrt{(x-a)^2} = \pm\sqrt{4a^2}$

$x - a = \pm 2a$

$x = a \pm 2a$

$x = 3a$ or $x = -a$

53. $(x+b)^2 = 16c^2$

$\sqrt{(x+b)^2} = \pm\sqrt{16c^2}$

$x + b = \pm 4c$

$x = -b \pm 4c$

55. $x^2 + 4x + 4 = 4$

$(x+2)^2 = 4$

$\sqrt{(x+2)^2} = \pm\sqrt{4}$

$x + 2 = \pm 2$

$x = -2 \pm 2$

$x = 0$ or $x = -4$

57. $9x^2 - 12x + 4 = 16$

$(3x-2)^2 = 16$

$\sqrt{(3x-2)^2} = \pm\sqrt{16}$

$3x - 2 = \pm 4$

$3x = 2 \pm 4$

$3x = 6$ **or** $3x = -2$

$x = 2$ $\qquad x = -\frac{2}{3}$

59. $4x^2 + 4x + 1 = 20$

$(2x+1)^2 = 20$

$\sqrt{(2x+1)^2} = \pm\sqrt{20}$

$2x + 1 = \pm\sqrt{4}\sqrt{5}$

$2x = -1 \pm 2\sqrt{5}$

$x = \dfrac{-1 \pm 2\sqrt{5}}{2}$

61. $6(x^2 - 1) = 4(x^2 + 3)$

$6x^2 - 6 = 4x^2 + 12$

$2x^2 = 18$

$x^2 = 9$

$\sqrt{x^2} = \pm\sqrt{9}$

$x = \pm 3$

63. $8(x^2 - 6) = 4(x^2 + 13)$

$8x^2 - 48 = 4x^2 + 52$

$4x^2 = 100$

$x^2 = 25$

$\sqrt{x^2} = \pm\sqrt{25}$

$x = \pm 5$

65. $5(x+1)^2 = (x+1)^2 + 32$

$4(x+1)^2 = 32$

$(x+1)^2 = 8$

$\sqrt{(x+1)^2} = \pm\sqrt{8}$

$x + 1 = \pm\sqrt{4}\sqrt{2}$

$x = -1 \pm 2\sqrt{2}$

67. Let x and $x + 2$ represent the integers.

$x(x+2) = 48$

$x^2 + 2x = 48$

$x^2 + 2x - 48 = 0$

$(x+8)(x-6) = 0$

$x + 8 = 0$ **or** $x - 6 = 0$

$x = -8$ $\qquad x = 6$

Since the integers must be positive, the solution must be $x = 6$, and the integers are 6 and 8.

69. Let x and $x + 2$ represent the integers.

$$x^2 + (x+2)^2 = 52$$
$$x^2 + x^2 + 4x + 4 = 52$$
$$2x^2 + 4x - 48 = 0$$
$$2(x^2 + 2x - 24) = 0$$
$$2(x+6)(x-4) = 0$$
$$x + 6 = 0 \quad \text{or} \quad x - 4 = 0$$
$$x = -6 \qquad\qquad x = 4$$

Since the integers must be negative, the solution must be $x = -6$, and the integers are -6 and -4.

71. Let w represent the width. Then $w + 4$ must represent the length.

$$\text{width} \cdot \text{length} = \text{area}$$
$$w(w + 4) = 32$$
$$w^2 + 4w = 32$$
$$w^2 + 4w - 32 = 0$$
$$(w+8)(w-4) = 0$$
$$w + 8 = 0 \quad \text{or} \quad w - 4 = 0$$
$$w = -8 \qquad\qquad w = 4$$

Since the width must be positive, the solution must be $w = 4$, and the dimensions are 4 ft by 8 ft.

73. $\dfrac{1}{2} \cdot \text{base} \cdot \text{height} = \text{area}$

$$\frac{1}{2}(x+4)(x) = 30$$
$$x(x+4) = 60$$
$$x^2 + 4x = 60$$
$$x^2 + 4x - 60 = 0$$
$$(x+10)(x-6) = 0$$
$$x + 10 = 0 \quad \text{or} \quad x - 6 = 0$$
$$x = -10 \qquad\qquad x = 6$$

Since the height must be positive, the solution must be $x = 6$ in.

75. Let $x = $ the uniform width. Refer to the figure to the right.

$$\text{width} \cdot \text{length} = \text{area}$$
$$(16 - 2x)(24 - 2x) = 180$$
$$384 - 32x - 48x + 4x^2 = 180$$
$$4x^2 - 80x + 204 = 0$$
$$4(x^2 - 20x + 51) = 0$$
$$4(x-3)(x-17) = 0$$
$$x - 3 = 0 \quad \text{or} \quad x - 17 = 0$$
$$x = 3 \qquad\qquad x = 17$$

The solution $x = 17$ does not make sense. The solution must be $x = 3$, and the dimensions are 10 ft by 18 ft.

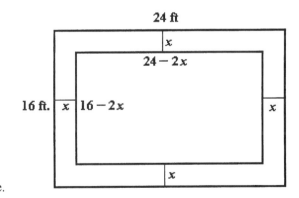

77.
$$s = 16t^2$$
$$1454 = 16t^2$$
$$90.875 = t^2$$
$$\sqrt{90.875} = t$$
$$9.5 \text{ sec} \approx t$$
(The negative square root was ignored.)

79. $h = -16t^2 + 144t$
$$= -16(3)^2 + 144(3)$$
$$= -16(9) + 432 = -144 + 432 = 288 \text{ ft}$$

81. Let $t = 4.5$ seconds.
$$h = -16t^2 + 144t$$
$$= -16(4.5)^2 + 144(4.5)$$
$$= 324 \text{ ft}$$

83. Set $h = 0$:
$$h = -16t^2 + 1088t$$
$$0 = -16t^2 + 1088t$$
$$0 = -16t(t - 68)$$
$$-16t = 0 \quad \textbf{or} \quad t - 68 = 0$$
$$t = 0 \qquad\qquad t = 68$$
It will hit the ground after 68 seconds.

85. Answers may vary.

87. When both sides are divided by x, it is assumed that $x \neq 0$, and yet $x = 0$ is a valid solution.

Exercise 9.2 (page 548)

1.
$$\frac{3t(2t + 1)}{2} + 6 = 3t^2$$
$$2\left(\frac{6t^2 + 3t}{2} + 6\right) = 2(3t^2)$$
$$6t^2 + 3t + 12 = 6t^2$$
$$3t = -12$$
$$t = -4$$

3.
$$\frac{2}{3x} - \frac{5}{9} = -\frac{1}{x}$$
$$9x\left(\frac{2}{3x} - \frac{5}{9}\right) = 9x\left(-\frac{1}{x}\right)$$
$$3(2) - x(5) = 9(-1)$$
$$6 - 5x = -9$$
$$-5x = -15$$
$$x = 3$$

5. completing

7. $(x + b)^2$

9. square; 8

11. $\frac{1}{2}(2) = 1; 1^2 = \mathbf{1}$
$$x^2 + 2x + \mathbf{1} = (x + 1)^2$$

13. $\frac{1}{2}(-4) = -2; (-2)^2 = \mathbf{4}$
$$x^2 - 4x + \mathbf{4} = (x - 2)^2$$

15. $\frac{1}{2}(7) = \frac{7}{2}; \left(\frac{7}{2}\right)^2 = \frac{\mathbf{49}}{\mathbf{4}}$
$$x^2 + 7x + \frac{\mathbf{49}}{\mathbf{4}} = \left(x + \frac{7}{2}\right)^2$$

17. $\frac{1}{2}(-3) = -\frac{3}{2}; \left(-\frac{3}{2}\right)^2 = \frac{\mathbf{9}}{\mathbf{4}}$
$$a^2 - 3a + \frac{\mathbf{9}}{\mathbf{4}} = \left(a - \frac{3}{2}\right)^2$$

19. $\dfrac{1}{2}\left(\dfrac{2}{3}\right) = \dfrac{1}{3}; \left(\dfrac{1}{3}\right)^2 = \dfrac{1}{9}$

$b^2 + \dfrac{2}{3}b + \dfrac{1}{9} = \left(b + \dfrac{1}{3}\right)^2$

21. $\dfrac{1}{2}\left(-\dfrac{5}{2}\right) = -\dfrac{5}{4}; \left(-\dfrac{5}{4}\right)^2 = \dfrac{25}{16}$

$c^2 - \dfrac{5}{2}c + \dfrac{25}{16} = \left(c - \dfrac{5}{4}\right)^2$

23. $x^2 + 6x + 8 = 0$

$x^2 + 6x = -8$

$x^2 + 6x + 9 = -8 + 9$

$(x+3)^2 = 1$

$\sqrt{(x+3)^2} = \pm\sqrt{1}$

$x + 3 = \pm 1$

$x = -3 \pm 1$

$x = -2 \text{ or } x = -4$

25. $x^2 - 8x + 12 = 0$

$x^2 - 8x = -12$

$x^2 - 8x + 16 = -12 + 16$

$(x-4)^2 = 4$

$\sqrt{(x-4)^2} = \pm\sqrt{4}$

$x - 4 = \pm 2$

$x = 4 \pm 2$

$x = 6 \text{ or } x = 2$

27. $x^2 - 2x - 15 = 0$

$x^2 - 2x = 15$

$x^2 - 2x + 1 = 15 + 1$

$(x-1)^2 = 16$

$\sqrt{(x-1)^2} = \pm\sqrt{16}$

$x - 1 = \pm 4$

$x = 1 \pm 4$

$x = 5 \text{ or } x = -3$

29. $x^2 - 7x + 12 = 0$

$x^2 - 7x = -12$

$x^2 - 7x + \dfrac{49}{4} = -12 + \dfrac{49}{4}$

$\left(x - \dfrac{7}{2}\right)^2 = -\dfrac{48}{4} + \dfrac{49}{4}$

$\left(x - \dfrac{7}{2}\right)^2 = \dfrac{1}{4}$

$\sqrt{\left(x - \dfrac{7}{2}\right)^2} = \pm\sqrt{\dfrac{1}{4}}$

$x - \dfrac{7}{2} = \pm\dfrac{1}{2}$

$x = \dfrac{7}{2} \pm \dfrac{1}{2}$

$x = \dfrac{8}{2} = 4 \text{ or } x = \dfrac{6}{2} = 3$

31. $x^2 + 5x - 6 = 0$

$x^2 + 5x = 6$

$x^2 + 5x + \dfrac{25}{4} = 6 + \dfrac{25}{4}$

$\left(x + \dfrac{5}{2}\right)^2 = \dfrac{24}{4} + \dfrac{25}{4}$

$\left(x + \dfrac{5}{2}\right)^2 = \dfrac{49}{4}$

$\sqrt{\left(x + \dfrac{5}{2}\right)^2} = \pm\sqrt{\dfrac{49}{4}}$

$x + \dfrac{5}{2} = \pm\dfrac{7}{2}$

$x = -\dfrac{5}{2} \pm \dfrac{7}{2}$

$x = \dfrac{2}{2} = 1 \text{ or } x = -\dfrac{12}{2} = -6$

33.
$$2x^2 = 4 - 2x$$
$$2x^2 + 2x = 4$$
$$x^2 + x = 2$$
$$x^2 + x + \frac{1}{4} = 2 + \frac{1}{4}$$
$$\left(x + \frac{1}{2}\right)^2 = \frac{8}{4} + \frac{1}{4}$$
$$\left(x + \frac{1}{2}\right)^2 = \frac{9}{4}$$
$$\sqrt{\left(x + \frac{1}{2}\right)^2} = \pm \sqrt{\frac{9}{4}}$$
$$x + \frac{1}{2} = \pm \frac{3}{2}$$
$$x = -\frac{1}{2} \pm \frac{3}{2}$$
$$x = \frac{2}{2} = 1 \text{ or } x = -\frac{4}{2} = -2$$

35.
$$3x^2 + 48 = -24x$$
$$3x^2 + 24x = -48$$
$$x^2 + 8x = -16$$
$$x^2 + 8x + 16 = -16 + 16$$
$$(x + 4)^2 = 0$$
$$\sqrt{(x + 4)^2} = \pm \sqrt{0}$$
$$x + 4 = \pm 0$$
$$x = -4 \pm 0$$
$$x = -4 \text{ or } x = -4$$

37.
$$2x^2 = 3x + 2$$
$$2x^2 - 3x = 2$$
$$x^2 - \frac{3}{2}x = 1$$
$$x^2 - \frac{3}{2}x + \frac{9}{16} = 1 + \frac{9}{16}$$
$$\left(x - \frac{3}{4}\right)^2 = \frac{16}{16} + \frac{9}{16}$$
$$\left(x - \frac{3}{4}\right)^2 = \frac{25}{16}$$
$$\sqrt{\left(x - \frac{3}{4}\right)^2} = \pm \sqrt{\frac{25}{16}}$$
$$x - \frac{3}{4} = \pm \frac{5}{4}$$
$$x = \frac{3}{4} \pm \frac{5}{4}$$
$$x = \frac{8}{4} = 2 \text{ or } x = -\frac{2}{4} = -\frac{1}{2}$$

39.
$$4x^2 = 2 - 7x$$
$$4x^2 + 7x = 2$$
$$x^2 + \frac{7}{4}x = \frac{2}{4}$$
$$x^2 + \frac{7}{4}x + \frac{49}{64} = \frac{1}{2} + \frac{49}{64}$$
$$\left(x + \frac{7}{8}\right)^2 = \frac{32}{64} + \frac{49}{64}$$
$$\left(x + \frac{7}{8}\right)^2 = \frac{81}{64}$$
$$\sqrt{\left(x + \frac{7}{8}\right)^2} = \pm \sqrt{\frac{81}{64}}$$
$$x + \frac{7}{8} = \pm \frac{9}{8}$$
$$x = -\frac{7}{8} \pm \frac{9}{8}$$
$$x = \frac{2}{8} = \frac{1}{4} \text{ or } x = -\frac{16}{8} = -2$$

41.
$$x^2 + 4x + 1 = 0$$
$$x^2 + 4x = -1$$
$$x^2 + 4x + 4 = -1 + 4$$
$$(x + 2)^2 = 3$$
$$\sqrt{(x + 2)^2} = \pm \sqrt{3}$$
$$x + 2 = \pm \sqrt{3}$$
$$x = -2 \pm \sqrt{3}$$

43.
$$x^2 - 2x - 4 = 0$$
$$x^2 - 2x = 4$$
$$x^2 - 2x + 1 = 4 + 1$$
$$(x - 1)^2 = 5$$
$$\sqrt{(x - 1)^2} = \pm \sqrt{5}$$
$$x - 1 = \pm \sqrt{5}$$
$$x = 1 \pm \sqrt{5}$$

45.
$$x^2 = 4x + 3$$
$$x^2 - 4x = 3$$
$$x^2 - 4x + 4 = 3 + 4$$
$$(x - 2)^2 = 7$$
$$\sqrt{(x - 2)^2} = \pm\sqrt{7}$$
$$x - 2 = \pm\sqrt{7}$$
$$x = 2 \pm \sqrt{7}$$

47.
$$2x^2 = 2 - 4x$$
$$2x^2 + 4x = 2$$
$$x^2 + 2x = 1$$
$$x^2 + 2x + 1 = 1 + 1$$
$$(x + 1)^2 = 2$$
$$\sqrt{(x + 1)^2} = \pm\sqrt{2}$$
$$x + 1 = \pm\sqrt{2}$$
$$x = -1 \pm \sqrt{2}$$

49.
$$2x(x + 3) = 8$$
$$2x^2 + 6x = 8$$
$$x^2 + 3x = 4$$
$$x^2 + 3x + \frac{9}{4} = 4 + \frac{9}{4}$$
$$\left(x + \frac{3}{2}\right)^2 = \frac{16}{4} + \frac{9}{4}$$
$$\left(x + \frac{3}{2}\right)^2 = \frac{25}{4}$$
$$\sqrt{\left(x + \frac{3}{2}\right)^2} = \pm\sqrt{\frac{25}{4}}$$
$$x + \frac{3}{2} = \pm\frac{5}{2}$$
$$x = -\frac{3}{2} \pm \frac{5}{2}$$
$$x = \tfrac{2}{2} = 1 \text{ or } x = -\tfrac{8}{2} = -4$$

51.
$$6\left(x^2 - 3\right) = 5x$$
$$6x^2 - 18 = 5x$$
$$6x^2 - 5x = 18$$
$$x^2 - \frac{5}{6}x = 3$$
$$x^2 - \frac{5}{6}x + \frac{25}{144} = 3 + \frac{25}{144}$$
$$\left(x - \frac{5}{12}\right)^2 = \frac{432}{144} + \frac{25}{144}$$
$$\left(x - \frac{5}{12}\right)^2 = \frac{457}{144}$$
$$\sqrt{\left(x - \frac{5}{12}\right)^2} = \pm\sqrt{\frac{457}{144}}$$
$$x - \frac{5}{12} = \pm\sqrt{\frac{457}{144}}$$
$$x = \frac{5}{12} \pm \sqrt{\frac{457}{144}}$$
$$x \approx 2.20 \text{ or } x \approx -1.36$$

53. **Answers may vary.**

55. $2x^2 + 6x \Rightarrow a = 2, b = 6$
$$4a\left(2x^2 + 6x\right) + b^2 = 4(2)\left(2x^2 + 6x\right) + 6^2 = 8\left(2x^2 + 6x\right) + 36 = 16x^2 + 48x + 36 = (4x + 6)^2$$

Exercise 9.3 (page 555)

1.
$$A = p + prt$$
$$A - p = prt$$
$$\frac{A - p}{pt} = \frac{prt}{pt}$$
$$\frac{A - p}{pt} = r$$

3.
$$y - y_1 = m(x - x_1)$$
$$y - 12 = \frac{3}{5}(x - 0)$$
$$y - 12 = \frac{3}{5}x$$
$$5(y - 12) = 5\left(\frac{3}{5}x\right)$$
$$5y - 60 = 3x$$
$$3x - 5y = -60$$

5. $\sqrt{80} = \sqrt{16}\sqrt{5} = 4\sqrt{5}$

7. $\dfrac{x}{\sqrt{7x}} = \dfrac{x\sqrt{7x}}{\sqrt{7x}\sqrt{7x}} = \dfrac{x\sqrt{7x}}{7x} = \dfrac{\sqrt{7x}}{7}$

9. $ax^2 + bx + c = 0$

11. 0

13. $-4; 8; 0$

15. $x^2 + 4x + 3 = 0$
$a = 1, b = 4, c = 3$

17. $3x^2 - 2x + 7 = 0$
$a = 3, b = -2, c = 7$

19.
$$4y^2 = 2y - 1$$
$$4y^2 - 2y + 1 = 0$$
$$a = 4, b = -2, c = 1$$

21.
$$x(3x - 5) = 2$$
$$3x^2 - 5x = 2$$
$$3x^2 - 5x - 2 = 0$$
$$a = 3, b = -5, c = -2$$

23.
$$7(x^2 + 3) = -14x$$
$$7x^2 + 21 = -14x$$
$$7x^2 + 14x + 21 = 0$$
$$a = 7, b = 14, c = 21$$

25.
$$(2q + 3)(q - 2) = (q + 1)(q - 1)$$
$$2q^2 - 4q + 3q - 6 = q^2 - q + q - 1$$
$$2q^2 - q - 6 = q^2 - 1$$
$$q^2 - q - 5 = 0$$
$$a = 1, b = -1, c = -5$$

27. $x^2 - 5x + 6 = 0 \Rightarrow a = 1, b = -5, c = 6$

$$x = \frac{-b \pm \sqrt{b^2 - 4ac}}{2a} = \frac{-(-5) \pm \sqrt{(-5)^2 - 4(1)(6)}}{2(1)} = \frac{5 \pm \sqrt{25 - 24}}{2} = \frac{5 \pm \sqrt{1}}{2} = \frac{5 \pm 1}{2}$$

$$x = \frac{6}{2} = 3 \text{ or } x = \frac{4}{2} = 2$$

29. $x^2 + 7x + 12 = 0 \Rightarrow a = 1, b = 7, c = 12$

$$x = \frac{-b \pm \sqrt{b^2 - 4ac}}{2a} = \frac{-7 \pm \sqrt{7^2 - 4(1)(12)}}{2(1)} = \frac{-7 \pm \sqrt{49 - 48}}{2} = \frac{-7 \pm \sqrt{1}}{2} = \frac{-7 \pm 1}{2}$$

$$x = \frac{-6}{2} = -3 \text{ or } x = \frac{-8}{2} = -4$$

31. $2x^2 - x - 1 = 0 \Rightarrow a = 2, b = -1, c = -1$

$$x = \frac{-b \pm \sqrt{b^2 - 4ac}}{2a} = \frac{-(-1) \pm \sqrt{(-1)^2 - 4(2)(-1)}}{2(2)} = \frac{1 \pm \sqrt{1 + 8}}{4} = \frac{1 \pm \sqrt{9}}{4} = \frac{1 \pm 3}{4}$$

$$x = \frac{-2}{4} = -\frac{1}{2} \text{ or } x = \frac{4}{4} = 1$$

33. $3x^2 + 5x + 2 = 0 \Rightarrow a = 3, b = 5, c = 2$

$$x = \frac{-b \pm \sqrt{b^2 - 4ac}}{2a} = \frac{-5 \pm \sqrt{5^2 - 4(3)(2)}}{2(3)} = \frac{-5 \pm \sqrt{25 - 24}}{6} = \frac{-5 \pm \sqrt{1}}{6} = \frac{-5 \pm 1}{6}$$

$$x = \frac{-4}{6} = -\frac{2}{3} \text{ or } x = \frac{-6}{6} = -1$$

35. $x^2 + 3x + 1 = 0 \Rightarrow a = 1, b = 3, c = 1$

$$x = \frac{-b \pm \sqrt{b^2 - 4ac}}{2a} = \frac{-3 \pm \sqrt{3^2 - 4(1)(1)}}{2(1)} = \frac{-3 \pm \sqrt{9 - 4}}{2} = \frac{-3 \pm \sqrt{5}}{2}$$

37. $x^2 + 5x - 3 = 0 \Rightarrow a = 1, b = 5, c = -3$

$$x = \frac{-b \pm \sqrt{b^2 - 4ac}}{2a} = \frac{-5 \pm \sqrt{5^2 - 4(1)(-3)}}{2(1)} = \frac{-5 \pm \sqrt{25 + 12}}{2} = \frac{-5 \pm \sqrt{37}}{2}$$

39. $4x^2 + 5x - 1 = 0 \Rightarrow a = 4, b = 5, c = -1$

$$x = \frac{-b \pm \sqrt{b^2 - 4ac}}{2a} = \frac{-5 \pm \sqrt{5^2 - 4(4)(-1)}}{2(4)} = \frac{-5 \pm \sqrt{25 + 16}}{8} = \frac{-5 \pm \sqrt{41}}{8}$$

41. $5x^2 - 8x - 1 = 0 \Rightarrow a = 5, b = -8, c = -1$

$$x = \frac{-b \pm \sqrt{b^2 - 4ac}}{2a} = \frac{-(-8) \pm \sqrt{(-8)^2 - 4(5)(-1)}}{2(5)} = \frac{8 \pm \sqrt{64 + 20}}{10} = \frac{8 \pm \sqrt{84}}{10}$$

$$= \frac{8 \pm 2\sqrt{21}}{10}$$

$$= \frac{4 \pm \sqrt{21}}{5}$$

43. $x^2 + 2x + 7 = 0 \Rightarrow a = 1, b = 2, c = 7$

$$x = \frac{-b \pm \sqrt{b^2 - 4ac}}{2a} = \frac{-2 \pm \sqrt{2^2 - 4(1)(7)}}{2(1)} = \frac{-2 \pm \sqrt{4 - 28}}{2} = \frac{-2 \pm \sqrt{-24}}{2}$$

The solutions are not real numbers.

45. $2x^2 + x = 5 \Rightarrow 2x^2 + x - 5 = 0 \Rightarrow a = 2, b = 1, c = -5$

$$x = \frac{-b \pm \sqrt{b^2 - 4ac}}{2a} = \frac{-1 \pm \sqrt{1^2 - 4(2)(-5)}}{2(2)} = \frac{-1 \pm \sqrt{1 + 40}}{4} = \frac{-1 \pm \sqrt{41}}{4}$$

47. $x^2 + 1 = -4x \Rightarrow x^2 + 4x + 1 = 0 \Rightarrow a = 1, b = 4, c = 1$

$$x = \frac{-b \pm \sqrt{b^2 - 4ac}}{2a} = \frac{-4 \pm \sqrt{4^2 - 4(1)(1)}}{2(1)} = \frac{-4 \pm \sqrt{16 - 4}}{2} = \frac{-4 \pm \sqrt{12}}{2}$$

$$= \frac{-4 \pm 2\sqrt{3}}{2}$$

$$= \frac{-2 \pm \sqrt{3}}{1} = -2 \pm \sqrt{3}$$

49. $x^2 + 5 = 2x \Rightarrow x^2 - 2x + 5 = 0 \Rightarrow a = 1, b = -2, c = 5$

$$x = \frac{-b \pm \sqrt{b^2 - 4ac}}{2a} = \frac{-(-2) \pm \sqrt{(-2)^2 - 4(1)(5)}}{2(1)} = \frac{2 \pm \sqrt{4 - 20}}{2} = \frac{2 \pm \sqrt{-16}}{2}$$

The solutions are not real numbers.

51. $x^2 = 1 - 2x \Rightarrow x^2 + 2x - 1 = 0 \Rightarrow a = 1, b = 2, c = -1$

$$x = \frac{-b \pm \sqrt{b^2 - 4ac}}{2a} = \frac{-2 \pm \sqrt{2^2 - 4(1)(-1)}}{2(1)} = \frac{-2 \pm \sqrt{4 + 4}}{2} = \frac{-2 \pm \sqrt{8}}{2}$$

$$= \frac{-2 \pm 2\sqrt{2}}{2}$$

$$= \frac{-1 \pm \sqrt{2}}{1} = -1 \pm \sqrt{2}$$

53. $3x^2 = 6x + 2 \Rightarrow 3x^2 - 6x - 2 = 0 \Rightarrow a = 3, b = -6, c = -2$

$$x = \frac{-b \pm \sqrt{b^2 - 4ac}}{2a} = \frac{-(-6) \pm \sqrt{(-6)^2 - 4(3)(-2)}}{2(3)} = \frac{6 \pm \sqrt{36 + 24}}{6} = \frac{6 \pm \sqrt{60}}{6}$$

$$= \frac{6 \pm 2\sqrt{15}}{6}$$

$$= \frac{3 \pm \sqrt{15}}{3}$$

55. $$h^2 + (h + 2)^2 = 10^2$$
$$h^2 + h^2 + 4h + 4 = 100$$
$$2h^2 + 4h - 96 = 0$$
$$a = 2, b = 4, c = -96$$
$$h = \frac{-b \pm \sqrt{b^2 - 4ac}}{2a} = \frac{-4 \pm \sqrt{4^2 - 4(2)(-96)}}{2(2)} = \frac{-4 \pm \sqrt{16 + 768}}{4} = \frac{-4 \pm \sqrt{784}}{4}$$

$$= \frac{-4 \pm 28}{4}$$

$h = \dfrac{24}{4} = 6$ or $h = \dfrac{-32}{4} = -8 \Rightarrow$ The dimensions are 6 in. by 8 in.

57. Let x and $x + 10$ represent the distances. $\qquad x^2 + (x + 10)^2 = 50^2$

$$x^2 + x^2 + 20x + 100 = 2500$$

$$2x^2 + 20x - 2400 = 0 \Rightarrow a = 2, b = 20, c = -2400$$

$$x = \frac{-b \pm \sqrt{b^2 - 4ac}}{2a} = \frac{-20 \pm \sqrt{20^2 - 4(2)(-2400)}}{2(2)} = \frac{-20 \pm \sqrt{400 + 19200}}{4}$$

$$= \frac{-20 \pm \sqrt{19600}}{4}$$

$$= \frac{-20 \pm 140}{4}$$

$$h = \frac{120}{4} = 30 \text{ or } h = \frac{-160}{4} = -40$$

The boats have sailed 30 and 40 nautical miles.

59. $\qquad A = P(1 + r)^2$

$$5724.50 = 5000(1 + r)^2$$

$$1.1449 = (1 + r)^2$$

$$1.1449 = r^2 + 2r + 1$$

$$0 = r^2 + 2r - 0.1449$$

$$a = 1, b = 2, c = -0.1449$$

$$r = \frac{-b \pm \sqrt{b^2 - 4ac}}{2a} = \frac{-2 \pm \sqrt{2^2 - 4(1)(-0.1449)}}{2(1)} = \frac{-2 \pm \sqrt{4 + 0.5796}}{2} = \frac{-2 \pm \sqrt{4.5796}}{2}$$

$$= \frac{-2 \pm 2.14}{2}$$

$$r = \frac{-2 + 2.14}{2} = \frac{0.14}{2} = 0.07 \Rightarrow \text{The rate needs to be 7\%.}$$

61. $R = -\dfrac{1}{6}x^2 + 450x = -\dfrac{1}{6}(600)^2 + 450(600) = -\dfrac{1}{6}(360000) + 270000 = -60000 + 270000$

$$= \$210000$$

63. $\qquad$ Length $\cdot$ Width = Area

$$(12 - 2x)(12 - 2x) = 64$$

$$144 - 24x - 24x + 4x^2 = 64$$

$$4x^2 - 48x + 80 = 0$$

$$x^2 - 12x + 20 = 0$$

$$x = \frac{-b \pm \sqrt{b^2 - 4ac}}{2a} = \frac{-(-12) \pm \sqrt{(-12)^2 - 4(1)(20)}}{2(1)} = \frac{12 \pm \sqrt{144 - 80}}{2} = \frac{12 \pm \sqrt{64}}{2}$$

$$= \frac{12 \pm 8}{2}$$

$$x = \frac{20}{2} = 10 \text{ or } x = \frac{4}{2} = 2. \quad \text{The solution } x = 10 \text{ does not make sense. The depth is 2 inches.}$$

65. Let x = hours needed for the second pipe to fill the tank alone.

$$\boxed{\begin{array}{c}\text{1st in}\\\text{1 hour}\end{array}} + \boxed{\begin{array}{c}\text{2nd in}\\\text{1 hour}\end{array}} = \boxed{\begin{array}{c}\text{Both in}\\\text{1 hour}\end{array}}$$

$$\frac{1}{4} + \frac{1}{x} = \frac{1}{x-2}$$

$$4x(x-2)\left(\frac{1}{4} + \frac{1}{x}\right) = 4x(x-2) \cdot \frac{1}{x-2}$$

$$x(x-2) + 4(x-2) = 4x$$

$$x^2 - 2x + 4x - 8 = 4x$$

$$x^2 - 2x - 8 = 0$$

$$(x-4)(x+2) = 0$$

$$x - 4 = 0 \quad \text{or} \quad x + 2 = 0$$

$$x = 4 \qquad\qquad x = -2$$

It takes the 2nd pipe 4 hours.

67. Answers may vary.

69. Answers may vary.

71. $x_1 + x_2 = \dfrac{-b + \sqrt{b^2 - 4ac}}{2a} + \dfrac{-b - \sqrt{b^2 - 4ac}}{2a} = \dfrac{-2b}{2a} = -\dfrac{b}{a}$

Exercise 9.4 (page 566)

1. $\sqrt{12} + \sqrt{27} = \sqrt{4}\sqrt{3} + \sqrt{9}\sqrt{3} = 2\sqrt{3} + 3\sqrt{3} = 5\sqrt{3}$

3. $\left(\sqrt{3} + 1\right)\left(\sqrt{3} - 1\right) = \sqrt{9} - \sqrt{3} + \sqrt{3} - 1 = 3 - 1 = 2$

5. quadratic

7. y-intercept

9. $a > 0; a < 0$

11. $(0, c)$

13. $y = x^2 + 1$

x	y
2	5
1	2
0	1
-1	2
-2	5

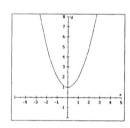

15. $y = -x^2$

x	y
2	-4
1	-1
0	0
-1	-1
-2	-4

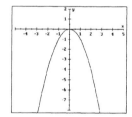

17. $y = x^2 + x$

x	y
2	6
1	2
0	0
-1	0
-2	2

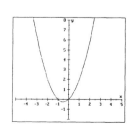

19. $y = -x^2 - 4x$

x	y
0	0
-1	3
-2	4
-3	3
-4	0

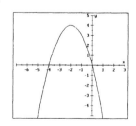

21. $y = x^2 + 4x + 4$

x	y
0	4
-1	1
-2	0
-3	1
-4	4

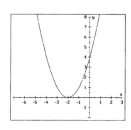

23. $y = x^2 - 4x + 6$

x	y
4	6
3	3
2	2
1	3
0	6

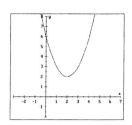

25. $y = -3(x-2)^2 + 4$
Vertex: $(2, 4)$

27. $y = 5(x+1)^2 - 5$
Vertex: $(-1, -5)$

29. $y = (x-1)^2$
Vertex: $(1, 0)$

31. $y = -7x^2 + 4$
$y = -7(x-0)^2 + 4$
Vertex: $(0, 4)$

33. $y = x^2 + 2x + 5$
$y = x^2 + 2x + 1 + 5 - 1$
$y = (x+1)^2 + 4$
Vertex: $(-1, 4)$

35. $y = x^2 - 6x - 12$
$y = x^2 - 6x + 9 - 12 - 9$
$y = (x-3)^2 - 21$
Vertex: $(3, -21)$

37. $y = x^2 - 4x + 4$
$y = (x-2)^2$
Vertex: $(2, 0)$

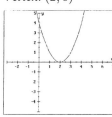

39. $y = -x^2 - 2x - 1$
$y = -(x^2 + 2x + 1)$
$y = -(x+1)^2$
Vertex: $(-1, 0)$

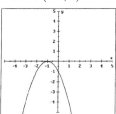

41. $y = x^2 + 2x - 3$
$y = x^2 + 2x + 1 - 3 - 1$
$y = (x+1)^2 - 4$
Vertex: $(-1, -4)$

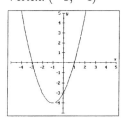

43. $y = -x^2 - 6x - 7$
$y = -(x^2 + 6x) - 7$
$y = -(x^2 + 6x + 9 - 9) - 7$
$y = -(x+3)^2 - (-9) - 7$
$y = -(x+3)^2 + 2$
Vertex: $(-3, 2)$

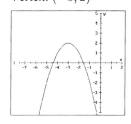

45. $y = 2x^2 + 8x + 6$
$y = 2(x^2 + 4x) + 6$
$y = 2(x^2 + 4x + 4 - 4) + 6$
$y = 2(x+2)^2 + 2(-4) + 6$
$y = 2(x+2)^2 - 2$
Vertex: $(-2, -2)$

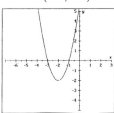

47. $y = -3x^2 + 6x - 2$
$y = -3(x^2 - 2x) - 2$
$y = -3(x^2 - 2x + 1 - 1) - 2$
$y = -3(x - 1)^2 - 3(-1) - 2$
$y = -3(x - 1)^2 + 1$
Vertex: $(1, 1)$

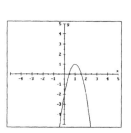

49. $y = x^2 - x - 2$
$a = 1, b = -1, c = -2$

$$x = \frac{-b}{2a} \qquad y = x^2 - x - 2$$

$$= \frac{-(-1)}{2(1)} \qquad y = \left(\frac{1}{2}\right)^2 - \frac{1}{2} - 2$$

$$= \frac{1}{2} \qquad y = \frac{1}{4} - \frac{1}{2} - 2 = -\frac{9}{4}$$

Vertex: $\left(\frac{1}{2}, -\frac{9}{4}\right)$

$x = 0$:
_____
$y = 0^2 - 0 - 2$
$y = -2$
$(0, -2)$

$y = 0$:
_____
$0 = x^2 - x - 2$
$0 = (x - 2)(x + 1)$
$x = 2$ or $x = -1$
$(2, 0), (-1, 0)$

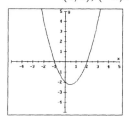

51. $y = -x^2 + 2x + 3$
$a = -1, b = 2, c = 3$

$$x = \frac{-b}{2a} \qquad y = -x^2 + 2x + 3$$

$$= \frac{-2}{2(-1)} \qquad y = -(1)^2 + 2(1) + 3$$

$$= \frac{-2}{-2} = 1 \qquad y = -1 + 2 + 3 = 4$$

Vertex: $(1, 4)$

$x = 0$:
_____
$y = -(0)^2 + 2(0) + 3$
$y = 3$
$(0, 3)$

$y = 0$:
_____
$0 = -x^2 + 2x + 3$
$0 = -(x^2 - 2x - 3)$
$0 = -(x - 3)(x + 1)$
$x = 3$ or $x = -1$
$(3, 0), (-1, 0)$

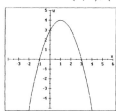

53. $y = 2x^2 + 3x - 2$
$a = 2, b = 3, c = -2$

$x = \dfrac{-b}{2a}$ $\quad y = 2x^2 + 3x - 2$

$= \dfrac{-3}{2(2)}$ $\quad y = 2\left(-\dfrac{3}{4}\right)^2 + 3\left(-\dfrac{3}{4}\right) - 2$

$= -\dfrac{3}{4}$ $\quad y = \dfrac{18}{16} - \dfrac{9}{4} - 2 = -\dfrac{25}{8}$

Vertex: $\left(-\dfrac{3}{4}, -\dfrac{25}{8}\right)$

$x = 0:$
$y = 2(0)^2 + 3(0) - 2$
$y = -2$
$(0, -2)$

$y = 0:$
$0 = 2x^2 + 3x - 2$
$0 = (2x - 1)(x + 2)$
$x = \dfrac{1}{2}$ or $x = -2$
$\left(\dfrac{1}{2}, 0\right), (-2, 0)$

55. Let $y =$ the revenue.
Revenue = $\boxed{\text{\# sold}} \cdot \boxed{\text{price}}$

$y = x\left(450 - \dfrac{1}{6}x\right)$

$y = 450x - \dfrac{1}{6}x^2$

$y = -\dfrac{1}{6}x^2 + 450x$

$a = -\dfrac{1}{6}, b = 450$

Find the vertex:

$x = \dfrac{-b}{2a} = \dfrac{-450}{2\left(-\frac{1}{6}\right)} = \dfrac{-450}{-\frac{1}{3}} = 1350$

The company must sell 1,350 TV's.

57. Let $y =$ the revenue.
Revenue = $\boxed{\text{\# sold}} \cdot \boxed{\text{price}}$

$y = n\left(150 - \dfrac{1}{10}n\right)$

$y = 150n - \dfrac{1}{10}n^2$

$y = -\dfrac{1}{10}n^2 + 150n$

$a = -\dfrac{1}{10}, b = 150$

Find the vertex:

$x = \dfrac{-b}{2a} = \dfrac{-150}{2\left(-\frac{1}{10}\right)} = \dfrac{-150}{-\frac{1}{5}} = 750$

The retailer must buy 750 players.

59. Let $y =$ the revenue.
Revenue = $\boxed{\text{\# sold}} \cdot \boxed{\text{price}}$

$y = x\left(1800 - \dfrac{3}{2}x\right)$

$y = 1800x - \dfrac{3}{2}x^2$

$y = -\dfrac{3}{2}x^2 + 1800x$

$a = -\dfrac{3}{2}, b = 1800$

Find the vertex:

$x = \dfrac{-b}{2a} = \dfrac{-1800}{2\left(-\frac{3}{2}\right)} = \dfrac{-1800}{-3} = 600$

Find the price:

Price $= 1800 - \dfrac{3}{2}x = 1800 - \dfrac{3}{2}(600) = 1800 - 900 = 900$

The price should be \$900.

61. $C(x) = 0.2x^2 - 10x + 650$

$a = 0.2, b = -10 \Rightarrow$ Find the vertex:

$x = \dfrac{-b}{2a} = \dfrac{-(-10)}{2(0.2)} = \dfrac{10}{0.4} = 25$

$y = 0.2x^2 - 10x + 650$

$\quad = 0.2(25)^2 - 10(25) + 650$

$\quad = 125 - 250 + 650 = 525$

25 vases should be made, for a minimum cost of $525.

63. Graph $y = x^2 - 5x + 6$ and find the x-coordinates where the graph crosses the x-axis.

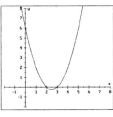

$x = 3, x = 2$

65. Graph $y = x^2 + 5x + 2$ and find the x-coordinates where the graph crosses the x-axis.

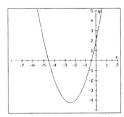

$x = -0.44, x = -4.56$

67. **Answers may vary.**

69. **Answers may vary.**

Chapter 9 Summary (page 570)

1. $x^2 + 2x = 0$

$x(x + 2) = 0$

$x = 0 \quad$ **or** $\quad x + 2 = 0$

$x = 0 \qquad\qquad x = -2$

2. $2x^2 - 6x = 0$

$2x(x - 3) = 0$

$2x = 0 \quad$ **or** $\quad x - 3 = 0$

$x = 0 \qquad\qquad x = 3$

3. $x^2 - 9 = 0$

$(x - 3)(x + 3) = 0$

$x - 3 = 0 \quad$ **or** $\quad x + 3 = 0$

$x = 3 \qquad\qquad x = -3$

4. $p^2 - 25 = 0$

$(p - 5)(p + 5) = 0$

$p - 5 = 0 \quad$ **or** $\quad p + 5 = 0$

$p = 5 \qquad\qquad p = -5$

5. $a^2 - 7a + 12 = 0$

$(a - 3)(a - 4) = 0$

$a - 3 = 0 \quad$ **or** $\quad a - 4 = 0$

$a = 3 \qquad\qquad a = 4$

6. $t^2 - 2t - 15 = 0$

$(t - 5)(t + 3) = 0$

$t - 5 = 0 \quad$ **or** $\quad t + 3 = 0$

$t = 5 \qquad\qquad t = -3$

7.
$$2x - x^2 + 24 = 0$$
$$-x^2 + 2x + 24 = 0$$
$$-(x^2 - 2x - 24) = 0$$
$$-(x - 6)(x + 4) = 0$$
$$x - 6 = 0 \quad \textbf{or} \quad x + 4 = 0$$
$$x = 6 \qquad\qquad x = -4$$

8.
$$2x^2 + x - 3 = 0$$
$$(2x + 3)(x - 1) = 0$$
$$2x + 3 = 0 \quad \textbf{or} \quad x - 1 = 0$$
$$2x = -3 \qquad\qquad x = 1$$
$$x = -\frac{3}{2} \qquad\qquad x = 1$$

9.
$$x^2 - 7x = -12$$
$$x^2 - 7x + 12 = 0$$
$$(x - 3)(x - 4) = 0$$
$$x - 3 = 0 \quad \textbf{or} \quad x - 4 = 0$$
$$x = 3 \qquad\qquad x = 4$$

10.
$$x^2 = 25$$
$$\sqrt{x^2} = \pm\sqrt{25}$$
$$x = \pm 5$$

11.
$$x^2 = 36$$
$$\sqrt{x^2} = \pm\sqrt{36}$$
$$x = \pm 6$$

12.
$$2x^2 = 18$$
$$x^2 = 9$$
$$\sqrt{x^2} = \pm\sqrt{9}$$
$$x = \pm 3$$

13.
$$4x^2 = 9$$
$$x^2 = \frac{9}{4}$$
$$\sqrt{x^2} = \pm\sqrt{\frac{9}{4}}$$
$$x = \pm\frac{3}{2}$$

14.
$$x^2 = 8$$
$$\sqrt{x^2} = \pm\sqrt{8}$$
$$x = \pm 2\sqrt{2}$$

15.
$$x^2 = 75$$
$$\sqrt{x^2} = \pm\sqrt{75}$$
$$x = \pm 5\sqrt{3}$$

16.
$$(x - 1)^2 = 25$$
$$\sqrt{(x - 1)^2} = \pm\sqrt{25}$$
$$x - 1 = \pm 5$$
$$x = 1 \pm 5$$
$$x = 6 \text{ or } x = -4$$

17.
$$(x + 3)^2 = 36$$
$$\sqrt{(x + 3)^2} = \pm\sqrt{36}$$
$$x + 3 = \pm 6$$
$$x = -3 \pm 6$$
$$x = 3 \text{ or } x = -9$$

18.
$$2(x + 1)^2 = 18$$
$$(x + 1)^2 = 9$$
$$\sqrt{(x + 1)^2} = \pm\sqrt{9}$$
$$x + 1 = \pm 3$$
$$x = -1 \pm 3$$
$$x = 2 \text{ or } x = -4$$

19.
$$4(x - 2)^2 = 9$$
$$(x - 2)^2 = \frac{9}{4}$$
$$\sqrt{(x - 2)^2} = \pm\sqrt{\frac{9}{4}}$$
$$x - 2 = \pm\frac{3}{2}$$
$$x = 2 \pm \frac{3}{2} = \frac{4}{2} \pm \frac{3}{2}$$
$$x = \tfrac{7}{2} \text{ or } x = \tfrac{1}{2}$$

20.
$$(x-8)^2 = 8$$
$$\sqrt{(x-8)^2} = \pm\sqrt{8}$$
$$x-8 = \pm 2\sqrt{2}$$
$$x = 8 \pm 2\sqrt{2}$$

21.
$$(x+5)^2 = 75$$
$$\sqrt{(x+5)^2} = \pm\sqrt{75}$$
$$x+5 = \pm 5\sqrt{3}$$
$$x = -5 \pm 5\sqrt{3}$$

22. Let h = the height. Then $2h + 3$ = the base of the panel.

$$\frac{1}{2} \cdot \text{base} \cdot \text{height} = \text{area}$$
$$\frac{1}{2}(2h+3)(h) = 45$$
$$h(2h+3) = 90$$
$$2h^2 + 3h - 90 = 0$$
$$(2h+15)(h-6) = 0$$
$$2h+15 = 0 \quad \text{or} \quad h-6=0$$
$$h = -\frac{15}{2} \qquad\qquad h = 6$$

The height must be positive, so $h = 6$ and the base is 15 feet.

23. Let $h = 0$:
$$h = 3000 + 40t - 16t^2$$
$$0 = -16t^2 + 40t + 3000$$
$$0 = -8(2t^2 - 5t - 375)$$
$$0 = -8(2t+25)(t-15)$$
$$2t+25 = 0 \quad \text{or} \quad t-15 = 0$$
$$t = -\frac{25}{2} \qquad\qquad t = 15$$

It will hit after 15 seconds.

24. Let w = the width and $2w + 3$ = the length.
$$\text{Width} \cdot \text{Length} = \text{Area}$$
$$w(2w+3) = 27$$
$$2w^2 + 3w = 27$$
$$2w^2 + 3w - 27 = 0$$
$$(2w+9)(w-3) = 0$$
$$2w+9 = 0 \quad \text{or} \quad w-3 = 0$$
$$w = -\frac{9}{2} \qquad\qquad w = 3$$

The dimensions are 3 ft by 9 ft.

25. Let w = the width and $w + 3$ = the length.
$$\text{Width} \cdot \text{Length} = 2 \cdot \text{Width} + 2 \cdot \text{Length}$$
$$w(w+3) = 2w + 2(w+3)$$
$$w^2 + 3w = 2w + 2w + 6$$
$$w^2 - w - 6 = 0$$
$$(w+2)(w-3) = 0$$
$$w+2 = 0 \quad \text{or} \quad w-3 = 0$$
$$w = -2 \qquad\qquad w = 3$$

The dimensions are 3 ft by 6 ft.

26. $\frac{1}{2}(4) = 2;\ 2^2 = \mathbf{4}$
$$x^2 + 4x + \mathbf{4} = (x+2)^2$$

27. $\frac{1}{2}(8) = 4;\ 4^2 = \mathbf{16}$
$$y^2 + 8y + \mathbf{16} = (y+4)^2$$

28. $\frac{1}{2}(-10) = -5;\ (-5)^2 = \mathbf{25}$
$$z^2 - 10z + \mathbf{25} = (z-5)^2$$

29. $\frac{1}{2}(-5) = -\frac{5}{2};\ \left(-\frac{5}{2}\right)^2 = \mathbf{\frac{25}{4}}$
$$t^2 - 5t + \mathbf{\frac{25}{4}} = \left(t - \frac{5}{2}\right)^2$$

30. $\frac{1}{2}\left(\frac{3}{4}\right) = \frac{3}{8};\ \left(\frac{3}{8}\right)^2 = \mathbf{\frac{9}{64}}$
$$a^2 + \frac{3}{4}a + \mathbf{\frac{9}{64}} = \left(a + \frac{3}{8}\right)^2$$

31. $\frac{1}{2}\left(-\frac{7}{3}\right) = -\frac{7}{6};\ \left(-\frac{7}{6}\right)^2 = \mathbf{\frac{49}{36}}$
$$c^2 - \frac{7}{3}c + \mathbf{\frac{49}{36}} = \left(x - \frac{7}{6}\right)^2$$

32. $x^2 + 5x - 14 = 0$

$$x^2 + 5x = 14$$

$$x^2 + 5x + \frac{25}{4} = 14 + \frac{25}{4}$$

$$\left(x + \frac{5}{2}\right)^2 = \frac{56}{4} + \frac{25}{4}$$

$$\left(x + \frac{5}{2}\right)^2 = \frac{81}{4}$$

$$\sqrt{\left(x + \frac{5}{2}\right)^2} = \pm\sqrt{\frac{81}{4}}$$

$$x + \frac{5}{2} = \pm\frac{9}{2}$$

$$x = -\frac{5}{2} \pm \frac{9}{2}$$

$$x = \tfrac{4}{2} = 2 \text{ or } x = -\tfrac{14}{2} = -7$$

33. $x^2 - 8x + 15 = 0$

$$x^2 - 8x = -15$$

$$x^2 - 8x + 16 = -15 + 16$$

$$(x - 4)^2 = 1$$

$$\sqrt{(x - 4)^2} = \pm\sqrt{1}$$

$$x - 4 = \pm 1$$

$$x = 4 \pm 1$$

$$x = 5 \text{ or } x = 3$$

34. $x^2 + 4x - 77 = 0$

$$x^2 + 4x = 77$$

$$x^2 + 4x + 4 = 77 + 4$$

$$(x + 2)^2 = 81$$

$$\sqrt{(x + 2)^2} = \pm\sqrt{81}$$

$$x + 2 = \pm 9$$

$$x = -2 \pm 9$$

$$x = 7 \text{ or } x = -11$$

35. $x^2 - 2x - 1 = 0$

$$x^2 - 2x = 1$$

$$x^2 - 2x + 1 = 1 + 1$$

$$(x - 1)^2 = 2$$

$$\sqrt{(x - 1)^2} = \pm\sqrt{2}$$

$$x - 1 = \pm\sqrt{2}$$

$$x = 1 \pm \sqrt{2}$$

36. $x^2 + 4x - 3 = 0$

$$x^2 + 4x = 3$$

$$x^2 + 4x + 4 = 3 + 4$$

$$(x + 2)^2 = 7$$

$$\sqrt{(x + 2)^2} = \pm\sqrt{7}$$

$$x + 2 = \pm\sqrt{7}$$

$$x = -2 \pm \sqrt{7}$$

37. $x^2 - 6x + 4 = 0$

$$x^2 - 6x = -4$$

$$x^2 - 6x + 9 = -4 + 9$$

$$(x - 3)^2 = 5$$

$$\sqrt{(x - 3)^2} = \pm\sqrt{5}$$

$$x - 3 = \pm\sqrt{5}$$

$$x = 3 \pm \sqrt{5}$$

38.

$$2x^2 + 5x - 3 = 0$$

$$2x^2 + 5x = 3$$

$$x^2 + \frac{5}{2}x = \frac{3}{2}$$

$$x^2 + \frac{5}{2}x + \frac{25}{16} = \frac{3}{2} + \frac{25}{16}$$

$$\left(x + \frac{5}{4}\right)^2 = \frac{49}{16}$$

$$\sqrt{\left(x + \frac{5}{4}\right)^2} = \pm\sqrt{\frac{49}{16}}$$

$$x + \frac{5}{4} = \pm\frac{7}{4}$$

$$x = -\frac{5}{4} \pm \frac{7}{4}$$

$$x = \tfrac{2}{4} = \tfrac{1}{2} \text{ or } x = \tfrac{-12}{4} = -3$$

39.

$$2x^2 - 2x - 1 = 0$$

$$2x^2 - 2x = 1$$

$$x^2 - x = \frac{1}{2}$$

$$x^2 - x + \frac{1}{4} = \frac{1}{2} + \frac{1}{4}$$

$$\left(x - \frac{1}{2}\right)^2 = \frac{3}{4}$$

$$\sqrt{\left(x - \frac{1}{2}\right)^2} = \pm\sqrt{\frac{3}{4}}$$

$$x - \frac{1}{2} = \pm\frac{\sqrt{3}}{2}$$

$$x = \frac{1}{2} \pm \frac{\sqrt{3}}{2} = \frac{1 \pm \sqrt{3}}{2}$$

40. $x^2 - 2x - 15 = 0 \Rightarrow a = 1, b = -2, c = -15$

$$x = \frac{-b \pm \sqrt{b^2 - 4ac}}{2a} = \frac{-(-2) \pm \sqrt{(-2)^2 - 4(1)(-15)}}{2(1)} = \frac{2 \pm \sqrt{4 + 60}}{2} = \frac{2 \pm \sqrt{64}}{2}$$

$x = \frac{2+8}{2} = \frac{10}{2} = 5 \text{ or } x = \frac{2-8}{2} = \frac{-6}{2} = -3$

41. $x^2 - 6x - 7 = 0 \Rightarrow a = 1, b = -6, c = -7$

$$x = \frac{-b \pm \sqrt{b^2 - 4ac}}{2a} = \frac{-(-6) \pm \sqrt{(-6)^2 - 4(1)(-7)}}{2(1)} = \frac{6 \pm \sqrt{36 + 28}}{2} = \frac{6 \pm \sqrt{64}}{2}$$

$x = \frac{6+8}{2} = \frac{14}{2} = 7 \text{ or } x = \frac{6-8}{2} = \frac{-2}{2} = -1$

42. $x^2 - 15x + 26 = 0 \Rightarrow a = 1, b = -15, c = 26$

$$x = \frac{-b \pm \sqrt{b^2 - 4ac}}{2a} = \frac{-(-15) \pm \sqrt{(-15)^2 - 4(1)(26)}}{2(1)} = \frac{15 \pm \sqrt{225 - 104}}{2} = \frac{15 \pm \sqrt{121}}{2}$$

$x = \frac{15+11}{2} = \frac{26}{2} = 13 \text{ or } x = \frac{15-11}{2} = \frac{4}{2} = 2$

43. $2x^2 - 7x + 3 = 0 \Rightarrow a = 2, b = -7, c = 3$

$$x = \frac{-b \pm \sqrt{b^2 - 4ac}}{2a} = \frac{-(-7) \pm \sqrt{(-7)^2 - 4(2)(3)}}{2(2)} = \frac{7 \pm \sqrt{49 - 24}}{4} = \frac{7 \pm \sqrt{25}}{4}$$

$x = \frac{7+5}{4} = \frac{12}{4} = 3 \text{ or } x = \frac{7-5}{4} = \frac{2}{4} = \frac{1}{2}$

44. $6x^2 - 7x - 3 = 0 \Rightarrow a = 6, b = -7, c = -3$

$$x = \frac{-b \pm \sqrt{b^2 - 4ac}}{2a} = \frac{-(-7) \pm \sqrt{(-7)^2 - 4(6)(-3)}}{2(6)} = \frac{7 \pm \sqrt{49 + 72}}{12} = \frac{7 \pm \sqrt{121}}{12}$$

$$x = \frac{7+11}{12} = \frac{18}{12} = \frac{3}{2} \text{ or } x = \frac{7-11}{12} = \frac{-4}{12} = -\frac{1}{3}$$

45. $x^2 + 4x + 1 = 0 \Rightarrow a = 1, b = 4, c = 1$

$$x = \frac{-b \pm \sqrt{b^2 - 4ac}}{2a} = \frac{-4 \pm \sqrt{4^2 - 4(1)(1)}}{2(1)} = \frac{-4 \pm \sqrt{16 - 4}}{2} = \frac{-4 \pm \sqrt{12}}{2} = \frac{-4 \pm 2\sqrt{3}}{2}$$

$$= -2 \pm \sqrt{3}$$

46. $x^2 - 6x + 7 = 0 \Rightarrow a = 1, b = -6, c = 7$

$$x = \frac{-b \pm \sqrt{b^2 - 4ac}}{2a} = \frac{-(-6) \pm \sqrt{(-6)^2 - 4(1)(7)}}{2(1)} = \frac{6 \pm \sqrt{36 - 28}}{2} = \frac{6 \pm \sqrt{8}}{2} = \frac{6 \pm 2\sqrt{2}}{2}$$

$$= 3 \pm \sqrt{2}$$

47. $x^2 + 3x = 0 \Rightarrow a = 1, b = 3, c = 0$

$$x = \frac{-b \pm \sqrt{b^2 - 4ac}}{2a} = \frac{-3 \pm \sqrt{3^2 - 4(1)(0)}}{2(1)} = \frac{-3 \pm \sqrt{9}}{2} = \frac{-3 \pm 3}{2}$$

$$x = \frac{0}{2} = 0 \text{ or } x = \frac{-6}{2} = -3$$

48.

$$w^2 + (w + 14)^2 = 26^2 \qquad\qquad 2(w^2 + 14w - 240) = 0$$
$$w^2 + w^2 + 28w + 196 = 676 \qquad\quad 2(w - 10)(w + 24) = 0$$
$$2w^2 + 28w - 480 = 0 \qquad\qquad w - 10 = 0 \quad \textbf{or} \quad w + 24 = 0$$
$$2(w^2 + 14w - 240) = 0 \qquad\qquad\quad w = 10 \qquad\qquad w = -24$$

The width is 10 cm and the perimeter is 68 cm.

49. $y = x^2 + 8x + 10$

$a = 1, b = 8, c = 10$

$$x = \frac{-b}{2a}$$

$$= \frac{-8}{2(1)}$$

$$= \frac{-8}{2} = -4$$

$y = x^2 + 8x + 10$

$y = (-4)^2 + 8(-4) + 10$

$y = 16 - 32 + 10 = -6$

Vertex: $(-4, -6)$

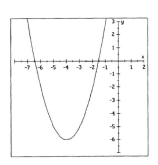

50. $y = -2x^2 - 4x - 6$

$a = -2, b = -4, c = -6$

$$x = \frac{-b}{2a}$$

$$= \frac{-(-4)}{2(-2)}$$

$$= \frac{4}{-4} = -1$$

Vertex: $(-1, -4)$

$y = -2x^2 - 4x - 6$

$y = -2(-1)^2 - 4(-1) - 6$

$y = -2 + 4 - 6 = -4$

51. $y = 5(x - 6)^2 + 7$

Vertex: $(6, 7)$

52. $y = 3(x + 3)^2 - 5$

Vertex: $(-3, -5)$

53. $y = 2x^2 - 4x + 7$

$a = 2, b = -4, c = 7$

$$x = \frac{-b}{2a} = \frac{-(-4)}{2(2)} = \frac{4}{4} = 1$$

$y = 2x^2 - 4x + 7 = 2(1)^2 - 4(1) + 7$

$$= 2 - 4 + 7 = 5$$

Vertex: $(1, 5)$

54. $y = -3x^2 + 18x - 11$

$a = -3, b = 18, c = -11$

$$x = \frac{-b}{2a} = \frac{-18}{2(-3)} = \frac{-18}{-6} = 3$$

$y = -3x^2 + 18x - 11$

$$= -3(3)^2 + 18(3) - 11$$

$$= -27 + 54 - 11 = 16$$

Vertex: $(3, 16)$

Chapter 9 Test (page 573)

1.
$$6x^2 + x - 1 = 0$$
$$(2x + 1)(3x - 1) = 0$$
$$2x + 1 = 0 \quad \text{or} \quad 3x - 1 = 0$$
$$x = -\tfrac{1}{2} \qquad\qquad x = \tfrac{1}{3}$$

2.
$$10x^2 + 43x = 9$$
$$10x^2 + 43x - 9 = 0$$
$$(5x - 1)(2x + 9) = 0$$
$$5x - 1 = 0 \quad \text{or} \quad 2x + 9 = 0$$
$$x = \tfrac{1}{5} \qquad\qquad x = -\tfrac{9}{2}$$

3.
$$x^2 = 16$$
$$\sqrt{x^2} = \pm\sqrt{16}$$
$$x = \pm 4$$

4.
$$(x - 2)^2 = 3$$
$$\sqrt{(x - 2)^2} = \pm\sqrt{3}$$
$$x - 2 = \pm\sqrt{3}$$
$$x = 2 \pm \sqrt{3}$$

5. $\frac{1}{2}(14) = 7; \; 7^2 = \mathbf{49}$

6. $\frac{1}{2}(-7) = \left(-\frac{7}{2}\right); \; \left(-\frac{7}{2}\right)^2 = \mathbf{\frac{49}{4}}$

7. $3a^2 + 6a - 12 = 0$ $\qquad (a+1)^2 = 5$

$\qquad a^2 + 2a - 4 = 0$ $\qquad \sqrt{(a+1)^2} = \pm\sqrt{5}$

$\qquad\quad a^2 + 2a = 4$ $\qquad\quad a + 1 = \pm\sqrt{5}$

$\qquad a^2 + 2a + 1 = 4 + 1$ $\qquad\qquad a = -1 \pm \sqrt{5}$

$\qquad\quad (a+1)^2 = 5$

8. $x = \dfrac{-b \pm \sqrt{b^2 - 4ac}}{2a}$

9. $x^2 + 3x - 10 = 0 \Rightarrow a = 1, b = 3, c = -10$

$\qquad x = \dfrac{-b \pm \sqrt{b^2 - 4ac}}{2a} = \dfrac{-3 \pm \sqrt{3^2 - 4(1)(-10)}}{2(1)} = \dfrac{-3 \pm \sqrt{9 + 40}}{2} = \dfrac{-3 \pm \sqrt{49}}{2} = \dfrac{-3 \pm 7}{2}$

$\qquad x = \frac{-3+7}{2} = \frac{4}{2} = 2$ or $x = \frac{-3-7}{2} = \frac{-10}{2} = -5$

10. $2x^2 - 5x = 12 \Rightarrow 2x^2 - 5x - 12 = 0 \Rightarrow a = 2, b = -5, c = -12$

$\qquad x = \dfrac{-b \pm \sqrt{b^2 - 4ac}}{2a} = \dfrac{-(-5) \pm \sqrt{(-5)^2 - 4(2)(-12)}}{2(2)} = \dfrac{5 \pm \sqrt{25 + 96}}{4} = \dfrac{5 \pm \sqrt{121}}{4}$

$\qquad x = \frac{5+11}{4} = \frac{16}{4} = 4$ or $x = \frac{5-11}{4} = \frac{-6}{4} = -\frac{3}{2}$

11. $2x^2 + 5x + 1 = 2 \Rightarrow 2x^2 + 5x - 1 = 0 \Rightarrow a = 2, b = 5, c = -1$

$\qquad x = \dfrac{-b \pm \sqrt{b^2 - 4ac}}{2a} = \dfrac{-5 \pm \sqrt{5^2 - 4(2)(-1)}}{2(2)} = \dfrac{-5 \pm \sqrt{25 + 8}}{4} = \dfrac{-5 \pm \sqrt{33}}{4}$

12. Let h = the height and $h + 2$ = the base. $\qquad h^2 + 2h - 80 = 0$

$\qquad \frac{1}{2} \cdot$ base $\cdot$ height = area $\qquad\qquad\qquad (h + 10)(h - 8) = 0$

$\qquad\quad \frac{1}{2}(h + 2)(h) = 40$ $\qquad\qquad\qquad h + 10 = 0$ $\quad$ **or** $\quad h - 8 = 0$

$\qquad\qquad h(h + 2) = 80$ $\qquad\qquad\qquad\qquad h = -10 \qquad\qquad h = 8$

$\qquad\quad h^2 + 2h - 80 = 0$ $\qquad\qquad$ The height must be positive, so $h = 8$

$\qquad\qquad\qquad\qquad\qquad\qquad\qquad\qquad$ and the base is 10 meters.

13. $y = -4(x + 5)^2 - 4$

Vertex: $(-5, -4)$

14. $y = x^2 + 4x + 2$

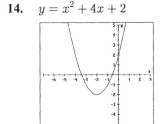

Sample Final Examination (page A-1)

1. Prime numbers between 20 and 30: 23, 29
Answer: b

2. $\dfrac{x+z}{y} = \dfrac{3+(-1)}{(-2)} = \dfrac{2}{-2} = -1$
Answer: b

3. $\dfrac{|x-z|}{|y|} = \dfrac{|-2-(-4)|}{|-3|} = \dfrac{|2|}{|-3|} = \dfrac{2}{3}$
Answer: c

4. Distributive property: $a(b+c) = ab + ac$
Answer: c

5. $7x - 4 = 24$
$7x = 28$
$x = 4$
Answer: a

6. $6z - (9 - 3z) = -3(z+2)$
$6z - 9 + 3z = -3z - 6$
$9z - 9 = -3z - 6$
$12z = 3$
$z = \dfrac{3}{12} = \dfrac{1}{4}$
Answer: e

7. $\dfrac{r}{5} - \dfrac{r-3}{10} = 0$
$10\left(\dfrac{r}{5} - \dfrac{r-3}{10}\right) = 10(0)$
$2r - (r - 3) = 0$
$2r - r + 3 = 0$
$r = -3$
Answer: a

8. $\dfrac{ax}{b} + c = 4$
$\dfrac{ax}{b} = 4 - c$
$b \cdot \dfrac{ax}{b} = b(4 - c)$
$ax = b(4 - c)$
$x = \dfrac{b(4 - c)}{a}$
Answer: c

9. Let $x =$ the number of 10¢ pencils. Then
$25 - x =$ the number of 15¢ pencils.
$10x + 15(25 - x) = 300$
$10x + 375 - 15x = 300$
$-5x = -75$
$x = 15$
He bought 15 of the 10¢ pencils.
Answer: c

10. $-3(x - 2) + 3 \geq 6$
$-3x + 6 + 3 \geq 6$
$-3x + 9 \geq 6$
$-3x \geq -3$
$\dfrac{-3x}{-3} \leq \dfrac{-3}{-3}$
$x \leq 1$
Answer: d

11. $x^2 x^3 x^7 = x^{2+3+7} = x^{12}$
Answer: b

12. $\dfrac{\left(x^2\right)^7}{x^3 x^4} = \dfrac{x^{2\cdot 7}}{x^{3+4}} = \dfrac{x^{14}}{x^7} = x^{14-7} = x^7$
Answer: c

13. $\dfrac{x^{-2} y^3}{xy^{-1}} = x^{-2-1} y^{3-(-1)} = x^{-3} y^4 = \dfrac{y^4}{x^3}$
Answer: d

14. $73{,}000{,}000 = 7.3 \times 10^7$
Answer: a

15. $f(-2) = 2(-2)^2 + 3(-2) - 4$
$= 2(4) - 6 - 4 = 8 - 6 - 4 = -2$
Answer: b

16. $2(y + 3) - 3(y - 2) = 2y + 6 - 3y + 6$
$= -y + 12$
Answer: d

17. $-3x^2y^2(2xy^3) = -6x^3y^5$
Answer: c

18. $2a^3b^2(3a^2b - 2ab^2) = 6a^5b^3 - 4a^4b^4$
Answer: b

19. $(x + 7)(2x - 3) = 2x^2 - 3x + 14x - 21$
$= 2x^2 + 11x - 21$
Answer: c

20.
$$\begin{array}{r} x - 2 \\ x + 7 \overline{\smash{\big)}\ x^2 + 5x - 14} \\ \underline{x^2 + 7x} \\ -2x - 14 \\ \underline{-2x - 14} \\ 0 \end{array}$$
Answer: b

21. $r^2h + r^2a = r^2(h + a)$
Answer: d

22. $m^2n^4 - 49 = (mn^2)^2 - 7^2$
$= (mn^2 + 7)(mn^2 - 7)$
Answer: d

23. $x^2 - 5x + 6 = (x - 2)(x - 3)$
Answer: d

24. $36x^2 + 12x + 1 = (6x + 1)(6x + 1)$
Answer: b

25. $2x^2 + 7xy + 6y^2 = (2x + 3y)(x + 2y)$
Answer: e

26. $8x^3 - 27 = (2x)^3 - 3^3$
$= (2x - 3)\big((2x)^2 + 2x(3) + 3^2\big)$
$= (2x - 3)\big(4x^2 + 6x + 9\big)$
Answer: d

27. $2x^2 + 2xy - 3x - 3y = 2x(x + y) - 3(x + y) = (x + y)(2x - 3)$
Answer: d

28. $x^2 + x - 6 = 0$
$(x - 2)(x + 3) = 0$
$x - 2 = 0 \quad \text{or} \quad x + 3 = 0$
$x = 2 \qquad\qquad x = -3$
Answer: b

29. $6x^2 - 7x - 3 = 0$
$(2x - 3)(3x + 1) = 0$
$2x - 3 = 0 \quad \text{or} \quad 3x + 1 = 0$
$x = \frac{3}{2} \qquad\qquad x = -\frac{1}{3}$
Answer: c

30. $\dfrac{x^2 - 16}{x^2 - 8x + 16} = \dfrac{(x + 4)(x - 4)}{(x - 4)(x - 4)}$
$= \dfrac{x + 4}{x - 4}$
Answer: a

31. $\dfrac{x^2 + 11x - 12}{x - 5} \cdot \dfrac{x^2 - 5x}{x - 1} = \dfrac{(x + 12)(x - 1)}{x - 5} \cdot \dfrac{x(x - 5)}{x - 1} = \dfrac{x(x + 12)}{1} = x(x + 12)$
Answer: b

32. $\dfrac{t^2 + 7t}{t^2 + 5t} \div \dfrac{t^2 + 4t - 21}{t - 3} = \dfrac{t^2 + 7t}{t^2 + 5t} \cdot \dfrac{t - 3}{t^2 + 4t - 21} = \dfrac{t(t + 7)}{t(t + 5)} \cdot \dfrac{t - 3}{(t + 7)(t - 3)} = \dfrac{1}{t + 5}$

Answer: a

33. $\dfrac{3x}{2} - \dfrac{x}{4} = \dfrac{3x(2)}{2(2)} - \dfrac{x}{4} = \dfrac{6x}{4} - \dfrac{x}{4} = \dfrac{5x}{4}$

Answer: d

34. $\dfrac{a + 3}{2a - 6} - \dfrac{2a + 3}{a^2 - 3a} + \dfrac{3}{4} = \dfrac{a + 3}{2(a - 3)} - \dfrac{2a + 3}{a(a - 3)} + \dfrac{3}{4}$

$$= \dfrac{(a + 3)(2a)}{2(a - 3)(2a)} - \dfrac{(2a + 3)(4)}{a(a - 3)(4)} + \dfrac{3(a)(a - 3)}{4(a)(a - 3)}$$

$$= \dfrac{2a^2 + 6a}{4a(a - 3)} - \dfrac{8a + 12}{4a(a - 3)} + \dfrac{3a^2 - 9a}{4a(a - 3)}$$

$$= \dfrac{2a^2 + 6a - 8a - 12 + 3a^2 - 9a}{4a(a - 3)}$$

$$= \dfrac{5a^2 - 11a - 12}{4a(a - 3)} = \dfrac{(5a + 4)(a - 3)}{4a(a - 3)} = \dfrac{5a + 4}{4a}$$

Answer: a

35. $\dfrac{x + \frac{1}{y}}{\frac{1}{x} + y} = \dfrac{\left(x + \frac{1}{y}\right)xy}{\left(\frac{1}{x} + y\right)xy} = \dfrac{x^2y + x}{y + xy^2}$

$$= \dfrac{x(xy + 1)}{y(1 + xy)} = \dfrac{x}{y}$$

Answer: d

36. $\dfrac{1}{x} + \dfrac{1}{2x} = \dfrac{1}{4}$

$$4x\left(\dfrac{1}{x} + \dfrac{1}{2x}\right) = 4x \cdot \dfrac{1}{4}$$

$$4 + 2 = x$$

$$6 = x$$

Answer: c

37. $\dfrac{2}{s + 1} + \dfrac{1 - s}{s} = \dfrac{1}{s^2 + s}$

$$\dfrac{2}{s + 1} + \dfrac{1 - s}{s} = \dfrac{1}{s(s + 1)}$$

$$s(s + 1)\left(\dfrac{2}{s + 1} + \dfrac{1 - s}{s}\right) = s(s + 1) \cdot \dfrac{1}{s(s + 1)}$$

$$2s + (s + 1)(1 - s) = 1$$

$$2s + s - s^2 + 1 - s = 1$$

$$-s^2 + 2s = 0$$

$$-s(s - 2) = 0$$

$$-s = 0 \quad \text{or} \quad s - 2 = 0$$

$$s = 0 \qquad\qquad s = 2$$

$s = 0$ does not check.

Answer: b

38. $3x - 4y = 12$
$3x - 4(0) = 12$
$3x = 12$
$x = 4 \Rightarrow (4, 0)$
Answer: c

39.

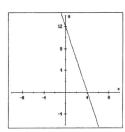

Answer: c

40. $m = \dfrac{y_2 - y_1}{x_2 - x_1} = \dfrac{-6 - 4}{8 - (-2)} = \dfrac{-10}{10} = -1$
Answer: a

41. $m = \dfrac{y_2 - y_1}{x_2 - x_1} = \dfrac{-6 - 4}{8 - (-2)} = \dfrac{-10}{10} = -1$
$y - y_1 = m(x - x_1)$
$y - (-6) = -(x - 8)$
$y + 6 = -x + 8$
$y = -x + 2$
Answer: a

42. $f(a + 2) = (a + 2)^2 - 2(a + 2) = a^2 + 4a + 4 - 2a - 4 = a^2 + 2a$
Answer: b

43. $2x - 5y = 5 \Rightarrow \times (-2) \quad -4x + 10y = -10$
$\underline{3x - 2y = -16 \Rightarrow \times (5)} \quad \underline{15x - 10y = -80}$
$ 11x = -90$
$ x = -\frac{90}{11}$
Answer: e

44. $8x - y = 29 8x - y = 29$
$\underline{2x + y = 11 \Rightarrow \times (-4)} \quad \underline{-8x - 4y = -44}$
$ -5y = -15$
$ y = 3$
Answer: c

45. $\sqrt{12} = \sqrt{4}\sqrt{3} = 2\sqrt{3}$
Answer: a

46. $\sqrt{\dfrac{3}{4}} = \dfrac{\sqrt{3}}{\sqrt{4}} = \dfrac{\sqrt{3}}{2}$
Answer: c

47. $\sqrt{75x^3} = \sqrt{25x^2}\sqrt{3x} = 5x\sqrt{3x}$
Answer: e

48. $3\sqrt{5} - \sqrt{20} = 3\sqrt{5} - \sqrt{4}\sqrt{5}$
$\phantom{3\sqrt{5} - \sqrt{20}} = 3\sqrt{5} - 2\sqrt{5} = \sqrt{5}$
Answer: d

49. $\dfrac{11}{\sqrt{11}} = \dfrac{11\sqrt{11}}{\sqrt{11}\sqrt{11}} = \dfrac{11\sqrt{11}}{11} = \sqrt{11}$
Answer: b

50. $\dfrac{7}{3-\sqrt{2}} = \dfrac{7\left(3+\sqrt{2}\right)}{\left(3-\sqrt{2}\right)\left(3+\sqrt{2}\right)} = \dfrac{7\left(3+\sqrt{2}\right)}{9-3\sqrt{2}+3\sqrt{2}-2} = \dfrac{7\left(3+\sqrt{2}\right)}{7} = 3+\sqrt{2}$

Answer: c

51.
$$\sqrt{\dfrac{3x-1}{5}} = 2$$
$$\left(\sqrt{\dfrac{3x-1}{5}}\right)^2 = 2^2$$
$$\dfrac{3x-1}{5} = 4$$
$$3x-1 = 20$$
$$3x = 21$$
$$x = 7$$

Answer: a

52.
$$3\sqrt{n}-1 = 1$$
$$3\sqrt{n} = 2$$
$$\left(3\sqrt{n}\right)^2 = 2^2$$
$$9n = 4$$
$$n = \dfrac{4}{9}$$

Answer: c

53. $\left(a^6 b^4\right)^{1/2} = \left(a^6\right)^{1/2}\left(b^4\right)^{1/2} = a^3 b^2$

Answer: b

54. $\left(\dfrac{8}{125}\right)^{2/3} = \left(\left(\dfrac{8}{125}\right)^{1/3}\right)^2 = \left(\dfrac{2}{5}\right)^2 = \dfrac{4}{25}$

Answer: a

55. $\dfrac{1}{2}(12) = 6; 6^2 = \mathbf{36}$

Answer: d

56. $x = \dfrac{-b \pm \sqrt{b^2-4ac}}{2a}$

Answer: b

57.
$$x^2 - 2x - 2 = 0$$
$$x^2 - 2x = 2$$
$$x^2 - 2x + 1 = 2 + 1$$
$$\left(x-1\right)^2 = 3$$
$$x - 1 = \pm\sqrt{3}$$
$$x = 1 \pm \sqrt{3}$$

Answer: c

58. $y = x^2 - 2x + 1 = \left(x-1\right)^2 + 0$

Vertex: $(1, 0)$

Answer: a

59.

Answer: e

60. $x = \dfrac{ky^2}{t}$

Answer: b

Exercise II.1 (page A-11)

1. $x^2 + 9 = 0$
$$x^2 = -9$$
$$\sqrt{x^2} = \pm\sqrt{-9}$$
$$x = \pm 3i$$

3. $3x^2 = -16$
$$x^2 = -\frac{16}{3}$$
$$\sqrt{x^2} = \pm\sqrt{-\frac{16}{3}}$$
$$x = \pm\frac{\sqrt{16}}{\sqrt{3}}i = \pm\frac{4\sqrt{3}}{3}i$$

5. $x^2 + 2x + 2 = 0 \Rightarrow a = 1, b = 2, c = 2$
$$x = \frac{-b \pm \sqrt{b^2 - 4ac}}{2a} = \frac{-2 \pm \sqrt{2^2 - 4(1)(2)}}{2(1)} = \frac{-2 \pm \sqrt{4 - 8}}{2} = \frac{-2 \pm \sqrt{-4}}{2} = \frac{-2 \pm 2i}{2}$$
$$= -1 \pm i$$

7. $2x^2 + x + 1 = 0 \Rightarrow a = 2, b = 1, c = 1$
$$x = \frac{-b \pm \sqrt{b^2 - 4ac}}{2a} = \frac{-1 \pm \sqrt{1^2 - 4(2)(1)}}{2(2)} = \frac{-1 \pm \sqrt{1 - 8}}{4} = \frac{-1 \pm \sqrt{-7}}{4} = \frac{-1 \pm i\sqrt{7}}{4}$$
$$= -\frac{1}{4} \pm \frac{\sqrt{7}}{4}i$$

9. $3x^2 - 4x + 2 = 0 \Rightarrow a = 3, b = -4, c = 2$
$$x = \frac{-b \pm \sqrt{b^2 - 4ac}}{2a} = \frac{4 \pm \sqrt{(-4)^2 - 4(3)(2)}}{2(3)} = \frac{4 \pm \sqrt{16 - 24}}{6} = \frac{4 \pm \sqrt{-8}}{6} = \frac{4 \pm 2i\sqrt{2}}{6}$$
$$= \frac{2}{3} \pm \frac{\sqrt{2}}{3}i$$

11. $i^{21} = i^{20}i^1 = \left(i^4\right)^5 i^1 = i^1 = i$

13. $i^{27} = i^{24}i^3 = \left(i^4\right)^6 i^3 = i^3 = -i$

15. $i^{100} = \left(i^4\right)^{25} = 1$

17. $i^{97} = i^{96}i^1 = \left(i^4\right)^{24} i^1 = i^1 = i$

19. $(3 + 4i) + (5 - 6i) = 3 + 4i + 5 - 6i$
$$= 8 - 2i$$

21. $(7 - 3i) - (4 + 2i) = 7 - 3i - 4 - 2i$
$$= 3 - 5i$$

23. $\left(8 + \sqrt{-25}\right) + \left(7 + \sqrt{-4}\right) = (8 + 5i) + (7 + 2i) = 8 + 5i + 7 + 2i = 15 + 7i$

25. $\left(-8 - \sqrt{-3}\right) - \left(7 - \sqrt{-27}\right) = \left(-8 - i\sqrt{3}\right) - \left(7 - 3i\sqrt{3}\right) = -8 - i\sqrt{3} - 7 + 3i\sqrt{3}$
$$= -15 + 2\sqrt{3}\, i$$

27. $3i(2 - i) = 6i - 3i^2 = 6i - 3(-1) = 6i + 3 = 3 + 6i$

29. $(2 + 3i)(3 - i) = 6 - 2i + 9i - 3i^2 = 6 + 7i - 3(-1) = 6 + 7i + 3 = 9 + 7i$

31. $(2-4i)(3+2i) = 6+4i-12i-8i^2 = 6-8i-8(-1) = 6-8i+8 = 14-8i$

33. $\left(2+\sqrt{-2}\right)\left(3-\sqrt{-2}\right) = \left(2+\sqrt{2}\,i\right)\left(3-\sqrt{2}\,i\right) = 6-2\sqrt{2}\,i+3\sqrt{2}\,i-2i^2$

$$= 6+\sqrt{2}\,i-2(-1)$$
$$= 6+\sqrt{2}\,i+2 = 8+\sqrt{2}\,i$$

35. $\left(-2-\sqrt{-16}\right)\left(1+\sqrt{-4}\right) = (-2-4i)(1+2i) = -2-4i-4i-8i^2 = -2-8i-8(-1)$

$$= -2-8i+8$$
$$= 6-8i$$

37. $\left(2+\sqrt{-3}\right)\left(3-\sqrt{-2}\right) = \left(2+i\sqrt{3}\right)\left(3-i\sqrt{2}\right) = 6-2i\sqrt{2}+3i\sqrt{3}-i^2\sqrt{6}$

$$= 6-2i\sqrt{2}+3i\sqrt{3}-(-1)\sqrt{6}$$
$$= 6+\sqrt{6}+\left(3\sqrt{3}-2\sqrt{2}\right)i$$

39. $\left(8-\sqrt{-5}\right)\left(-2-\sqrt{-7}\right) = \left(8-i\sqrt{5}\right)\left(-2-i\sqrt{7}\right) = -16-8i\sqrt{7}+2i\sqrt{5}+i^2\sqrt{35}$

$$= -16-8i\sqrt{7}+2i\sqrt{5}+(-1)\sqrt{35}$$
$$= -16-\sqrt{35}+\left(2\sqrt{5}-8\sqrt{7}\right)i$$

41. $\dfrac{1}{i} = \dfrac{1i}{ii} = \dfrac{1i}{i^2} = \dfrac{i}{-1} = -i = 0-i$

43. $\dfrac{4}{5i^3} = \dfrac{4i}{5i^3i} = \dfrac{4i}{5i^4} = \dfrac{4i}{5} = \dfrac{4}{5}i = 0+\dfrac{4}{5}i$

45. $\dfrac{3i}{8\sqrt{-9}} = \dfrac{3i}{8(3i)} = \dfrac{1}{8} = \dfrac{1}{8}+0i$

47. $\dfrac{-3}{5i^5} = \dfrac{-3}{5i^4i^1} = \dfrac{-3}{5i} = \dfrac{-3i}{5ii} = \dfrac{-3i}{5i^2} = \dfrac{-3i}{-5} = \dfrac{3}{5}i = 0+\dfrac{3}{5}i$

49. $\dfrac{-6}{\sqrt{-32}} = \dfrac{-6}{4i\sqrt{2}} = \dfrac{-3}{2i\sqrt{2}}\cdot\dfrac{i\sqrt{2}}{i\sqrt{2}} = \dfrac{-3i\sqrt{2}}{2i^2(2)} = \dfrac{-3i\sqrt{2}}{-4} = \dfrac{3\sqrt{2}}{4}i = 0+\dfrac{3\sqrt{2}}{4}i$

51. $\dfrac{3}{5+i} = \dfrac{3(5-i)}{(5+i)(5-i)} = \dfrac{3(5-i)}{25-5i+5i-i^2} = \dfrac{3(5-i)}{25-(-1)} = \dfrac{3(5-i)}{26} = \dfrac{15-3i}{26} = \dfrac{15}{26}-\dfrac{3}{26}i$

53. $\dfrac{-12}{7-\sqrt{-1}} = \dfrac{-12}{7-i} = \dfrac{-12(7+i)}{(7-i)(7+i)} = \dfrac{-12(7+i)}{49+7i-7i-i^2} = \dfrac{-12(7+i)}{49-(-1)} = \dfrac{-12(7+i)}{50}$

$$= \dfrac{-6(7+i)}{25}$$
$$= \dfrac{-42-6i}{25}$$
$$= -\dfrac{42}{25}-\dfrac{6}{25}i$$

55. $\dfrac{5i}{6+2i} = \dfrac{5i(6-2i)}{(6+2i)(6-2i)} = \dfrac{30i-10i^2}{36-12i+12i-4i^2} = \dfrac{30i-10(-1)}{36-4(-1)} = \dfrac{10+30i}{40} = \dfrac{10}{40} + \dfrac{30}{40}i$

$$= \dfrac{1}{4} + \dfrac{3}{4}i$$

57. $\dfrac{3-2i}{3+2i} = \dfrac{(3-2i)(3-2i)}{(3+2i)(3-2i)} = \dfrac{9-6i-6i+4i^2}{9-6i+6i-4i^2} = \dfrac{9-12i+4(-1)}{9-4(-1)} = \dfrac{5-12i}{13} = \dfrac{5}{13} - \dfrac{12}{13}i$

59. $\dfrac{3+\sqrt{-2}}{2+\sqrt{-5}} = \dfrac{3+i\sqrt{2}}{2+i\sqrt{5}} = \dfrac{\left(3+i\sqrt{2}\right)\left(2-i\sqrt{5}\right)}{\left(2+i\sqrt{5}\right)\left(2-i\sqrt{5}\right)} = \dfrac{6-3i\sqrt{5}+2i\sqrt{2}+\sqrt{10}}{4-(-1)(5)}$

$$= \dfrac{6+\sqrt{10}+\left(2\sqrt{2}-3\sqrt{5}\right)i}{9}$$

$$= \dfrac{6+\sqrt{10}}{9} + \dfrac{2\sqrt{2}-3\sqrt{5}}{9}i$$

61. $|6+8i| = \sqrt{6^2+8^2} = \sqrt{36+64}$

$$= \sqrt{100} = 10$$

63. $|12-5i| = \sqrt{12^2+(-5)^2} = \sqrt{144+25}$

$$= \sqrt{169} = 13$$

65. $|5+7i| = \sqrt{5^2+7^2} = \sqrt{25+49} = \sqrt{74}$

67. $\left|4+\sqrt{-2}\right| = \left|4+i\sqrt{2}\right| = \sqrt{4^2+\left(\sqrt{2}\right)^2} = \sqrt{16+2} = \sqrt{18} = 3\sqrt{2}$

69. $\left|8+\sqrt{-5}\right| = \left|8+i\sqrt{5}\right| = \sqrt{8^2+\left(\sqrt{5}\right)^2} = \sqrt{64+5} = \sqrt{69}$

71. $|5-0i| = \sqrt{5^2+(-0)^2} = \sqrt{25+0}$

$$= \sqrt{25} = 5$$

73. **Answers may vary.**